D1072520

# Handbook of Giant Magnetostrictive Materials

ACADEMIC PRESS SERIES IN ELECTROMAGNETISM

Electromagnetism is a classical area of physics and engineering that still plays a very important role in the development of new technology. Electromagnetism often serves as a link between electrical engineers, material scientists, and applied physicists. This series presents volumes on those aspects of applied and theoretical electromagnetism that are becoming increasingly important in modern and rapidly developing technology. Its objective is to meet the needs of researchers, students, and practicing engineers.

*This is a volume in*
ELECTROMAGNETISM

ISAAK MAYERGOYZ, SERIES EDITOR
UNIVERSITY OF MARYLAND, COLLEGE PARK, MARYLAND

# Handbook of Giant Magnetostrictive Materials

**Göran Engdahl**

*Royal Institute of Technology*
*Stockholm, Sweden*

## ACADEMIC PRESS

A Harcourt Science and Technology Company

San Diego   San Francisco   New York
Boston   London   Sydney   Tokyo

ACADEMIC PRESS
*A Harcourt Science and Technology Company*
525 B Street, Suite 1900, San Diego, CA 92101-4495, USA
http://www.apnet.com

Academic Press
24–28 Oval Road, London NW1 7DX, UK
http://www.hbuk.co.uk/ap/

**Library of Congress Catalog Card Number:** 99-65058
**International Standard Book Number:** 0-12-238640-X

Printed in the United States of America
99 00 01 02 03 IP 9 8 7 6 5 4 3 2 1

To Helena, my wife, and our children August, Samuel, Julia, Gabriella, and Johannes for their support and patience.

# Academic Press Series in Electromagnetism

<small>EDITED BY</small>
Isaak Mayergoyz, University of Maryland, College Park, Maryland

<small>BOOKS PUBLISHED IN THE SERIES</small>
Georgio Bertotti, *Hysterisis in Magnetism: For Physicists, Material Scientists, and Engineers*
Alain Bossavit, *Computational Electromagnetism: Variational Formulations, Complementarity, Edge Elements*
M.V.K. Chari and S.J. Salon, *Numerical Methods in Electromagnetics*
Göran Engdahl, *Handbook of Giant Magnetostrictive Materials*
John C. Mallinson, *Magneto-Resistive Heads: Fundamentals and Applications*
Isaak Mayergoyz, *Nonlinear Diffusion of Electromagnetic Fields*
Shan X. Wang and Alexander M. Taratorin, *Magnetic Information Storage Technology*

<small>RELATED BOOKS</small>
John C. Mallison, *The Foundations of Magnetic Recording, Second Edition*
Reinaldo Perez, *Handbook of Electromagnetic Compatibility*

# CONTENTS

# FOREWORD

This volume in the Academic Press Electromagnetism series presents an in-depth, comprehensive and self-contained treatment of giant magneto-constrictive materials and their applications. These materials have unique magneto-elastic properties. For this reason, they hold promise for many novel applications. Currently, there does not exist any book that covers the physics of giant magnetostrictive materials, their manufacture and characterization, as well as their modeling and device applications. This book represents the first and successful attempt to give synthetic expositon of all these issues within one volume.

This book contains the contributions of the well known experts in the field of giant magnetostrictive material technology. These contributions were assembled and edited by Dr. Göran Engdahl, who has been a pioneer in the area of practical utilization of these materials. The book is intended for readers without extensive experience in the area of giant magnetostrictive materials. It reviews the physical origin of giant magnetostriction and the manufacturing processes for bulk composite giant magnetostrictive materials. It covers in depth the modeling and characterization issues for these materials as well as the current and potential device applications. The final chapter of the book deals with the emerging magnetostrictive thin film technologies.

I maintain that this book will be a valuable reference for both experts and beginners in the field. Electrical and mechanical engineers, physicists and material scientists, experienced researchers and graduate students will find this book to be a valuable source of new facts, fundamental concepts, and penetrating insights. This book will enrich its readers with a better understanding of the physics of giant magnetostrictive materials and their promise for device applications.

Issak Mayergoyz, Series Editor

# PREFACE

The creation of this book is a result of a number of lucky circumstances and coincidences. The first such event occurred in 1986 when I was at Asea Brown Boveri in Västerås, and was asked to determine the magnetomechanical coupling factor of a piece of deformed metal-like material called TERFENOL-D.

This led to a number of feasibility studies, and some years later, I had established a research group at the Royal Institute of Technology in Stockholm for characterizing and modeling giant magnetostrictive materials, i.e., TERFENOL-D and related materials.

One of my ambitions was to create realistic models usable for design that included magnetomechanical hysteresis effects. These efforts led to collaborations with Isaak Mayergoyz, who also has been functioning as opponent during the dissertations of two of my students (Lars Kvarnsjö, 1993 and Anders Bergqvist, 1984). In fact, it was Prof. Isaak Mayergoyz that encouraged and convinced me to accept an offer from Academic Press (at that time AIP) to function as an editor for this handbook on giant magnetostrictive materials.

One other important factor has been my wife, Helena, who said that I *of course* should accept the offer from Academic Press. In fact, Helena influenced the process much earlier by encouraging me to initiate my research activities at the Royal Institute of Technology. Her firm conviction has always been that in the long run I should be active as a relaxed professor at Royal Institute of Technology rather than a stressed project manager at ABB.

At that time, in 1995, the situation indeed was not that relaxed. At ABB I was engaged as project manager in a development project that represented a full-time workload and at the Royal Institute, I functioned as a research leader for a group of five persons.

The first contribution from Arthur Clark and Kristl Hathaway given to me was then really encouraging and led me to believe that this book project was possible to perform. When Dale McMasters delivered his metallurgy contribution some time later, I considered that the die was cast. I knew that there really should be a book.

So here it is finally. My personal engagement in the writing of this book proved to be much more comprehensive than I ever had imagined when I accepted. One fortunate circumstance was that during the last years I have had the opportunity to spend concentrated time on this book project.

This book is the first handbook of giant magnetostrictive materials. The intention is for it to function as a source of information for electrical and mechanical engineers who consider using giant magnetostrictive materials in their constructions. My ambition, then, has been to give a comprehensive scope of the book. Irrespective of the background of the reader, there will always be some parts of the book that can be easily assimilated.

The chapters of the book are organized so it is possible to read separate chapters in arbitrarily order. I hope that this feature of the book will inspire you as a reader to go on reading chapter after chapter, even those that are out of your professional discipline. It is my belief that going outside ones professional discipline promotes new ideas, concepts inventions etc. This supports my aspirations that the book also encourages progress in the development of the giant magnetostrictive materials technology.

The following is a brief review of the contents of the six chapters in the book.

## CHAPTER 1

The first chapter enlightens the reader about the quantum mechanical origin of giant magnetostriction. Magnetic and magneto-elastic concepts are explained and magnetostriction phenomena are described. Giant magnetostriction in crystalline rare earth alloys and the option to design materials with implications for applications are described. The chapter also includes a section in which metallurgical structure and composition of the giant magnetostrictive materials are covered. Then it is described what happens microscopically when the material is magnetized, giving the reader a basic understanding of what is occurring in the material during the magneto-mechanical transduction process. In the end of the chapter the manufacturing processes and the making of current giant magnetostrictive materials are described, including manufacture methods for bulk and composite giant magnetostrictive materials.

## CHAPTER 2

The second chapter describes current linear models of magnetostrictive behavior and nonlinear methods of modeling. The models described comprise analytical methods, equivalent circuits, dynamic simulation, finite difference, and FE methods. The chapter also gives information regarding different current computational tools concerning their features, capabilities, limitations, etc. Topics such as e.g., longitudial waves, eddy currents, magnetic circuits, hysteresis, systems interaction, power supply, and control are covered. Numerical values of magnetostrictive constants used in the linear models are also given.

## CHAPTER 3

The third chapter covers different general aspects of magnetostrictive design that always appear during development of device applications and the general characteristics and physical limitations of the giant magnetostrictive materials. In this chapter general design experience in the field is covered. Important topics such as magnetic and mechanical operation ranges, magnetic, electrical, mechanical and thermal design, electromechanical response, and mechanical transmission are covered. The chapter intends to give the reader an insight of a general design methodology that can be applied by the reader.

## CHAPTER 4

The fourth chapter first covers time domain experimental methodologies for characterization of the magnetoelastic behavior of the magnetostrictive materials and actuators by quasi-static characterization and dynamic measurements. Frequency domain methodologies such as the resonance/antiresonance and impedance locus methods are also presented. The concepts of maximal and potential efficiency of magnetostrictive actuators are also presented and explained.

## CHAPTER 5

The fifth chapter first gives some insight into the background and driving forces regarding the development of magnetostrictive materials and their applications. A structured inventory is presented, which covers a broad spectrum of current and potential device applications. In this inventory, the applications are described with technical and commercial details. The intention is to demonstrate the huge diversity and potential of the technology and to give inspiration for a continued development of device applications based on giant magnetostrictive materials.

## CHAPTER 6

The sixth chapter gives a brief orientation regarding emerging magneto-strictive thin film technologies. Material features, production methods, measurement methodologies, and some application case studies are covered.

The field of giant magnetostrictive materials and their applications comprises more topics and details than is covered by this book. To attain a reasonable volume of the book and to make the material manageable, the material has been limited to those topics that can be primarily related to the design process. This means, for example, that topics related to physical material characterization are omitted. Specific and detailed technical information regarding material manufacturing and device applications is also omitted.

Readers interested in achieving a more detailed knowledge are encouraged to study the appropriate publications in the supplied reference lists.

As a complement, I have also assembled a condensed market inventory (Appendix D) that hopefully will save valuable time if you consider initiating the development of new device applications based on giant magnetostrictive materials.

Finally, I wish you a pleasant and hopefully informative reading of this book.

## ACKNOWLEDGMENTS

This book has been created through the teamwork of all the contributors. First of all, I want to thank all the co-writers for accepting to participate in this book project and all their writing efforts. This collaboration has been really stimulating and during the work I have gained substantial new experiences on various aspects of the topic  of the book "giant magnetostrictive materials."

I also want to thank my former students and colleagues who supplied figures and gave assistance during the writing. I especially want to thank Dr. Lars Kvarnsjö, who has given me numerous pictures, Dr. Anders Bergqvist for faithful discussions, and Dr. Frederik Stillesjö for his support.

At ABB I want to thank Birger Drugge who, in the mid 1980's, gave me the opportunity to work with giant magnetostrictive materials and encouraged me to undertake research and development in this field. I also want to thank Prof. Ronald Eriksson, the head of the department of Electric Power Engineering at KTH where I work now, for his continuous support regarding the buildup of our research group on magnetic and magnetostrictive materials.

There are several more people I would like to thank for their indirect contributions to this book. These include all colleagues and students at ABB and KTH that have supplied a friendly and creative atmosphere in all of my research and development projects over the years. I remember, for example all the efforts done by the experimental workshops, trouble-shooting regarding measurement systems and computer codes, and lots more at ABB Corporate Research and KTH. I also want to thank all persons related to my co-writers that have assisted them during their writing.

Finally, I want to thank my wife, Helena, for her continuous encouragement and support and all of my family for their heroic patience during my work on this book.


Göran Engdahl

Täby, August 1999

# CONTRIBUTORS

*Numbers in parentheses indicate the pages on which the authors' contributions begin.*

*Charles B. Bright* (207, 287), ETREMA Products, Inc. 2500 North Loop Drive, Ames, Iowa 50010

*Tord Cedell* (107), Lund University, Department of Mechanical Engineering, P.O. Box 118, 221 00 Lund, Sweden

*Frank Claeyssen* (353), CEDRAT RECHERCHE SA, AMA Department, Zirst, 38246 MEYLAN Cedex, France

*Arthur Clark* (1), Clark Associates, 10421 Floral Dr., Hyattsville, Maryland 20783

*Göran Engdahl* (127, 207, 265, 287, 345, 349), Kungliga Tekniska Högskolan, Electric Power Engineering, 100 44 Stockholm, Sweden

*Kristl Hathaway* (1), Naval Research Laboratory, 45555 Overlook Ave., SW, Washington, D.C. 20375

*Don Lord* ( 69), Joule Physics Laboratory, School of Sciences, University of Salford, Salford M5 4WT, England

*Dale McMasters* (52, 95), ETREMA Products, Inc., 2500 North Loop Drive, Ames, Iowa 50010

*Eckhard Quandt* (323), Stiflung caesar, Center of Advanced European Studies and Research, Friedensplatz 16, D-53111 Bonn, Germany

# CHAPTER 1

# Physics of Giant Magnetostriction*

## PHYSICAL ORIGIN OF GIANT MAGNETOSTRICTION

Magnetostriction [a] can be described most generally as the deformation of a body in response to a change in its magnetization (magnetic moment per unit volume). The change in magnetization can be brought about either by a change in temperature or by the application of a magnetic field. For example, cooling a body into a ferromagnetic state produces a volume expansion. Alternatively, two manifestations of magnetostriction due to a magnetic field are shown in Fig. 1.1: (i) a linear deformation with increasing field magnitude and (ii) a rotating deformation caused by changing the direction of a field of fixed magnitude. All magnetic materials exhibit magnetostriction to some degree; however, giant magnetostriction occurs in a small number of materials containing rare earth elements. It is one manifestation of the strong magnetoelastic coupling or magnetoelasticity in these materials. Another manifestation of strong magnetoelastic coupling is a large $\Delta E$ *effect*, a change in elastic moduli accompanying a change in magnetization (1) (Fig. 1.2).

Magnetoelasticity is the coupling between the *classical* properties of elasticity and strain and the intrinsically *quantum mechanical* and *relativistic* phenomena of magnetism. All materials are elastic. The strain response to stress in a solid derives from the energies associated with stretching, compressing, or bending the bonds between atoms. On the other hand, only a small percentage of materials are ferromagnetic: nine elements (all metals), the metallic alloys and compounds of these elements, and a very few nonmetallic compounds. Magnetism occurs because of an unusual imbalance in the magnetic moments of a material's electrons. When this imbalance occurs the electrons can order in such a

*Please refer to the table of contents and contributor's list for the authors of the various sections in this chapter.

**1**

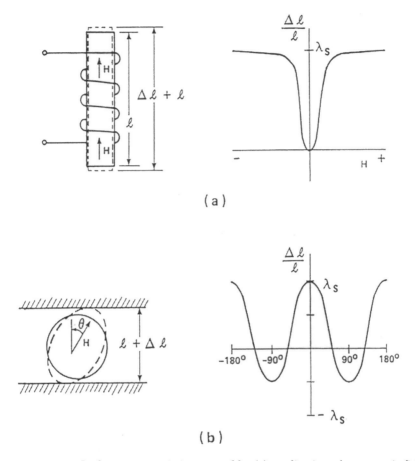

**FIGURE 1.1.**   Joule magnetostriction caused by (a) application of a magnetic field and (b) rotation of a saturating magnetic field.

way that the net magnetic moment is in a particular direction, lowering the crystal symmetry and producing new properties, one of which is magnetostriction. In this section, some simplified descriptions of magnetism and magnetostriction that will form a base for understanding the potential and limitations of materials exhibiting giant magnetostriction are discussed. The discussion is more specialized than in a general magnetism text in that topics not essential to understanding magnetostriction have been omitted, and the chemical elements chosen as examples

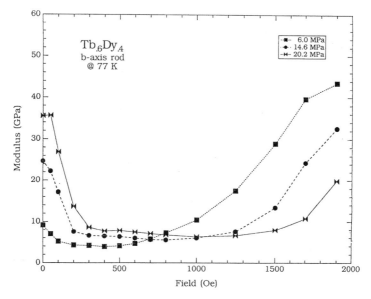

**FIGURE 1.2.** Young's modulus vs magnetic field for $b$-axis rod of $Tb_{0.6}Dy_{0.4}$ at 77 K taken under compressive stresses of 6.0, 14.6, and 20.2 MPa (*1*).

to illustrate fundamental concepts are chosen from those comprising giant magnetostriction materials. [Text shown in brackets is intended to compensate for the oversimplification with more rigorous descriptions.]

## Magnetism

### Atomic Magnetism

The Bohr model of the atom consists of a small heavy nucleus made up of positively charged protons and neutral neutrons, encircled by negatively charged electrons. In the classical theory of electromagnetism, any moving charge produces a magnetic field and changes its motion in response to an external magnetic field. Therefore, it is tempting to think of these orbiting electrons as small circular currents forming magnetic dipoles, which combine to give a magnetic moment for the atom. In fact, a net atomic dipole moment can arise from these orbital dipoles only when they are unbalanced. Consider the case of two electrons moving in the same orbital path but in opposite directions. This case of balanced or compensated angular momenta produces a net magnetic dipole of zero. On the other hand, an unbalanced situation with electrons moving in different orbits produces a nonzero magnetic dipole.

Quantum mechanics (2) tells us which electron orbits or states are allowed in a given atom and in what order these states are preferentially occupied by the available electrons. One can readily see that any atom whose electron orbital momenta do not cancel should be magnetic. In fact, although the electron orbits do contribute to the magnetic moments for all atoms except those for which the orbital momenta completely cancel ("closed-shell atoms"), this contribution is usually only a part of the total moment.

Each electron also carries with it an intrinsic angular moment called "spin," which is responsible for the second part of an atom's magnetic moment. [The protons and neutrons also have spin moments but they are smaller than those of the electrons by a factor of $10^4$ and can be neglected.] The electron spin moment has two possible states, "up" ($+$) and "down" ($-$), and according to quantum mechanics a given atomic orbit can be populated by two electrons, one in each spin state. If every orbit is populated by both a ($+$) and ($-$) spin electron the net spin moment for the atom is zero. However, if, as is usually the case, there are $n_+$ orbits with ($+$) spin electrons and $n_-$ orbits with ($-$) spin electrons the net spin moment for the atom is given by

$$\mu_S = \mu_B(n_+ - n_-) \tag{1.1}$$

where $\mu_B$ is the magnetic moment of a single electron spin.

The net orbital moment, $\mu_L$, and the total (spin + orbital) moment of an atom, $\mu_J$, can be calculated by similar procedures, which are governed by the rules of quantum mechanics for combining angular momenta. The electron states of atoms are grouped into "shells" with numbers corresponding to the energy with which an electron in a shell is bound to the nucleus. Moving to the right and down in the periodic table, as the number of electrons per atom increases the shells are generally filled by first placing a ($+$) electron in each orbital state in the shell and then adding the ($-$) electrons. In a pictorial sense the "shape" of an atom depends on which of these orbital states are filled: If every orbital state is filled equally, either with a single ($+$) electron or with a pair of ($+$) and ($-$) electrons, the atom is said to have a half-filled or filled shell and its shape is spherical. This interplay between the filling of orbitals, the net moment and the atomic shape will prove to be crucial to understanding magnetostriction in rare earth elements and compounds.

## Magnetism in Solids

Since most atoms have partially filled electron shells, why are so few solids magnetic? The answer is that when atoms come together and bond to form a solid, they share electrons in ways that eliminate the spin and orbital imbalance required for magnetism. Electrons in molecules travel in "molecular orbitals," and in solids they occupy even more spatially extended states known as "band states." The atomic orbitals which have the largest radius (and the smallest energy binding them to the nucleus) overlap in space with similar orbitals on neighboring atoms to produce the most extended states in the solid. These states provide the bonding between the atoms and determine the solid's elastic stiffness. If the original atomic orbital states had unbalanced populations of electron spins, these will usually be redistributed in the extended states to nearly balance. One unbalanced spin in an atom can produce significant magnetism, whereas one unbalanced spin in a solid is negligible. Thus, the magnetic moments which are preserved in the solid are more characteristic of the *ionic* electron configuration, i.e., the atom with the bonding electrons removed, or with sufficient bonding electrons added to complete the shell. Consequently, the only groups of elements in the periodic table which exhibit magnetic moments in solids are those in which the unbalanced electron populations occur in an *inner* shell, namely the transition metals (3d, 4d, and 5d) and the rare earths and actinides. Table 1.1 shows calculated atomic and ionic moments, and moments observed in solids at high temperatures, for some representative elements with and without partially filled inner shells.

## Rare Earth Magnetism

In one interesting group, the rare earths, the unbalanced spins are sheltered in an inner orbital shell which has essentially no overlap with orbitals on neighboring atoms and no participation in the bonding. This occurs because in the rare earths the 4f orbital shell is filled "out of order." Two or three valence electrons occupy the 6s and 5d states before the more

**Table 1.1.** Magnetic Moments (in $\mu_B$) of Representative Elements with and without Partially Filled Inner Shells

| Element | Atomic moment | Ion | Ionic moment | Predicted paramagnetic moment[a] | Observed paramagnetic moment |
|---|---|---|---|---|---|
| Carbon | 0.0 | $C^{4+}$ | 0.0 | 0.0 | 0.0 |
| Aluminum | 0.3 | $Al^{3+}$ | 0.0 | 0.0 | 0.0 |
| Sulfur | 3.0 | $S^{2-}$ | 0.0 | 0.0 | 0.0 |
| Iron | 6.0 | $Fe^{2+}$ | 6.0 | 6.7 (4.9) | 4.8 |
| | | $Fe^{3+}$ | 5.0 | 5.9 (5.9) | 5.4 |
| Copper | 1.0 | $Cu^{1+}$ | 0.0 | 0.0 | 0.0 |
| | | $Cu^{2+}$ | 3.0 | 3.6 (1.7) | 1.9 |
| Neodymium | 2.4 | $Nd^{3+}$ | 3.3 | 3.6 | 3.5 |
| Terbium | 12.0 | $Tb^{3+}$ | 9.0 | 9.7 | 9.5 |
| Bismuth | 3.0 | $Bi^{3-}$ | 0.0 | 0.0 | 0.0 |
| Uranium | 4.3 | $U^{3+}$ | 3.3 | 3.6 | 3.1 |

[a] The predicted paramagnetic moments (observed at temperatures above the ordering temperature) are slightly higher than the ionic moments because of contributions from non-ground state configurations poplulated at elevated temperatures. Moments in parentheses are calculated for spin moments only (assuming orbital moments quenched for 3d transition metals)

tightly bound 4f shell is filled. The typical rare earth atomic configuration is [Xe]$4f^n 5d^{(1\,or\,0)}6s^2$.] The 4f subshell has seven orbital states which can hold a total of 14 electrons. For simplicity, we consider only 3 + ions. Gd, in the middle of the rare earth series, has 7 4f electrons which exactly half-fill the available states ($n_+ = 7$ and $n_- = 0$) giving it the largest of the rare earth spin moments ($\mu_s = 7\mu_B$). The light rare earths (Ce–Eu) have fewer than 7 4f electrons. For example, Sm has five, giving it a spin moment of $5\mu_B$. The heavy rare earths (Gd–Yb) have a full component of 7 (+) electrons plus some (−) electrons which act to cancel part of the spin moment. For example, Tb has $n_+ = 7$ and $n_- = 1$, so $\mu_s = 6\mu_B$.

In addition to the spin imbalance, the partially filled 4f shell has an orbital imbalance which gives both an orbital contribution to the magnetic moment and an anisotropic physical shape to the ion. Gd, with each 4f orbital state occupied by one electron, is spherical (as is Lu with two electrons per 4*f* orbital). Sm, with five 4*f* orbital states occupied by a single electron, is prolate, and Tb, with one 4*f* orbital doubly occupied and six singly occupied, is oblate. Figure 1.3 shows the calculated ionic shapes for the rare earth series(3). The orbital contribution to the magnetic moment cancels part of the spin moment for the light rare earths and adds to the spin moment for the heavy rare earths. Table 1.2 gives the ionic spin moments, the predicted total ionic moments, and the experimenally observed moments for the rare earth elements. These atomic properties carry over to the solid because the moments and shapes are derived from electrons which do not participate significantly in the bonding with other atoms. Rare earth atoms retain essentially the same magnetic moments and physical shape, independent of their local environment—in the elements, in alloys, compounds, and oxides, and even in amorphous materials. [The actinide series (Ac–Lu) also has an unfilled *f* shell (5*f*), but this shell is less well shielded from its environment by the outer electrons than is the case for the 4*f* shell in the rare earths. Uranium, for example, is not ferromagnetic in elemental form but does have ferromagnetic compounds; see Table 1.1.]

## Transition Metal Magnetism

The 3*d* transition metal series (Sc–Zn) also has an "out of sequence" occupation of the electron orbitals. The 4*s* and 4*p* orbitals are partially

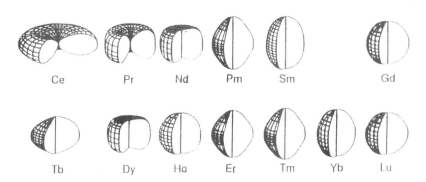

**FIGURE 1.3.** 4*f* electron charge cloud density for the rare earth elements taken from Ref. (3).

**Table 1.2.** Magnetic and Structural Properties of Rare Earths

| Property | La | Pr | Nd | Sm | Gd | Tb | Dy | Ho | Er | Tm | Lu |
|---|---|---|---|---|---|---|---|---|---|---|---|
| (3+) Ion configuration | $4f^0$ | $4f^2$ | $4f^3$ | $4f^5$ | $4f^7$ | $4f^8$ | $4f^9$ | $4f^{10}$ | $4f^{11}$ | $4f^{12}$ | $4f^{14}$ |
| $n^+$ | 0 | 2 | 3 | 5 | 7 | 7 | 7 | 7 | 7 | 7 | 7 |
| $n^-$ | 0 | 0 | 0 | 0 | 0 | 1 | 2 | 3 | 4 | 5 | 7 |
| Ion spin moment ($\mu_B$) | 0 | 2 | 3 | 5 | 7 | 6 | 5 | 4 | 3 | 2 | 0 |
| Ion moment ($\mu_B$) | 0 | 3.2 | 3.3 | 0.71 | 7.0 | 9.0 | 10.0 | 10.0 | 9.0 | 7.0 | 0.0 |
| Experimental saturation moment ($\mu_B$) | | 2.7 | 2.2 | 0.5 | 7.6 | 9.3 | 10.2 | 10.3 | 9.1 | 7.1 | |
| Curie temperature (K) | | | | | 293 | 220 | 89 | 20 | 20 | 32 | |
| Neel temperature (K) | | | 20 | 109 | | 230 | 179 | 132 | 85 | 58 | |
| Magnetic easy axis | | $a$ | $b$ | $a$ | $c+30°$ | $b$ | $a$ | $b$ | $c+30$ | $c$ | |
| Anisotropy $(10^6 \text{ ergs/cm}^2)$ $T \sim 0K$; $K_2$ | | | | | −2.1 | 900 | 870 | 660 | −97 | −300 | |
| $K^6_6$ | | 0.03 | 20 | | 0.006 | 2 | −11 | 27 | −21 | 35 | |
| Magnetostriction $(10^{-3})$ $T \sim 0K$; $\lambda^{\alpha 1}$ | | | | | 0.14 | −2.6 | | | | | |
| $\lambda^{\alpha 2}$ | | | | | −0.13 | 9.0 | | | | | |
| $\lambda^\gamma$ | | | | | 0.11 | 8.7 | 9.4 | 2.5 | −5.1 | | |

**Table 1.2.** *Continued*

| Property | La | Pr | Nd | Sm | Gd | Tb | Dy | Ho | Er | Tm | Lu |
|---|---|---|---|---|---|---|---|---|---|---|---|
| $\chi^{\epsilon}$ | | | | | 0.02 | 15.0 | 5.5 | | | | |
| Structure | dhcp | dhcp | dhcp | rhom | hcp | hcp | hcp | hcp | hcp | hcp | hcp |
| Lattice constant (Å) | | | | | | | | | | | |
| $c$ | 12.17 | 11.83 | 11.80 | 26.21 | 5.78 | 5.70 | 5.65 | 5.62 | 5.59 | 5.55 | 5.55 |
| $a$ | 3.77 | 3.67 | 3.66 | 3.63 | 3.63 | 3.61 | 3.59 | 3.58 | 3.56 | 3.54 | 3.51 |
| Density (g/cm³) | 6.15 | 6.77 | 7.01 | 7.52 | 7.90 | 8.23 | 8.55 | 8.80 | 9.07 | 9.32 | 9.84 |
| Melting temperature (°C) | 918 | 931 | 1021 | 1074 | 1313 | 1356 | 1412 | 1474 | 1529 | 1545 | 1663 |
| Elastic moduli ($T = 300$ K) | | | | | | | | | | | |
| Young's (GPa) | 36.6 | 37.3 | 41.4 | 49.7 | 54.8 | 55.7 | 61.4 | 64.8 | 69.9 | 74.0 | 68.6 |
| Bulk (GPa) | 27.9 | 28.8 | 31.8 | 37.8 | 37.9 | 38.7 | 40.5 | 40.2 | 44.4 | 44.5 | 47.6 |
| Shear (GPa) | 14.3 | 14.8 | 16.3 | 19.5 | 21.8 | 22.1 | 24.7 | 26.3 | 28.3 | 30.5 | 27.2 |
| Poisson's ratio | 0.280 | 0.281 | 0.281 | 0.274 | 0.259 | 0.261 | 0.247 | 0.231 | 0.237 | 0.213 | 0.261 |
| Yield strength (0.2%) (MPa) | 126 | 73 | 71 | 68 | 15 | | 43 | | 60 | | |
| Tensile strength (MPa) | 130 | 147 | 164 | 156 | 118 | 139 | 139 | | 136 | | |

*Note.* References used: "Handbook on the Physics and Chemistry of Rare Earths," Cumulative Index, Vols, 1–15 (K. A. Gschneidner, Jr., and L. Eyring, Eds.). North-Holland, Amsterdam, 1993; S. Legvold, Rare earth metals and alloys, in "Ferromagnetic Materials," (E. P. Wolfarth, Ed.), Vol. 1. North-Holland, Amsterdam, 1980.

occupied before the $3d$ subshell is filled. However, unlike the $4f$ subshell in the rare earths, the $3d$ subshell is not so tightly bound to the nucleus, and some of the $3d$ orbitals are spread out enough that the electrons in them participate (along with the $4s$ and $4p$ electrons) in bonding to other atoms. Which $3d$ orbitals are bonding orbitals and which are not depends very strongly on where the neighboring atoms sit. The result of this is that the magnetic properties and atomic shapes of the transition metals depend very much on crystal structure and chemistry. In most solids containing transition metals the orbital moment is nearly eliminated (or "quenched") by the environment, leaving only the ion spin moment. Iron, in its 2 + ionic state has six electrons filling the five $3d$ states. Their occupancy should be $n_+ = 5$ and $n_- = 1$ to give an ionic spin moment of 4 $\mu_B$. [Accurate quantum calculations predict spin moments near 4 $\mu_B$ for iron atoms in environments in which the bonding is weak or there are very few neighboring atoms.] However, in bulk body-centered cubic iron the measured spin moment is reduced by bonding to about 2.2 $\mu_B$ /atom. In oxides such as $Fe_3O_4$, iron atoms on different sites have different moments. Cobalt and nickel are similar to iron, with their bulk spin moments (1.7 and 0.6) somewhat reduced from the predicted 2 + ionic values (3 and 2). Table 1.3 shows the magnetic and structural properties for the three transition metals Fe, Co, and Ni, which are *ferromagnetic*, i.e., which have a net magnetic moment in the solid (4). Cr and Mn are not ferromagnetic in their bulk forms, but they do have *antiferromagnetic* states in which individual ions have magnetic moments, but these are oriented antiparallel to their neighbors giving zero net macroscopic moment. Others of the $3d$ transition metals and some of the $4d$ transition metals have an enhanced susceptibility to forming a spin moment when placed in an external magnetic field but are neither ferromagnetic nor antiferromagnetic.

## Magnetic Order and the Effect of Temperature

The last issue which determines whether a material will have an observable and usable magnetization is the nature and temperature dependence of the magnetic order. Simply having a magnetic moment on individual ions is not very useful if all the moments in a solid are oriented in different directions. In the simplest possible picture, the degree of extension and overlap of electron orbitals, which tends to make the moments in a solid less than in free atoms, also determines how much the moments on neighboring atoms "see" each other and interact. This

**Table 1.3.** Magnetic and Structural Properties of $3d$ Transition Metals

| Property | Cr | Mn-$\alpha$ | Fe | Co | Ni | Cu |
|---|---|---|---|---|---|---|
| (2 +) Ion Configuration | $3d^4$ | $3d^5$ | $3d^6$ | $3d^7$ | $3d^8$ | $3d^{10}$ (1 + ion) |
| $n^+$ | 4 | 5 | 5 | 5 | 5 | 5 |
| $n^-$ | 0 | 0 | 1 | 2 | 3 | 5 |
| Ion spin moment ($\mu_B$) | 4 | 5 | 4 | 3 | 2 | 0 |
| Experimental ion moment ($\mu_B$) | 0.5 | | 2.22 | 1.72 | 0.62 | 0 |
| Curie temperature (K) | 312[a] | 95[a] | 1044 | 1388 | 627 | |
| Magnetic easy axis | | | [100] | $c$ | [111] | |
| $K_4$ ($10^6$ erg/cm$^3$) at RT | | | 0.48 | 4.1 | 0.05 | |
| $K_6$ ($10^6$ erg/cm$^3$) at RT | | | 0.001 | 1.4 | 0.025 | |
| $\lambda_{100}(10^{-6})$ at RT | | | 24 | $\lambda^{\alpha 1} = 31$ | $-66$ | |
| $\lambda_{111}(10^{-6})$ at RT | | | $-23$ | $\lambda^{\alpha 2} = -136$ | $-29$ | |
| | | | | $\lambda^{\gamma} = -248$ | | |
| | | | | $\lambda^{\epsilon} = 57$ | | |
| Structure (Å) | bcc | cubic | bcc | hcp | fcc | fcc |
| Lattice Constant | | | | | | |
| $a$ | 2.88 | | 2.87 | 2.51 | 3.52 | 3.61 |
| $c$ | | | | 4.07 | | |
| Density (g/cm$^3$)[b] | 7.22 | 7.21 | 7.92 | 8.67 | 9.04 | 8.95 |
| Melting temperature (°C) | 1857 | 1244 | 1535 | 1495 | 1453 | 1083 |
| Elastic moduli (GPa) | | | | | | |
| E | | | | 210 | 210 | 120 |
| G | | | | | 3 | 50 |
| Yield strength (MPa) | | 240 | 250 | | 140 | 35 |
| Tensile strength (MPa) | | 500 | 400 | 237 | 500 | 220 |

[a] Antiferromagnet-Néel temperature.
[b] Calculated from X-ray data.

interaction property is called *magnetic exchange*. In those materials in which the magnetic orbitals are extended and participate in bonding (e.g., Fe, Co, and Ni) the magnetic exchange is strong, the atomic moments align themselves even in the presence of high temperatures which tend to disorder them. In the rare earth materials, in which the $4f$ orbital electrons are nearly unaware of their neighbors, the magnetic exchange is weaker and the moments align only at temperatures below room temperature. In a *ferromagnet* the atomic moments align parallel to produce a net moment, and the temperature at which ordering occurs is called the Curie temperature, $T_c$. The measured saturation magnetization, $M_S$, will be large at low temperatures and then decrease slowly with increasing temperature up to the Curie temperature, at which it decreases rapidly to zero. [Above $T_c$, even though the material is magnetically disordered in zero field, local magnetic moments may be aligned by a large applied magnetic field producing a *paramagnetic* magnetization.]

In *antiferromagnets*, such as Cr and Mn, the atomic moments align antiparallel below the ordering temperature (called the Néel temperature) and there is no net moment and thus no observable magnetization. Antiparallel ordering of atomic moments in a compound such as $TbFe_2$ produces a *ferrimagnetic* state in which the Tb moments are oriented in the opposite direction to the Fe moments but, having different magnitudes, still give a net moment leading to a macroscopic magnetization. [Because the magnetic exchange coupling between pairs of atomic moments in a solid can vary in both sign and strength, there exist other more complicated magnetic structures than the ferromagnets, antiferromagnets, and ferrimagnets discussed here.]

Other properties which depend on the magnetization, such as the magnetic anisotropy and magnetostriction described later, decrease with temperature as a function of the magnetization and also go to zero at $T_c$. The magnetostriction is predicted by theory to vary as $M^3(T)$ at low temperatures and as $M^2(T)$ at temperatures approaching $T_c$ (5). Figure 1.4 shows the temperature dependence of the single crystal magnetostriction constant, $\lambda^\gamma$, of Dy compared to the theoretical prediction (6). In Fig. 1.5, the magnetostriction of $TbFe_2$ is shown to be a good fit to the behavior predicted from the temperature dependence of the magnetization of the Tb *sublattice* rather than the net magnetization of the compound (7). In the following discussion, it is assumed that the temperature is sufficiently below $T_c$ so that its effect can be included by making the relevant coefficients temperature dependent. At a fixed temperature, $T$, the magnetization of a perfectly ordered ferromagnet is called the saturation magnetization, $M_s(T)$.

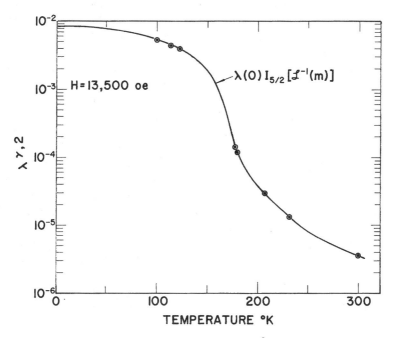

**FIGURE 1.4.**   The magnetostriction coefficient, $\lambda^{\gamma,2}$, of Dy as a function of temperature. The solid line is the temperature dependence predicted by single-ion theory (6).

## Magnetic Anisotropy and Magnetostriction

Since the discussion in the previous section explained that partially filled orbital shells lead to both magnetic moments and nonspherical atomic shapes, one may suspect that this might provide a coupling between magnetism and elasticity. Since a large part of the magnetic moment is the spin moment, this will occur only if there is a strong coupling between the *direction* of the spin moment of an atom and the *orientation* of its anisotropically shaped electron charge cloud. Such a coupling exists at the individual electron level and is called "spin-orbit coupling." Spin-orbit coupling is one of the smallest energies used to describe the state of an atom because it derives from the relativistic aspects of the electron motion (how fast the electron is traveling in its orbit relative to the speed of light) (8). For the purpose of this discussion it is sufficient to note that heavy atoms, having more positively charged protons in the nucleus and thus stronger centripetal forces, cause their electrons to move faster in a given

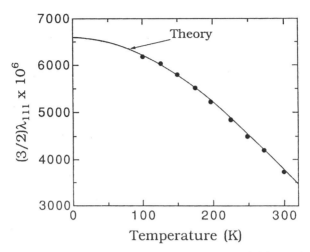

**FIGURE 1.5.** Single crystal magnetostriction, $\lambda_\parallel - \lambda_\perp = (3/2)\lambda_{111}$, for TbFe$_2$ for temperatures between 100 and 300 K. The theoretical curve is calculated from the Tb sublattice magnetization using the molecular field model, $\lambda(T) = \lambda(0)I_{5/2}[L^{-1}(m_R)]$, where $I_{5/2}$ is a normalized hyperbolic Bessel function and $L^{-1}(m_R)$ is the inverse Langevin function of the normalized Tb sublattice magnetization, $m_R$ (5, 7).

orbit than do light atoms. In rare earths the spin directions of the rapidly moving 4$f$ electrons are strongly coupled to the orientation of their orbits. This strong spin-orbit coupling of the indivual electrons leads to a strong coupling between the total spin moment and the total electron density. Thus, in the rare earths the spin moment can be envisioned as rigidly attached to the anisotropically shaped electron charge cloud.

With this coupling established the magnetic (or magnetocrystalline) anisotropy for rare earths can now be defined: It is the tendency of a magnetic moment to point in a particular crystalline direction because of the electrical attraction or repulsion between its attached electronic charge cloud and the neighboring charged ions (Fig. 1.6a). [The energy of this interaction is often called the crystal field energy.] Quantitatively, the anisotropy energy is that energy required to rotate the magnetic moment away from its preferred direction.

The magnetoelastic coupling is defined as the tendency of neighboring ions to shift their positions in response to the rotation of the magnetic moment and its rigidly attached anisotropic charge cloud (Fig. 1.6b). Quantitatively this is given by the change in elastic energy

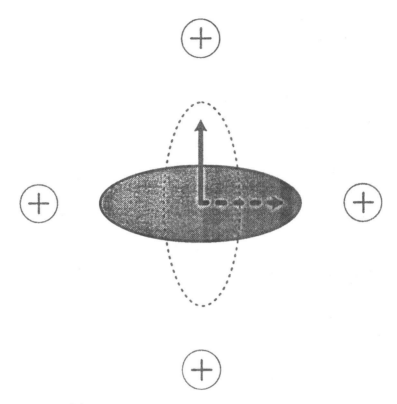

**FIGURE 1.6a.** Schematic of oblate $4f$ charge density of a rare earth element with + nearest neighbors, such as Tb, rotating in a magnetic field.

associated with a specified rotation of the magnetic moment or, conversely, the change in magnetic anisotropy energy due to a specified strain. This magnetoelastic coupling produces *Joule magnetostriction*, which is an anisotropic change in length due to the application of a magnetic field. [The coupling between the magnetic moment and the volume, which leads to *volume magnetostriction*, is usually small and has a different physical origin. It will not be considered here.]

In the magnetic $3d$ transition metals the spin-orbit coupling is weaker by approximately an order of magnitude and the local atomic picture is far too oversimplified. However, the operational definitions for magnetic anisotropy and magnetostriction remain the same.

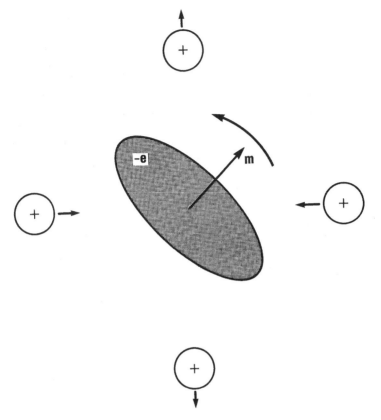

**FIGURE 1.6b.**   Schematic of the 4*f* charge density rotating in a magnetic field.

## Planar Magnet

An understanding of magnetostriction and magnetocrystalline anisotropy in single crystals has come from the phenomenological approach
(9–11) discussed in the next three sections. The magnetic anisotropy and
magnetoelastic coupling energies are given in terms of products of
powers of magnetization and strain components, with the forms of the
allowed terms dictated by the atomic symmetry of the crystal[1] (similar to
the standard formulation of elastic energies as products of stress and
strain of appropriate symmetries). The magnitudes of the coefficients of

the energy terms are unknown but may be obtained by comparisons with experimental data. [There has been some progress in calculating these coefficients from first principles (12) but this field is still in its infancy.] The symmetry-allowed energy terms may be thought of as elements of an expansion, and throughout this chapter we consider only the *lowest order* terms in the magnetic anisotropy and magnetoelastic coupling energies, i.e., those with the lowest powers of strain and magnetization. Also, all the energy terms discussed in this section are actually energy densities, which must be integrated over the volume of the sample. This integration is trivial only in the case in which the magnetizations and strains are uniform over the entire sample.

For purposes of illustration, the simplest possible system is a fictitious two-dimensional planar magnet (Fig. 1.7), elastically isotropic in the $x$–$y$ plane, with uniaxial magnetic anisotropy such that the magnetization with magnitude $M_s$ prefers to lie along the $x$-axis. [This model is a simplified version of the model used to describe the magnetoelastic response of a field-annealed metallic glass ribbon in Spano *et al.* (13)].

*Anisotropy.* For this simple system the uniaxial magnetic anisotropy energy has the form

$$E_{anis} = -K\alpha_x^2 \qquad (1.2)$$

where $\alpha_x$ is the normalized component of magnetization in the $x$ direction, $M_x/M_s$, and $K$ is the energy difference between $M$ along the $x$-axis and $M$ along the $y$-axis.

*Magnetoelastic Coupling.* The magnetoelastic coupling between the magnetization direction and the elastic strain, is assumed to be isotropic in the $x$–$y$ plane. It has the form

$$E_{me} = -b(\varepsilon_{xx}\alpha_x^2 + \varepsilon_{yy}\alpha_y^2 + \varepsilon_{xy}\alpha_x\alpha_y) \qquad (1.3)$$

---

[1]The allowed energy terms of a crystal must be invariant under all the symmetry operations of the crystal, and these terms may be worked out using group theory. For example, the proper expression for the magnetostriction for a particular symmetry can be determined by taking the direct product of the basis functions of the same irreducible representation, one function in the strain space and the other in the space of the magnetization. This method gives energy terms which not only reflect the crystal symmetry but also are orthogonal. Using orthogonal functions has the advantage that the coefficients of the lower order terms do not change as higher order terms are added to the expansion. In addition, the coefficients of these terms possess characteristic temperature dependences. More detailed discussions are given in Callen and Callen (5), Akulov (9), Becker and Döring (10), and Clark (11).

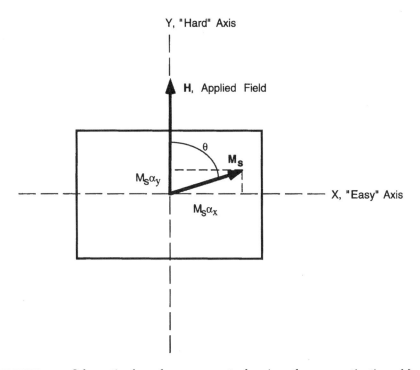

**FIGURE 1.7.** Schematic for planar magnet showing the magnetization, $M_s$, rotating in the $x$–$y$ plane away from the easy $x$-axis toward the field, $H$, applied along the $y$-axis.

where $b$ is the magnitude of the coupling.

   To reiterate, a large magnetic anisotropy derives from strong spin-orbit coupling combined with a strong interaction between a local anisotropic ionic charge cloud and the anisotropic electric field produced by its neighbors. A large magnetoelastic coupling comes from strong spin-orbit coupling plus a strong interaction between an ion charge cloud and *displacements* of its surrounding ions. Since magnetoelastic coupling energy is the strain derivative of the anisotropy, it may be large even if the anisotropy itself is small or zero.

*Magnetization Response to an Applied Magnetic Field (without Magnetoelastic Coupling).* We need to know how a material responds to an applied magnetic field in the absence of magnetoelastic coupling. The Zeeman or magnetic field energy of a body with a classical magnetization, *M*, in an applied field *H* is the dot product of these vectors:

$$E_H = \boldsymbol{M} \cdot \boldsymbol{H} \qquad (1.4)$$

[The dynamical behavior corresponding to this expression is that a field *H* applied at an angle to a magnetic moment *M* will cause the moment to precess around the field direction with no energy change. Eventual relaxation of *M* to its lowest energy state parallel to *H* is accomplished by non-energy-conserving interactions.] When a field is applied to a body the magnetization in that body will reorient its direction to lie as much along the field direction as possible.

Magnetic boundary conditions which make it unfavorable for the magnetization to point perpendicular to a surface (giving uncompensated magnetic poles on the surface) tend to break up a uniform magnetization into domains. The reorientation of magnetization in a field then can occur in two ways: rotation of the magnetization vector toward the field direction and/or expansion of domains with magnetization direction along the field. Knowing the relative contributions of these to the magnetization process is crucial to designing the optimum microstructure and macroscopic shape of a magnetostrictive material. Note that in either case, once the local magnetization vector is everywhere along the field direction, the total magnetization cannot be increased further, and it is said to be *saturated*.

In the planar magnet illustrative example the behavior of the magnetization with a magnetic field applied in the *y*-axis perpendicular to the magnetically preferred *x*-axis is given by minimizing the sum of energies

$$E_{\text{Total}} = E_{\text{anis}} + E_H = -K\alpha_x{}^2 - M_s H \alpha_y \qquad (1.5)$$

where $\alpha_y = M_y/M_s$. When *K* is positive the magnetization lies along the *x*-axis in zero field (the "easy axis"). Setting $\alpha_x = \sin\theta$ and $\alpha_y = \cos\theta$, and requiring the derivative of $E_{\text{Total}}$ with respect to $\theta$ to equal zero, gives

$$HM_s \sin\theta = 2K \sin\theta \cos\theta \qquad (1.6)$$

This equation is solved by two values of $\theta$ for each value of *H*, and the correct solution is the one that gives the lowest energy when substituted into Eq. (1.5). The correct solution at low fields is given by

$$\cos\theta = \frac{HM_s}{2K} \tag{1.7}$$

which describes the rotation of the magnetization away from the easy $x$-axis toward the applied magnetic field direction ($y$-axis). Equation (1.7) is the low-energy solution for fields up to $H = M_s/2K$. At this field the energies of the two solutions are equal, and for larger fields the solution given by $\sin\theta = 0$ is the low-energy solution, i.e., the magnetization measured along the field direction $(M_y = M_s\cos\theta = M_s)$ is a constant independent of field (saturated).

*Strain Response to an Applied Stress (without Magnetoelastic Coupling).* The purely elastic response of a material to an applied stress can be obtained by minimizing the energy as a function of strain.[2] The energy due to strain is given in terms of elastic constants, $C_{ij}$. [In a magnetoelastic material these non-magnetic elastic constants can only be measured in very high magnetic fields. The effective elastic constants measured at low or zero applied magnetic fields are softened by the magnetoelastic interaction as will be shown in the following.] For the elastically isotropic planar system, with no $z$ strain components, there are only two independent elastic constants, $C_{11}$ and $C_{12}$ [$C_{22} = C_{11}$, $C_{66} = (C_{11} - C_{12})/2$, and $C_{16} = C_{26} = 0$]. The energy due to the intrinsic elastic stiffness is given by

$$E_{elas} = \frac{1}{2}C_{11}\varepsilon_{xx}^2 + \frac{1}{2}C_{11}\varepsilon_{yy}^2 + C_{12}\varepsilon_{xx}\varepsilon_{yy} + \frac{1}{2}\left(\frac{C_{11} - C_{12}}{2}\right)\varepsilon_{xy}^2 \tag{1.8}$$

The coupling energy between strain and applied stress, $\sigma$ (in this case compression or tension applied along the $y$-axis), is

$$E_{stress} = -\sigma\varepsilon_{yy} \tag{1.9}$$

Requiring the derivatives of the sum $E_{elas} + E_{stress}$ with respect to each of the strains to equal zero gives

---

[2]The general tensor form for the elastic energy of a crystal is given by

$$E_{elas} = \frac{1}{2}\sum_{i,j,k,l=x,y,z} \varepsilon_{ij}C_{ijkl}\varepsilon_{kl}$$

The indices on the elastic constants are usually simplified according to the rule: $xx = 1$, $yy = 2$, $zz = 3$, $yz = 4$, $zx = 5$, and $xy = 6$, and the off-diagonal strains are redefined such that $\varepsilon_{ij} = 2\varepsilon_{ij}$ in the previous equation. The symmetry of the crystal determines how many of the elastic constants are independent, i.e., certain of the elastic constants may be required to be equal or zero.

$$\varepsilon_{yy} = \left(\frac{C_{11}}{C_{11}{}^2 - C_{12}{}^2}\right)\sigma = Y^{-1}\sigma$$

$$\varepsilon_{xx} = -\left(\frac{C_{12}}{C_{11}{}^2 - C_{12}{}^2}\right)\sigma \qquad (1.10)$$

$$\varepsilon_{xy} = 0$$

where $Y$ is the Young's modulus in this case.

[In this and the crystalline systems discussed in the following sections, we neglect terms in the elastic energy greater than second order in the strain because even giant magnetostrictions correspond to strains less than $10^{-2}$. This linear elastic response is sufficient to describe most materials in the small strain regime. For high strains, higher order terms are required; for materials near a structural phase transition, an expansion in powers of strain may not be sufficient. Moreover, finite elastic boundary conditions may lead to ferroelastic domains similar to magnetic domains discussed earlier. If such a material is strongly magnetoelastic the magnetic and elastic domain structures will be strongly coupled.]

*Magnetoelastic Coupled Strain and Magnetization Response to Applied Stress and Field.* The previous two sections discussed magnetic and elastic responses that were independent of each other because of the absence of magnetoelastic coupling. If we now add the magnetoelastic coupling in the form of Eq. (1.3), the magnetization response to a magnetic field will depend on the strain, and the strain response will depend on the magnetization. We will derive these coupled behaviors by considering the response of the planar magnet to applied stress or magnetic field. Gathering all the energy terms for the planar magnet, we have

$$E_{\text{Total}} = E_H + E_{\text{anis}} + E_{\text{me}} + E_{\text{elas}} + E_{\text{stress}}$$

$$E_{\text{Total}} = -HM_s\alpha_y - K\alpha_x{}^2 - b\left(\varepsilon_{xx}\alpha_x{}^2 + \varepsilon_{yy}\alpha_y{}^2 + \varepsilon_{xy}\alpha_x\alpha_y\right)$$

$$+\frac{1}{2}C_{11}\varepsilon_{xx}{}^2 + \frac{1}{2}C_{11}\varepsilon_{yy}{}^2 + C_{12}\varepsilon_{xx}\varepsilon_{yy} + \frac{1}{2}\left(\frac{C_{11} - C_{12}}{2}\right)\varepsilon_{xy}{}^2 - \sigma\varepsilon_{yy} \qquad (1.11)$$

It is obvious that if the magnetoelastic coupling did not exist ($b = 0$) the planar magnet would respond independently to field and stress, as in Eqs. (1.7) and (1.10). It is also fairly obvious that if the material is clamped (all $\varepsilon'$s $= 0$) it will respond to a field as in Eq. (1.7). It is perhaps less obvious that if the magnetization is prevented from rotating by a very large anisotropy or magnetic field the response to an applied stress will be purely elastic as in Eq. (1.10). The effect of the magnetoelastic coupling, $b$, is to allow energy from the elastic system to be transferred to the magnetic system and vice versa. It will be shown that the coupled system responds to stress with effective or renormalized elastic constants containing terms in $b$. Also, the magnetization responds to an applied field with an effective anisotropy constant, again containing the term(s) in $b$.

To obtain the coupled response we simultaneously minimize the total energy with respect to strain and the magnetization direction ($\alpha$). Differentiating Eq. (1.11) with respect to each of the strains yields the *equilibrium* strains in terms of the $\alpha'$s:

$$C_{11}\varepsilon_{xx} + C_{12}\varepsilon_{yy} = b\alpha_x{}^2$$
$$C_{11}\varepsilon_{yy} + C_{12}\varepsilon_{xx} = b\alpha_y{}^2 + \sigma \qquad (1.12)$$
$$\frac{1}{2}(C_{11} - C_{12})\varepsilon_{xy} = b\alpha_x\alpha_y$$

The *magnetostrictive strain* is the part of the strain that depends on $b$. It should be noted that these strains will not change if the direction of magnetization is reversed. (Reversing the signs of all of the $\alpha$'s does not change the equations.) The magnetostrictive strain is thus a function of $M^2$. Figure 1.8 shows the relationship between the magnetostriction and magnetization of $Tb_{0.27}Dy_{0.73}Fe_{1.95}$ *(14)*.

The *magnetostriction* is commonly determined by measuring the change in length along some direction as the magnetization rotates from perpendicular to parallel to that direction at a constant stress. If we consider the difference in $\varepsilon_{yy}$ for $M$ along $y$ ($\alpha_y = 1$, $\alpha_x = 0$) and $M$ along $x$ ($\alpha_y = 0$, $\alpha_x = 1$), we get

$$\frac{\Delta l}{l} = \Delta\varepsilon_{yy} = \frac{b}{C_{11} - C_{12}} \equiv \lambda \qquad (1.13)$$

The *magnetostriction constant*, $\lambda$, is here defined as the change in length along the magnetization direction that would be observed if the sample was initially magnetized in a perpendicular direction. From Eq. (1.13) it is clear that magnetostriction can be increased by increasing the magnetoelastic coupling, $b$, or by decreasing the elastic constants, $C = C_{11} - C_{12}$. However, the performance of magnetostrictive materials is often measured in terms of force, $\sim \lambda C$, or energy, $\sim \lambda^2 C$, which makes a reasonable elastic constant desirable.

We now determine how the magnetoelastically coupled planar magnet responds to applied stress and magnetic field and derive the *effective* elastic constants and magnetic anisotropy. In order to determine how the magnetization rotates with applied field and stress, we substitute $\alpha_x = \sin\theta$ and $\alpha_y = \cos\theta$, and set the derivative of $E_{\text{Total}}$ with respect to $\theta$ equal to zero, giving

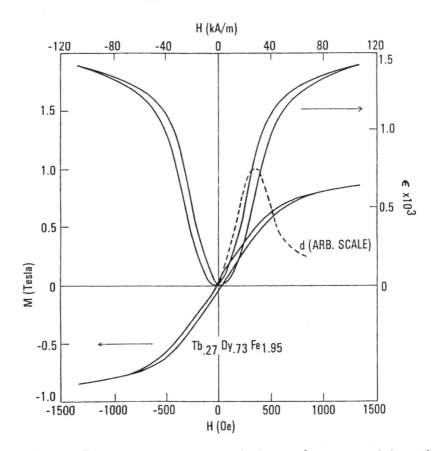

**FIGURE 1.8.** Room temperature magnetization and magnetostriction of $Tb_{0.27}Dy_{0.73}Fe_{1.95}$ at 12.4 MPa. The dotted line illustrates the magnitude of the $d$-constant which is defined by slope of the magnetostriction vs magnetic field (14).

$$0 = HM_s \sin\theta - 2K \sin\theta\cos\theta - b\left(2\left(\varepsilon_{xx} - \varepsilon_{yy}\right)\sin\theta\cos\theta + \varepsilon_{xy}\left(\cos^2\theta - \sin^2\theta\right)\right)$$

$$(1.14)$$

Inserting the equilibrium strains yields

$$0 = HM_s \sin\theta - 2K \sin\theta\cos\theta + \frac{2b\sigma}{C_{11} - C_{12}}\sin\theta\cos\theta \qquad (1.15)$$

The following two points are important. First the magnetoelastic coupling, $b$, gives a contribution to the anisotropy due to the stress. If the magnetoelastic coupling were not isotropic, it would produce an effective anisotropy even in the absence of an applied stress, as will be discussed in the Section "Magnetostriction in Cubic Materials." Again, the solutions of Eq. (1.15) are $\sin \theta = 0$ at large field or, for lower fields,

$$\cos \theta = \frac{HM_s}{2K - \frac{2b\sigma}{C_{11}-C_{12}}} = \frac{HM_s}{2K\left(1 - \frac{\lambda\sigma}{K}\right)} \tag{1.16}$$

That is, the magnetization rotates away from the easy axis toward the field direction, but now the rotation is opposed by the anisotropy, $K$, plus the magnetoelastic anisotropy, which can be either positive or negative. Comparing Eq. (1.16) with Eq. (1.7) shows that the total magnetic anisotropy is now $K' = K - b\sigma/(C_{11} - C_{12})$, with the second term being the additional anisotropy due to stress. The stress-induced magnetoelastic anisotropy is proportional to the magnetostriction times the stress. The amount of rotation of magnetization produced by a given stress increases as the ratio of magnetostriction. $\lambda$, to anisotropy, $K$. Figure 1.9a shows the magnetization along the field direction $(M_y = M_s \cos \theta)$ for a metallic glass ribbon that has been annealed to give it a uniaxial anisotropy (a system which approximates the planar magnet) and is subjected to stress (13). The data agree with the linear behavior predicted by Eq. (1.16) and show a reasonably sharp transition to the saturated state $(\sin \theta = 0)$.

It can also be seen from Eq. (1.16) that the longitudinal susceptibility, $\chi$, is proportional to the inverse of the effective anisotropy

$$\chi = \frac{\partial M}{\partial H} = M_s \frac{\partial \cos \theta}{\partial H} = \frac{M_s^2}{2K\left(1 - \frac{\lambda\sigma}{K}\right)} \tag{1.17}$$

Figure 1.9b shows the inverse susceptibility for the same metallic glass ribbon under different stresses. The inverse susceptibility decreases linearly with stress, following the decrease of the effective anisotropy, until it reaches a small limiting value, probably due to local inhomogeneities.

Second, magnetoelastic coupling also changes the effective elastic constants. Calculating $\varepsilon_{yy}$ from Eq. (1.12) gives

$$\varepsilon_{yy} = \varepsilon_{yy}^0 + \frac{C_{11}}{C_{11}^2 - C_{12}^2}\sigma - \frac{b}{C_{11} - C_{12}}\cos^2 \theta \tag{1.18}$$

[The first term is an artifact of our choice of the zero-stress state. Because it is independent of stress and field it can be neglected.] The total Young's

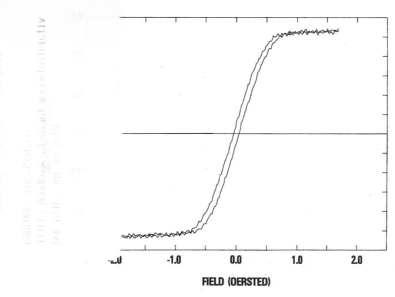

**FIGURE 1.9a.** Magnetization vs field for metallic glass ribbon at room temperature (*13*).

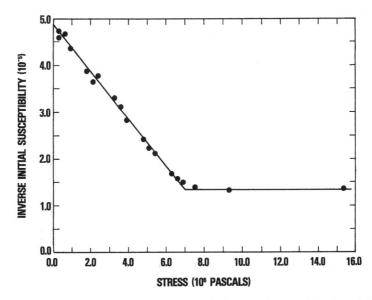

**FIGURE 1.9b.** Inverse susceptibility vs applied stress for metallic glass ribbon at room tempertature (*13*).

modulus, $Y_{tot}$, for the planar magnet model is defined by the derivative of $\varepsilon_{yy}$ with respect to the stress:

$$Y_{tot}^{-1} = \frac{C_{11}}{C_{11}^2 - C_{12}^2} - \left(\frac{HM_s}{2K}\right)^2 \frac{2\lambda^2}{K} \frac{1}{\left(1 - \frac{\lambda\sigma}{K}\right)^3} \tag{1.19}$$

The magnetoelastic contribution to the modulus depends on the applied stress (i.e, contributes to a nonlinear elastic response) and may be increased by increasing the magnetostriction or decreasing the anisotropy. In the linear elastic approximation, using the definition of the purely elastic Young's modulus, $Y$, from Eq. (1.10),

$$Y_{tot} = Y\left[1 - \left(\frac{HM_s}{2K}\right)^2 \frac{2Y\lambda^2}{K}\right]^{-1} \tag{1.20}$$

The field dependence of the elastic modulus of $Tb_{0.6}Fe_{0.4}$ is shown in Fig. 1.2. The magnetoelastic contributions to the moduli also change the sound velocity as illustrated in Fig. 1.10 for a $Tb_{0.3}Dy_{0.7}Fe_2$ polycrystal (15).

The general dependencies of the magnetoelastic properties derived here for the simple planar model remain the same for the giant magnetostriction materials discussed later.

## Isotropic Materials

Three-dimensional isotropic materials, which might be approximated by polycrystals or amorphous materials, are very similar to the planar magnet, but with $K = 0$. There is no preferred direction for the magnetization until a stress or field is applied. The elastic and magnetoelastic energies are given by

$$E_{elas} = \frac{1}{2}C_{11}\sum_{i=x,y,z}\varepsilon_{ii}^2 + C_{12}\sum_{i,j=x,y,z}\varepsilon_{ii}\varepsilon_{jj} + \frac{1}{2}(C_{11} - C_{12})\sum_{i,j=x,y,z}\varepsilon_{ij}^2$$

$$E_{me} = -b\left[\sum_{i=x,y,z}\varepsilon_{ii}\alpha_i^2 + \sum_{i,j=x,y,z}\varepsilon_{ij}\alpha_i\alpha_j\right] \tag{1.21}$$

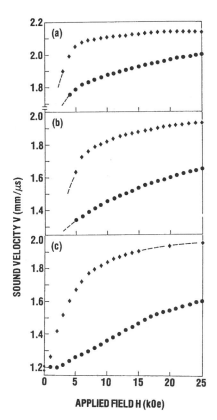

**FIGURE 1.10.**  Sound velocity of shear waves as a function of applied field $H$ for $H$ parallel ($\bullet$) and $H$ perpendicular ($\blacklozenge$) to the shear polarization in (a) $Tb_{0.3}Dy_{0.7}Fe_2$, (b) $Sm_{0.88}Ho_{0.12}Fe_2$, and (c) $Sm_{0.7}Ho_{0.3}Fe_2$ (15).

It can easily be shown, using the procedures of the preceding section, that along any measuring direction the magnetostrictive strain is given by

$$\frac{\Delta l}{l} = \frac{b}{C_{11} - C_{12}}\left(\alpha^2 - \frac{1}{3}\right) \equiv \frac{3}{2}\lambda_s\left(\alpha^2 - \frac{1}{3}\right) \qquad (1.22)$$

where $\alpha$ is the cosine of the angle between the magnetization and the measuring direction. $\lambda_s$ is the isotropic saturation magnetostriction constant, which is defined as the change of length as the material is magnetized starting from a demagnetized (random domain) state. [Inclusion of the 1/3 in Eq. (1.22) sets the zero of strain to occur at this demagnetized state, where the average of $\alpha^2$ over all domains is 1/3.]

## Designing Joule Magnetostriction Materials

The recent history of the development of magnetic materials consists in large part of efforts to combine the two groups of magnetic elements (rare earths and transition metals) together, and with other elements, to get the desired properties. One can make simple guesses as to which materials are best for a given application by examining Tables 1.2 and 1.3. Permanent magnets require high magnetic anisotropies (rare earths) but need to remain ordered well above room temperature (transition metals). The current highest performance permanent magnets combine rare earths and transition metals, e.g., $SmCo_5$ and $Nd_2Fe_{14}B$. Cost also plays a role. Since the rare earths and cobalt are expensive, permanent magnets for household use are usually iron oxides. Magnetic materials for high-frequency applications (e.g., radar) cannot tolerate the eddy current losses experienced by metals and are also usually Fe-based oxides. For applications requiring "soft" magnetic materials that readily change their direction of magnetization, low anisotropy is desired. Si-doped iron or even amorphous Fe obtained by rapid quenching with glass-forming alloying elements have minimum anisotropies. Since magnetoelastic coupling can contribute to the anisotropy, soft magnets usually require very low magnetostriction as well.

Although magnetostriction was discovered in the mid-nineteenth century (16), it was not exploited for applications until the 1900s. The intentional development of magnetic materials with large magnetostrictions began less than 30 years ago. From the discussions of the previous sections it is clear that the following are important requirements for useful magnetostriction:

1. The intrinsic magnetoelastic coupling must be large.

2. The magnetic ordering temperature (Curie temperature) must be well above the desired operating temperatures.

3. It must be possible to drive the material with a magnetic field. Since the energy change produced by a change in magnetic field is proportional to the magnetization, the larger the magnetization, the more energy may be transferred from the magnetic to elastic system.

4. One would like to change the direction of the magnetization with relatively small applied fields; thus, the coercive fields should be small. In practice, this usually requires a small magnetic anisotropy.

The tables of properties for the rare earths and transition metals indicate that rare earths provide the first and third properties but only at low temperatures. The transition metals have larger moments at room temperature and above but small magnetoelastic coupling. Thus, rare earth elements and alloys may be useful for magnetostriction at low temperatures, but for room temperature applications an alloy of rare earth and transition metals is the best choice. Because the magnetostriction of the rare earth ions is associated with the inner $4f$ shell electrons, rare earths exhibit large magnetostrictive effects in compounds with other magnetic materials and even with nonmagnetic materials. In a compound with other magnetic materials, if the exchange coupling is strong, the large magnetostrictive effects due to the rare earths will be exhibited up to high temperatures. The fourth requirement for low anisotropy may be met by adjustments in the rare earth compositions of the alloys, as discussed later. In the following sections, we discuss the development of giant magnetostriction in both the rare earth alloys and the rare earth compounds with transition metals. The magnetostrictions of some polycrystalline materials at room temperature are shown in Table 1.4 (11).

## Giant Magnetostriction in Crystalline Rare Earth Alloys

In bulk crystalline materials the form of the anisotropy and elastic energies and the magnetoelastic coupling are determined by the symmetry of the crystal structure. Once the forms of these energy terms are specified all the manipulations performed previously for the planar magnet can be performed to obtain the saturation magnetostriction and the magnetoelastic contributions to the anisotropy and elastic constants.

### Giant Magnetostriction in Hexagonal Materials

The largest known magnetostrictions in any materials are those observed in the rare earth metals at cryogenic temperatures. The magnetostrictive

**Table 1.4.** Magnetostriction of Some Polycrystalline Materials at Room Temperature

| Material | $\frac{3}{2}\lambda_s(10^{-6})$ | Material | $\frac{3}{2}\lambda_s(10^{-6})$ |
|---|---|---|---|
| Fe | $-14$ | $Ho_6Fe_{23}$ | 87 |
| Co | $-93$ | $Er_6Fe_{23}$ | $-54$ |
| Ni | $-50$ | $Tm_6Fe_{23}$ | $-38$ |
| $Fe_{0.4}Co_{0.6}$ | 102 | $Sm_2Fe_{17}$ | $-95$ |
| $Fe_{0.4}Ni_{0.6}$ | 38 | $Tb_2Fe_{17}$ | $-21$ |
| $YFe_2$ | 3 | $Dy_2Fe_{17}$ | $-90$ |
| $SmFe_2$ | $-2340$ | $Ho_2Fe_{17}$ | $-159$ |
| $GdFe_2$ | 59 | $Er_2Fe_{17}$ | $-83$ |
| $TbFe_2$ | 2630 | $Tm_2Fe_{17}$ | $-44$ |
| $TbFe_2$ (amorphous) | 462 | $YCo_3$ | 1 |
| $DyFe_2$ | 650 | $TbCo_3$ | 98 |
| $DyFe_2$ (amorphous) | 57 | $Y_2Co_{17}$ | 120 |
| $HoFe_2$ | 120 | $Pr_2Co_{17}$ | 504 |
| $ErFe_2$ | $-449$ | $Tb_2Co_{17}$ | 311 |
| $TmFe_2$ | $-185$ | $Dy_2Co_{17}$ | 110 |
| $SmFe_3$ | $-317$ | $Er_2Co_{17}$ | 42 |
| $TbFe_3$ | 1040 | $Tb_2Ni_{17}$ | $-6$ |
| $DyFe_3$ | 528 | $Fe_3O_4$ | 60 |
| $HoFe_3$ | 86 | $CoFe_2O_4$ | $-165$ |
| $ErFe_3$ | $-104$ | $NiFe_2O_4$ | $-39$ |
| $TmFe_3$ | $-64$ | $Y_3Fe_5O_{12}$ | $-3$ |

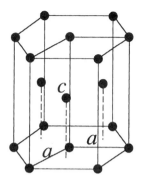

**FIGURE 1.11.**   Hexagonal unit cell.

rare earth metals crystallize in the hexagonal close-packed structure (Fig. 1.11). The four orthogonal elastic modes for cylindrical symmetry (appropriate for the hexagonal structure) are shown in Fig. 1.12.

Following the same phenomenological approach as in the planar magnet example discussed previously, each independent elastic mode is associated with an independent magnetoelastic coupling constant, with

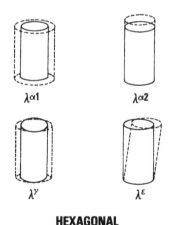

$\lambda^{\alpha 1}$          $\lambda^{\alpha 2}$

$\lambda^{\gamma}$           $\lambda^{\varepsilon}$

**HEXAGONAL**

**FIGURE 1.12.**   The four magnetostriction modes for cystals of hexagonal symmetry.

magnitudes to be determined by experiment. The magnetoelastic energy may then be written as

$$E_{me} = b_{12}\left(\alpha_z^2 - \frac{1}{3}\right)(\varepsilon_{xx} + \varepsilon_{yy}) + b_{22}\left(\alpha_z^2 - \frac{1}{3}\right)\varepsilon_{zz} + b_3\left[\frac{1}{2}(\alpha_x^2 - \alpha_y^2)(\varepsilon_{xx} - \varepsilon_{yy})\right.$$

$$\left. + \alpha_x\alpha_y\varepsilon_{xy}\right] + b_4(\alpha_z\alpha_x\varepsilon_{zx} + \alpha_y\alpha_z\varepsilon_{yz})$$

(1.23)

where $z$ is the hexagonal axis and the $\alpha_i = M_i/M_s$ are the usual Cartesian components of the magnetization. The elastic energy for a hexagonal crystal is given by

$$E_{elas} = \frac{1}{2}C_{11}(\varepsilon_{xx}^2 + \varepsilon_{yy}^2) + C_{12}\varepsilon_{xx}\varepsilon_{yy} + \frac{1}{2}C_{33}\varepsilon_{zz}^2 + C_{13}(\varepsilon_{xx}\varepsilon_{zz} + \varepsilon_{yy}\varepsilon_{zz})$$

$$+ \frac{1}{2}C_{44}(\varepsilon_{yz}^2 + \varepsilon_{zx}^2) + \frac{1}{4}(C_{11} - C_{12})\varepsilon_{xy}^2$$

(1.24)

and the anisotropy energy is

$$E_{anis} = K_2\alpha_z^2 + K_4\alpha_z^4 + K_6^0\alpha_z^6 + K_6^6\left[\left(\alpha_x^2 - \alpha_y^2\right)^3 - 12\alpha_x^2\alpha_y^2\left(\alpha_x^2 - \alpha_y^2\right)\right]$$

(1.25)

The equilibrium strains obtained by minimizing $E_{me} + E_{elas}$ are

$$\varepsilon_{xx} = \lambda^{\alpha 1}\left(\alpha_z^2 - \frac{1}{3}\right) + \lambda^{\gamma}\frac{1}{2}\left(\alpha_x^2 - \alpha_y^2\right)$$

$$\varepsilon_{yy} = \lambda^{\alpha 1}\left(\alpha_z^2 - \frac{1}{3}\right) - \lambda^{\gamma}\frac{1}{2}\left(\alpha_x^2 - \alpha_y^2\right)$$

$$\varepsilon_{zz} = \lambda^{\alpha 2}\left(\alpha_z^2 - \frac{1}{3}\right)$$

$$\varepsilon_{xy} = \lambda^{\gamma}2\alpha_x\alpha_y$$

$$\varepsilon_{yz} = \lambda^{\varepsilon}\alpha_y\alpha_z$$

$$\varepsilon_{xz} = \lambda^{\varepsilon}\alpha_x\alpha_z$$

(1.26)

where the magnetostriction constants are labeled by symmetry operation superscripts, and are defined by

$$\lambda^{\alpha 1} \equiv \frac{-b_{21}C_{33} + b_{22}C_{13}}{C_{33}(C_{11} + C_{12}) - 2C_{13}^2}$$

$$\lambda^{\alpha 2} \equiv \frac{2b_{21}C_{13} - b_{22}(C_{11} + C_{12})}{C_{33}(C_{11} + C_{12}) - 2C_{13}^2}$$

$$\lambda^{\gamma} \equiv \frac{-b_3}{(C_{11} - C_{12})} \tag{1.27}$$

$$\lambda^{\varepsilon} \equiv \frac{-b_4}{2C_{44}}$$

For example, consider an experiment which measures the change in length of a single-crystal sample along its hexagonal ($z$) axis.

$$\frac{\Delta l}{l} = \varepsilon_{zz} = \lambda^{\alpha 2}\left(\alpha_z^2 - \frac{1}{3}\right) \tag{1.28}$$

For a rotation of the magnetization from the $x$-axis in the basal plane ($\alpha_z = 0$) to the $z$-axis ($\alpha_z = 1$), we get

$$\frac{\Delta l}{l} = \lambda^{\alpha 2} \tag{1.29}$$

If, in a different experiment, the length change is measured along the $x$-axis,

$$\frac{\Delta l}{l} = \varepsilon_{xx} = \lambda^{\alpha 1}\left(\alpha_z^2 - \frac{1}{3}\right) + \lambda^{\gamma}(\alpha_x^2 - \alpha_y^2) \tag{1.30}$$

For a rotation of $M$ from $z$ to $x$,

$$\frac{\Delta l}{l} = \left[\lambda^{\alpha 1}\left(-\frac{1}{3}\right) + \lambda^{\gamma}\frac{1}{2}(1)\right] - \lambda^{\alpha 1}\left(\frac{2}{3}\right)$$

$$= -\lambda^{\alpha 1} + \frac{1}{2}\lambda^{\gamma} \tag{1.31}$$

and for a rotation from $y$ to $x$,

$$\frac{\Delta l}{l} = \lambda^{\gamma} \tag{1.32}$$

A general expression for the change in length along an arbitrary direction represented by the unit vector, $\beta$, can be obtained by substituting the equilibrium strains given by Eq. (1.26) into

$$\frac{\Delta l}{l} = \sum_{i,j} \varepsilon_{ij}\beta_i\beta_j \tag{1.33}$$

giving

$$
\frac{\Delta l}{l} = \lambda^{\alpha 1}\left(\alpha_z^{\,2} - \frac{1}{3}\right)(\beta_x^{\,2} + \beta_y^{\,2}) + \lambda^{\alpha 2}\left(\alpha_z^{\,2} - \frac{1}{3}\right)(\beta_z^{\,2})
$$

$$
+ \lambda^{\gamma}\left[\frac{1}{2}(\alpha_x^{\,2} - \alpha_y^{\,2})(\beta_x^{\,2} - \beta_y^{\,2}) + 2\alpha_x\alpha_y\beta_x\beta_y\right] + 2\alpha^{\varepsilon}(\alpha_y\alpha_z\beta_y\beta_z + \alpha_z\alpha_x\beta_z\beta_x)
$$

$$(1.34)$$

In order to measure all the fundamental magnetostriction constants in a hexagonal crystal, one must perform a series of experiments measuring the length changes along the crystal symmetry axes produced by various rotations of the magnetization, as indicated in Eqs. (1.28)–(1.32). However, the anisotropies of the rare earth elements are so huge, requiring such large fields to effect the rotations, that only a few of the single-crystal magnetostriction constants have been measured. The measured $\lambda^{\gamma}$ strains of single-crystal Dy in the hexagonal plane are shown as a function of temperature in Fig. 1.4 (6). At very low temperatures the magnetostrictive strains may be large enough to dominate the thermal expansion as shown for Dy in Fig. 1.13 (17).

The large anisotropies exhibited by the rare earth metals are undesirable for applications because they usually require very large fields to change the direction of the magnetization. [In samples that are chemically and structurally nearly perfect, in which there are few defects to impede the motion of domain walls, the magnetization direction may be changed easily by domain wall motion at relatively low fields, but large anisotropies still correlate with large hysteresis.] A more robust approach to eliminate the need for high driving fields is to develop rare earth alloys with low magnetic anisotropies which retain large magnetostrictions. Since magnetostriction is the strain derivative of the anisotropy, it is usual for large anisotropy and large magnetostriction to occur together in materials with large spin-orbit coupling. However, if the *order* (power of $\alpha$) of the magnetostrictive and anisotropy energy terms for a given crystal structure are different, there exists the possibility to cancel one and not the other.[3] Both these energies are related to the interaction of the anisotropic electron charge cloud of the rare-earth ion with its environment. Stevens

---

[3]The order of a function with respect to $\alpha$ is equivalent to the value of the angular momentum index, $l$, which indicates the number of nodes (or zeroes) of the function if the unit vector $\mathbf{a}$ experiences a full 360-degree rotation. The lowest order magnetostriction is then associated with a function of $l = 2$, as is the lowest order hexagonal anisotropy. The lowest order cubic anisotropy as discussed in the next section is $l = 4$.

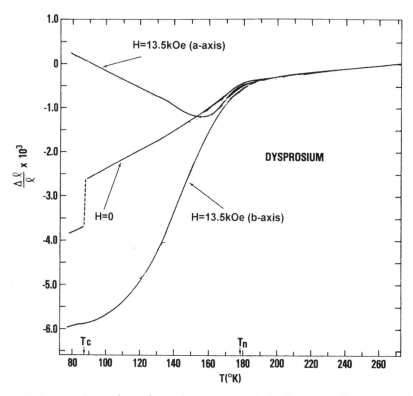

**FIGURE 1.13.** Anomalous thermal expansion of single-crystal dysprosium for applied magnetic fields of 0.0 and 13.5 kOe, applied parallel to either the $a$- or $b$-axis in the hexagonal plane (17).

(18) has shown that because of this the *signs* of the anisotropies and magnetostrictions follow well-defined relative sequences, as shown in Table 1.5, as the rare-earth species is varied in a given crystal lattice. This predictable relationship between the rare earths allows the exciting possibility of tailoring materials with desired and unusual anisotropies and/or magnetostrictions. In hexagonal materials the lowest order term in the anisotropy energy is second order in $\alpha$, and the lowest order magnetostriction constant is also second order in $\alpha$. Therefore, there is no possibility of canceling the $c$-axis anisotropy without also reducing the magnetostriction; the $c$-axis magnetostriction, $\lambda^{\alpha}$, may be achieved only by very large applied fields. However, for rare earths with $K_2 < 0$, in which

**Table 1.5.** Predicted Relative Signs of Magnetostriction and Anisotropy Constants

|  | Pr | Nd | Sm | Gd | Tb | Dy | Ho | Er | Tm |
|---|---|---|---|---|---|---|---|---|---|
| $\lambda, K_2$ | + | + | − | 0 | + | + | + | − | − |
| $K_4$ | + | + | − | 0 | − | + | + | − | − |
| $K_6, K_6^6$ | − | + | + | 0 | + | − | + | − | + |

the preferred magnetization direction is in the plane perpendicular to the hexagonal axis, the anisotropy *in the basal plane* is determined by the sixth order (in $\alpha_x$ and $\alpha_y$) term, $K_6^6$. Choosing rare earths with different signs of $K_6^6$ and the same sign of $\lambda$ (second order in $\alpha$) will produce an alloy with $K_6^6 \sim 0$, in which the magnetization direction will rotate easily in the plane under the action of a small applied field. For example, both Dy and Tb are easy-plane magnets, but the easy axis in the plane is the *a*-axis ($11\bar{2}0$) for Dy and the *b*-axis ($10\bar{1}0$) for Tb. This can be seen from the opposite signs of the sixth order anisotropy constants given in Table 1.5.

It has been shown that it is possible to alloy Tb and Dy to obtain $K_6^6 \sim 0$ *(19)* while at the same time retaining a large basal plane magnetostriction. Since Tb and Dy have different Curie temperatures, the magnitudes of $K_6^6$ vary with temperature differently for the two elements. Thus, the $Tb_xDy_{1-x}$ alloy composition required to achieve anisotropy compensation ($K_6^6 = 0$) will be different for different temperatures. In practice, one chooses the desired operating temperature and adjusts the alloy composition so that $K_6^6 = 0$ near that temperature. The alloy $Tb_{0.6}Dy_{0.4}$ has an anisotropy compensation temperature near liquid nitrogen (77 K). In addition, its anisotropy is sufficiently low for a range of temperatures around this compensation temperature to make this alloy technologically useful. The choice of Tb–Dy was also made to maximize the magnetostrictions because $\lambda$ has the same sign in the two elements. (Low-anisotropy alloys could be formed from other pairs of rare earths shown in Table 1.5 which have different signs of anisotropy, but those with opposite signs for magnetostriction will have a reduced magnetostriction due to the partial cancellation.) Figure 1.14 shows that the basal plane magnetostriction, $\lambda^\gamma$, of $Tb_xDy_{1-x}$ at low temperatures changes very little with alloy composition *(19)*. Single crystal rods of $Tb_{0.6}Dy_{0.4}$ alloy with the rod axis along the basal plane *a*-axis achieve $\lambda \sim 6.4 \times 10^{-3}$ at 77 K *(20)* (Fig. 1.15). Although the rare earth alloys exhibit giant magnetostriction at low

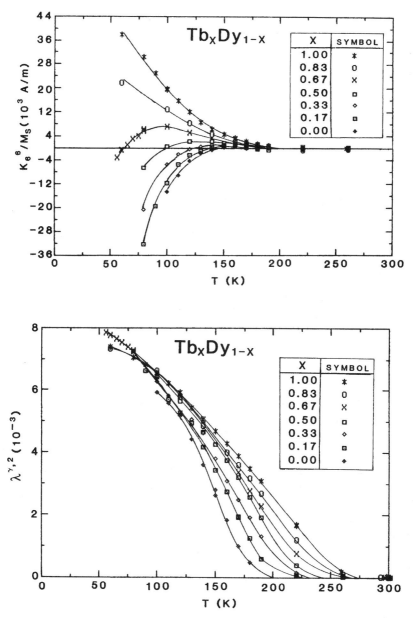

**FIGURE 1.14.** (a) Magnetocrystalline anisotropy to magnetization ratio, $K_6^6/M_s$, vs temperature for $Tb_xDy_{1-x}$ from 50 to 300 K and (b) Magnetostriction constant, $\lambda^{\gamma,2}$, vs temperature for $Tb_xDy_{1-x}$ over the same temperature range. (*19*).

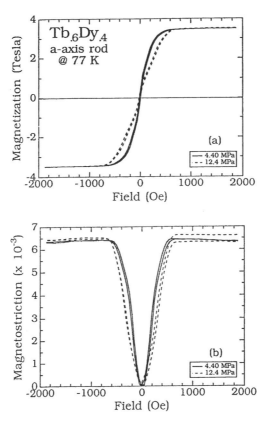

**FIGURE 1.15.** Magnetization and magnetostriction of single-crystal $Tb_{0.6}Dy_{0.4}$ along the orthohexagonal *a*-axis (20).

temperatures, they are not ideal for applications requiring very high forces because of their low yield strengths.

## Giant Magnetostriction in Cubic Materials

Following the measurement of the low-temperature magnetostrictions of the rare earth elements in the 1960s, a search was begun for alloys or compounds with high magnetostriction at room temperature, culminating in the measurement of strains exceeding $10^{-3}$ in $TbFe_2$ in 1971 (*21*). The magnetic transition metals (Fe, Co, and Ni) all have strong exchange interactions leading to high Curie temperatures and alloys of these elements with rare earths were investigated for magnetostriction at room temperature. Intermetallic compounds combining rare earths with Co or Ni behave as expected, with the transition metal-rich compounds having both the highest Curie temperatures and the lowest magnetostrictions. In the RFe compounds, however, the highest Curie temperature is achieved, surprisingly, for the most rare earth-rich composition, the $RFe_2$, as shown in Fig. 1.16 for the Tb–Fe compounds.

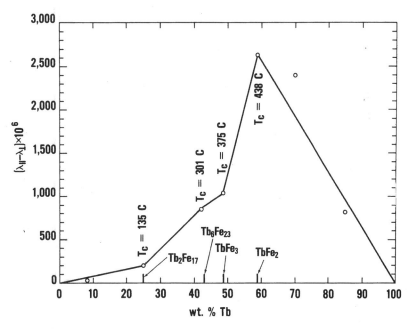

**FIGURE 1.16.** Room-temperature magnetostriction, $\lambda_\parallel - \lambda_\perp$, of polycrystalline Tb–Fe alloys for a magnetic field of 25 kOe (*21*).

**FIGURE 1.17.**    Laves phase crystal structure.

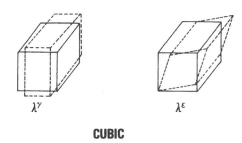

## CUBIC

**FIGURE 1.18.**   The two volume-conserving magnetostriction modes of cystals with cubic symmetry.

These $RFe_2$'s crystallize in the cubic Laves phase ($C_{15}$) structure (Fig. 1.17). The normal elastic modes for a cubic crystal are shown in Fig. 1.18 (*11*). The high cubic symmetry requires only two independent magnetoelastic coefficients. The magnetoelastic, anisotropy, and elastic energies are

$$
E_{me} = b_1\left(\alpha_x^2\varepsilon_{xx} + \alpha_y^2\varepsilon_{yy} + \alpha_z^2\varepsilon_{zz}\right) + b_2\left(\alpha_x\alpha_y\varepsilon_{xy} + \alpha_y\alpha_z\varepsilon_{yz} + \alpha_z\alpha_x\varepsilon_{zx}\right)
$$

$$
E_{anis} = K_4\left(\alpha_x^2\alpha_y^2 + \alpha_y^2\alpha_z^2 + \alpha_z^2\alpha_x^2\right) + K_6\alpha_x^2\alpha_y^2\alpha_z^2
$$

$$
E_{elas} = \frac{1}{2}C_{11}\left(\varepsilon_{xx}^2 + \varepsilon_{yy}^2 + \varepsilon_{zz}^2\right) + C_{12}\left(\varepsilon_{xx}\varepsilon_{yy} + \varepsilon_{yy}\varepsilon_{zz} + \varepsilon_{zz}\varepsilon_{xx}\right)
$$

$$
+ \frac{1}{2}C_{44}\left(\varepsilon_{xy}^2 + \varepsilon_{yz}^2 + \varepsilon_{zx}^2\right) \tag{1.35}
$$

The equilibrium strains in terms of the magnetostriction constants are

$$
\varepsilon_{ii} = \frac{3}{2}\lambda_{100}\alpha_i^2
$$
$$
\varepsilon_{ij} = 3\lambda_{111}\alpha_i\alpha_j \tag{1.36}
$$

where

$$
\lambda_{100} \equiv \frac{-2b_1}{3(C_{11} - C_{12})}
$$
$$
\lambda_{111} \equiv \frac{-b_2}{3C_{44}} \tag{1.37}
$$

The change in length along an arbitrary axis, $\beta$, is given by

$$
\frac{\Delta l}{l} = \frac{3}{2}\lambda_{100}\left(\alpha_x^2\beta_x^2 + \alpha_y^2\beta_y^2 + \alpha_z^2\beta_z^2 - \frac{1}{3}\right)
$$
$$
+ 3\lambda_{111}\left(\alpha_x\alpha_y\beta_x\beta_y + \alpha_y\alpha_z\beta_y\beta_z + \alpha_z\alpha_x\beta_z\beta_x\right) \tag{1.38}
$$

The magnetostrictions as a function of applied field for various polycrystalline $RFe_2$ compounds are shown in Fig. 1.19 (*11*). Large magnetostrictions persist in these compounds up to and higher than room temperature for two reasons. First, as noted earlier, the rare earths retain their strongly magnetoelastic behaviors in compounds because the magnetoelasticity is associated with the inner shell electrons which do not participate in the bonding. Second, in the $RFe_2$ compounds the rare earth ions are strongly exchange coupled to the Fe ions so that the rare earth sublattice remains ordered, with a large sublattice magnetization up to high temperatures. In Fig. 1.5 the magnetostriction of $TbFe_2$ was shown to decrease with temperature as the third power of the Tb sublattice magnetization, as predicted theoretically for temperatures well below the Curie temperature.[4]

In Fig. 1.19, full saturation of the magnetostriction is not achieved in DyFe$_2$ because $\lambda_{111}$ is much greater than $\lambda_{100}$, and at low fields the magnetization in DyFe$_2$ lies along the easy <100> cube edge directions with low magnetostriction. As the field is increased the magnetization rotates increasingly toward the hard <111> cube diagonal direction with larger magnetostriction. Such a difference in the magnetostriction constants of a single material is not unusual. Because these constants arise from independent strain derivatives of the magnetic anisotropy, their magnitudes may differ greatly or they may be nearly equivalent. In many cases, even the signs are different. [In the transition metals, in which the magnetism is associated with delocalized electrons, body-centered cubic Fe has opposite signs for $\lambda_{111}$ and $\lambda_{100}$, face-centered cubic Ni has $\lambda_{100}$ and $\lambda_{111}$ with the same sign but magnitudes differing by a factor of approximately 2/3, and hexagonal cobalt has $\lambda$'s with a mixture of signs and magnitudes.] Table 1.6 shows the measured single-crystal magnetostriction constants for the Laves phase compounds of rare earths with the magnetic 3$d$ elements and Al. For TbFe$_2$ and the other rare earth cubic Laves phase compounds with Fe, $\lambda_{111}$ is always much larger than $\lambda_{100}$ (11).

The sequences for the rare earth series worked out by Stevens, based on the anisotropy of the rare earth ion electron charge clouds, can be used to predict the relative *magnitudes* of anisotropies and magnetostrictions (in addition to their signs given in Table 1.5). As the rare earth species is varied within the RFe$_2$ series of compounds, the predicted magnitudes of $\lambda_{111}$ are compared with data in Table 1.7. It is clear that the Stevens sequence gives a good *qualitative* prediction of the behavior of magneto striction.

The field required to rotate the magnetization to produce the magnetostrictive strains is (again) usually large when the anisotropy is large. In cubic materials, however, the lowest order anisotropy term is fourth order, whereas the lowest order magnetostriction constant is still second order. Recall that when the order of the relevant anisotropy constant differs from that of the relevant magnetostrictive constants it

---

[4]In rare earth magnets the anisotropy and magnetostriction are usually assumed to be associated with the magnetic ordering of the individual rare earth ion moments (rather than, for example, a pair-ordering interaction), and this is referred to as a single ion mechanism. In this case, the temperature dependences of the magnetostriction and anisotropy can be related to the temperature dependence of the magnetization; they vary as a well-defined power of the magnetization which is determined by their order (in $\alpha$). The order of the function is equivalent to the quantum mechanical angular momentum index, $l$ (see footnote 3). The power is determined by taking quantum statistical averages of the appropriate function and is given in terms of $l$ by $l(l+1)/2$. Thus, the magnetostriction ($l = 2$) is predicted to vary as the third power of the rare earth magnetization at low temperatures.

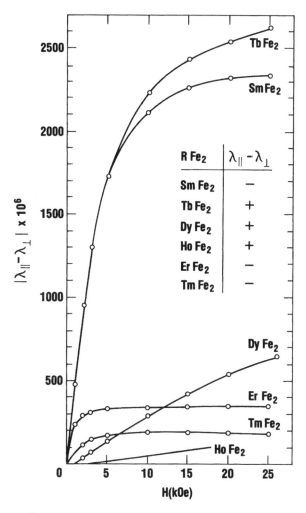

**FIGURE 1.19.** Room-temperature magnetostriction, $\lambda_{\parallel} - \lambda_{\perp}$, of rare earth–$Fe_2$ polycrystals vs magnetic field (*11*).

becomes possible to tailor the anisotropy by alloying the rare earth component. Using Table 1.5, it is clear that compounds containing both Tb and Dy, or Sm and Er, will result in a material with large (same-sign) magnetostrictions and a completely or partially canceled (different signs)

**Table 1.6.** Magnetostriction of Cubic Laves Phase Compounds with Rare Earths at $T = 0$

| Compound | $\lambda_{111}(10^{-6})$ | $\lambda_{100}(10^{-6})$ | $T_c$ (K) |
|---|---|---|---|
| $NdAl_2$ | — | $-700$ | 61 |
| $TbAl_2$ | $-3000$ | — | 114 |
| $DyAl_2$ | — | $-1700$ | 68 |
| $TbMn_2$ | $-3000$ | — | 40 |
| $TbFe_2$ | 4000, 4500 | — | 711 |
| $DyFe_2$ | — | $-70$ | 635 |
| $HoFe_2$ | — | $-750$ | 612 |
| $TmFe_2$ | $-3500, -2600$ | — | 610 |
| $TbCo_2$ | 4400 | — | 256 |
| $DyCo_2$ | — | $-2000$ | 159 |
| $HoCo_2$ | — | $-2200$ | 85 |
| $ErCo_2$ | $-2500$ | — | 36 |
| $TbNi_2$ | 1500 | — | 45 |
| $DyNi_2$ | — | $-1300$ | 30 |
| $HoNi_2$ | — | $-1000$ | 22 |

**Table 1.7.** Magnitudes of Single-Crystal Magnetostriction in Rare Earth–$Fe_2$ Compounds

| Compound | $\frac{3}{2}\lambda_{111}(10^{-6})$ (calculated at 0 K) | $\frac{3}{2}\lambda_{111}(10^{-6})$ (measured at room temperature) | $T_c$ |
|---|---|---|---|
| $SmFe_2$ | $-4800$ | $-3150$ | 676 |
| $TbFe_2$ | 6600 | 3690 | 697, 711 |
| $DyFe_2$ | 6300 | 1890 | 635 |
| $HoFe_2$ | 2400 | 288 | 606 |
| $ErFe_2$ | $-2250$ | $-450$ | 590, 597 |
| $TmFe_2$ | $-5550$ | $-315$ | 560 |

anisotropy. Alloys such as $Tb_{1-x}Dy_xFe_2$(TERFENOL-D) and $Sm_{1-x}Er_xFe_2$ (Samfenol-E) have been shown to have the desired high magnetostriction-to-anisotropy ratio. TERFENOL-D with $0.70 < x < 0.73$ has an anisotropy compensation temperature approximately at room temperature. [It is also possible to add a third rare earth to eliminate higher order anisotropy constants, as was shown by adding Ho to TERFENOL-D (22).]

Anisotropy compensated compounds of this type, when they also have highly unequal magnetostriction constants (e.g., opposite signs of $\lambda_{100}$ and $\lambda_{111}$), may exhibit spectacular temperature dependences. Unusual temperature behavior for magnetostrictrions and thermal expansions are shown in Figs. 1.20 (23) and 1.21 (24). For TERFENOL-D, the <100> axes are easy at low temperatures (below the anisotropy compensation temperature) and the <111> axes are easy at higher temperatures. Since $\lambda_{111} \gg \lambda_{100}$, the magnetostriction increases at the compensation temperature and then decreases in the normal way with increasing temperature, giving the unusual peaks in magnetostriction shown in Fig. 1.20. Note also that the magnetostriction is large near room temperature at low applied fields. Because the strong magnetic coupling persists to high temperatures, the magnetostrictions also remain large well above room temperature (Fig. 1.22) (25).

Unequal magnetostriction constants allow a magnetoelastic contribution to the magnetocrystalline anisotropy, even in the absence of stress. For cubic materials such as TERFENOL-D, the contribution to $K_4$ can be written as

$$K_4' = K_4 + \frac{9}{2}\left[\lambda_{100}{}^2\left(\frac{C_{11} - C_{12}}{2}\right) - \lambda_{111}{}^2 C_{44}\right] \qquad (1.39)$$

In the case of isotropic magnetostriction ($\lambda_{111} = \lambda_{100}$) and elasticity ($(C_{11} - C_{12})/2 = C_{44}$), the term in the square brackets equals zero, just as it did for the planar magnet example. However, for TERFENOL-D, where $\lambda_{111} \gg \lambda_{100}$ and $K_4$ is small near the compensation temperature, the magnetoelastic anisotropy is a large part of the anisotropy. [In hexagonal systems the lowest order anisotropy constant is not affected by the magnetoelastic coupling.]

Large differences in the magnetostriction constants also produce unusual behavior in polycrystals. Texturing becomes crucially important,

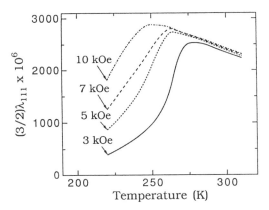

**FIGURE 1.20.**   Magnetostriction, $\lambda_{\parallel} - \lambda_{\perp} = (3/2)\lambda_{111}$, of $Tb_{0.27}Dy_{0.73}Fe_2$ at various magnetic fields from 220 to 300 K (23).

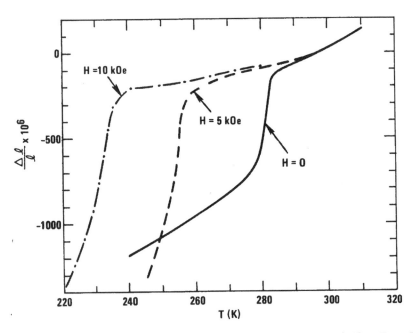

**FIGURE 1.21.**   Thermal expansion along [111] in single-crystal $Tb_{0.27}Dy_{0.73}Fe_2$ (24).

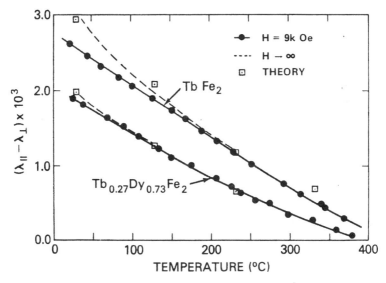

**FIGURE 1.22.** Magnetostriction, $\lambda_\parallel - \lambda_\perp$, of TbFe$_2$ and Tb$_{0.27}$Dy$_{0.73}$Fe$_2$ from 20 to 380°C (25).

and when stresses are applied along nonprincipal axes (in individual crystallites), the observed strains are not collinear with the stress.

A class of cubic compounds with large magnetostrictions at cryogenic temperatures, comparable to the hexagonal rare earths, are the intermetallic RZn compounds in which R = Tb or Dy or a combination (26). These (Tb$_x$Dy$_{1-x}$)Zn compounds have a simple chemical structure (CsCl) and are easy to grow in highly magnetostrictive single-crystal orientations, but they have a complicated magnetic phase diagram which is still not complete (Fig. 1.23). In contrast to the RFe$_2$ compounds in which $\lambda_{111} > \lambda_{100}$, in the RZn compounds $\lambda_{100} > \lambda_{111}$. Since $\lambda_{100}$ is large, it is desirable to operate in the <100> easy part of the phase diagram. The magnetostrictive properties under stress, which are important for transduction, are shown in Fig. 1.24 for TbZn at 77 K (27). Note that the saturation magnetostriction is comparable to that of the TbDy hexagonal alloy at this temperature. This magnetostriction combined with a reasonable elastic modulus makes this material a candidate for transduction at low temperatures. An advantage of the RZn compounds compared to the TbDy alloys is their higher yield strength. TbZn has been operated up to 100 MPa without degradation.

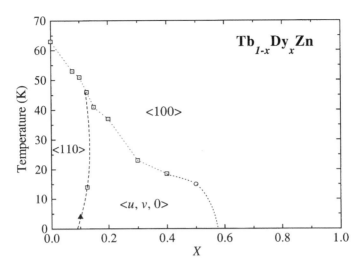

**FIGURE 1.23.** Easy axis magnetic phase diagram for $Tb_{1-x}Dy_xZn$. The open squares represent the lowest points of the sharp dips in the modulus vs temperature; the open circle represents the results from magnetization vs temperature data; and the solid triangle is estimated from the drop in the elastic modulus for $x = 0.1$ (*26*).

## Materials of Other Symmetries

Most highly magnetostrictive materials investigated to date crystallize in either the cubic or the hexagonal structures. In addition to those discussed previously, these also include $RFe_3$, which is hexagonal, and $R_6Fe_{23}$, which is cubic. The $R_2Fe_{17}$ compounds may be either hexagonal or rhombohedral. The fabrication of magnetic materials as epitaxial thin films on substrates often leads to strains due to the mismatch of lattice constants in the film and substrate. When these strains are large the magnetic film should be considered as a material with a lower symmetry. For example, a nominally cubic material grown as a (100) or (111) film will have tetragonal or trigonal symmetry, as discussed in Cullen *et al.* (*28*).

## Implications for Applications of Giant Magnetostriction Materials

The fundamental physical mechanisms underlying magnetostriction, as discussed in the preceding sections, give rise to useful properties for

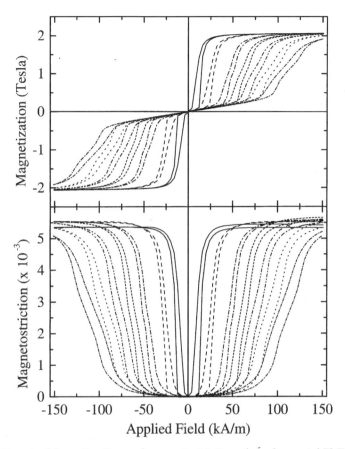

**FIGURE 1.24.** Magnetization and magnetostriction of single-crystal TbZn at 77 K along the [100] axis for compressive stresses from 5.3 to 50.5 Mpa (27).

applications that will be considered in other chapters. A brief summary of the materials with the largest magnetostrictions and their temperature ranges of operation are shown in Table 1.8. Here, we summarize how the fundamental properties of magnetostrictive materials impact their application:

1. Driving a magnetostrictive material with an applied magnetic field produces a rotation (or domain alignment) of the magnetization which saturates at some large field, resulting in a saturation Joule magnetostriction that cannot be exceeded by applying additional field.

2. The magnetostriction is independent of the sign of the applied field. A piezomagnetic response, i.e., a strain that changes sign with reversal of the applied field, can be achieved by applying a static bias field which is larger than the driving field.

3. Currently-developed magnetostrictive materials are not poled, as is done with piezoelectrics, to eliminate the need for a bias field. Piezoelectric ceramics make use of free charge to stabilize a domain configuration during poling, but there are no equivalent free magnetic poles. The very large coercivities which allow poling of permanent magnets derive from large anisiotropies which are undesirable in magnetostrictive materials.

4. The fields required to achieve large magnetostrictions may be minimized by compensating the magnetic anisotropy, but only over a limited range of temperatures. Different rare earth compositions may be optimized for different operating temperatures. Operation over a wide range of temperatures may require higher operating fields than that near the compensation temperature.

5. Materials with large magnetoelastic coupling also exhibit the reciprocal effect to magnetostriction—a large change in the magnetization due to stress.

6. The magnetoelastic coupling contributes to both the magnetic anisotropy and the elastic constants because energy is transferred between the magnetic and elastic systems due to the coupling.

**Table 1.8.**   Giant Magnetostriction

| Material | Temperature range | Magnetostriction (%) |
|---|:---:|---|
| $Tb_{1-x}Dy_x$, $(Tb_{1-x}Dy_x)Zn$ | 0–10 K | 0.9 |
| | 70–100 K | 0.6 |
| $RFe_2$ | −30–80 C | 0.2–0.3 |
| | 150–250 C | 0.1 |

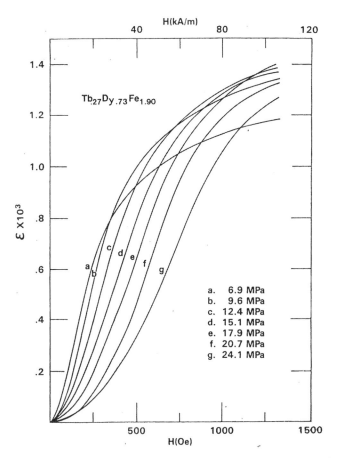

**FIGURE 1.25.** Magnetostriction of $Tb_{0.27}Dy_{0.73}Fe_2$ vs magnetic field under compressive stresses from 6.9 to 24.1 MPa (*14*).

7. In applications, a compressive bias stress is often applied. When work is performed by the magnetostrictive system, the magnetostrictive material must function under a compressive load. As the compressive load is increased, more mechanical energy is stored, a higher magnetic energy is needed, and thus a higher magnetic field must be applied. Figure 1.25 illustrates how the strain vs field curves depend on stress for TERFENOL-D (*14*).

## METALLURGY AND MICROSTRUCTURES OF GIANT MAGNESTOSTRICTIVE MATERIALS

The ideal room-temperature giant magnetostrictive transducer element would be a defect-free, single-crystal of TERFENOL-D ($Tb_{0.3}$ $Dy_{0.7}$ $Fe_2$) with the [111] crystallographic direction of the crystal aligned along the drive axis of the transducer element. This definition is the result of extensive research beginning in the 1960s with the measurement of the exceptionally high magnetostrictions of Tb and Dy at low temperatures [reported by Clark (29)]. This section deals with the metallurgy of the alloy systems used to develop the processes that yield these giant magnetostrictive materials in a practical form which approaches this ideal description and is useful for room-temperature transduction.

A major advance in magnetostrictive technology was accomplished with the low-temperature measurements of the basal plane magnetos-trictions of hexagonal (Tb) and dysprosium (Dy) rare earth metals. The basal plane projection is shown in Fig. 1.26. Strains approaching 1% (10,000 ppm) were observed which, in comparison to the 40- to 70-ppm room-temperature strains of the magnetic transition metals Ni, Co, and Fe, gave rise to the giant descriptive for these magnetostrictive materials. A second major advance occurred in the 1970s as a result of a systematic study to determine the room-temperature magnetostrictions of the intermetallic compounds formed between these heavy rare earths (Tb and Dy) and the magnetic transition metals (Ni, Co, and Fe). The highest magnetostrictions were found in the $RFe_2$ compounds, which crystallize in the Laves phase face-centered cubic (FCC) C15 structure type.[5] Of these $RFe_2$ compounds, $TbFe_2$ and $SmFe_2$ were found to possess the largest room-temperature magnetostrictions, $|8| > 2000$ ppm, positive and nega-tive, respectively. Figure 1.26 shows the results of these studies for the Tb–Fe system. Both compounds have high Curie temperatures ($T_c \sim 700$ K) and were considered to be prime candidates for use in devices operating over a wide temperature range including ambient.

### TERFENOL-Based Transducer Element Development

$TbFe_2$, TERFENOL,[6] became the overwhelming choice for development for use as drive elements in transducers in the 1980s and 1990s. The

---

[5]Strukurbricht nomenclature for this family of intermetallic compounds that crystallize C15, $MgCu_2$-type structure.

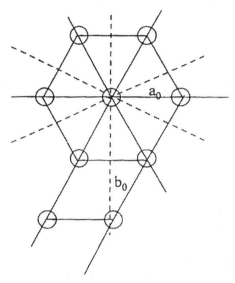

**FIGURE 1.26.** Schematic of the basal plane projection for hexagonal-close-packed (HCP) c-axis perpendicular to the plane of paper. HCP unit cell is extended to redefine crystal structure in the lower symmetry orthorhombic system; $a_0 \neq b_0 \neq c_0$.

largest positive magnetostrains at room temperature occur when Tb is a major constituent. It was recognized in earlier studies that two huge anisotropies exist in $TbFe_2$. The first is the anisotropy in the magnetization, which means that large, impractical fields are required to realize the huge magnetostrictive strains in $TbFe_2$.

## Magnetization Anisotropy

Magnetization anisotropy can be totally compensated by substitution of other rare earths with same-sign magnetostriction and opposite-sign anisotropy constants of those for $TbFe_2$ (see Table 1.5). The candidates to accomplish this anisotropy minimization in TERFENOL are Ce, Pr, Nd, Dy, and Ho. The selection of which candidate to use has a significant impact on the solidification of the final alloy.

[6]TERFENOL is an acronym: Ter, terbium; Fe, iron; and nol, Naval Ordinance Laboratory (now NSWCCD, Naval Surface Center Carderock Division).

Dy was found to be the most effective rare earth (R) to reduce the magnetization anisotropy in TERFENOL while producing a minimal reduction in strain. Consequently, the psuedobinary family of alloys TERFENOL-D (D = dysprosium), $Tb_x Dy_{1-x} Fe_2$, became the giant magnetostrictive material for further development in the 1980s and 1990s. For $x = 0.27$, the anisotropy is minimized and the sacrifice in strain, for example, $\frac{3}{2}\lambda_{111} = 3690$ ppm for Tb Fe$_2$ and $\frac{3}{2}\lambda_{111} = 2460$ ppm for $Tb_{0.27} Dy_{0.73}$ Fe$_2$, is justified since the fields required to produce these strains are reduced from the impractical 25 kOe for TbFe$_2$ to a practical drive field of <2 kOe for $Tb_{0.27}$ Dy$_{0.73}$ Fe$_2$.

Dy is the immediate neighbor of Tb in the rare earth series and is nearly of the same atomic size. Tb and Dy form a complete solid solution with a linear relationship between both the melting points and the high-temperature allotropic HCP to body-centered cubic transformations in these metals (Fig. 1.27) (30). TERFENOL-D can be considered a true psuedobinary intermetallic compound for all values of $x$ in $Tb_x Dy_{1-x} Fe_2$ and can be represented as RFe$_2$. RFe$_2$ has predictable metallurgical properties, including melting points, modes of formation, and phase relationships based on the phase diagrams for Tb–Fe (Fig. 1.28) (31) and Dy–Fe (Fig. 1.29) (32) binary alloy systems. Reasonably accurate psuedo-binary phase diagrams can be constructed for these R–Fe systems, where $R = Tb_x Dy_{1-x}$.

In both the Tb–Fe and Dy–Fe binary systems, TbFe$_2$ and DyFe$_2$ form by peritectic reactions as shown in Fig. 1.29 for DyFe$_2$. The three-phase microstructure for TbFe$_{1.99}$ shown in Fig. 1.30 is typical for this peritectic reaction as shown in Fig. 1.28. Upon cooling from the liquid, TbFe$_3$ dendrites freeze first, TbFe$_2$ freezes next in contact with TbFe$_3$, and the darker eutectic phase freezes last. To obtain single-phase TbFe$_2$, a long-term heat treatment is necessary to convert the TbFe$_3$ phase to TbFe$_2$ by reaction with the eutectic phase. Single-phase and single-crystal TbFe$_2$ can be prepared from the melt but the conditions for obtaining these materials are unreasonable for consideration for large-volume production. An example is the single-crystal of TbFe$_2$ shown in Fig. 1.31 which was grown in a cold crucible (containerless) apparatus with growth rates of about 1 mm/hr (33). For DyFe$_2$ the peritectic reaction is close to the liquidus temperature and the two-phase, liquid plus RFe$_3$ region is narrow in this area of the diagram (Fig. 1.29). Under proper conditions, DyFe$_2$ forms as if it were a congruent melting compound. Since TERFENOL-D, $(Tb_x Dy_{1-})Fe_2$ (where $x = 0.27$-0.3) was the target stoichiometry for commercial development, the processes to obtain single-phase materials are feasible since the solidification is more like that

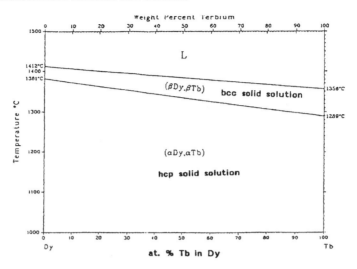

**FIGURE 1.27.** Dysprosium–terbium (Dy–Tb) phase diagram based on information from Ref.(2), not determined experimentally.

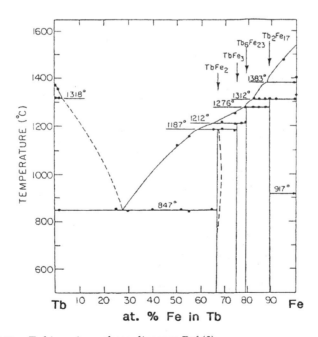

**FIGURE 1.28.** Terbium–iron phase diagram Ref.(3).

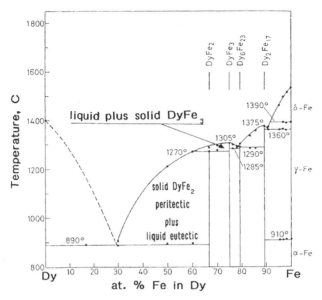

**FIGURE 1.29.** Dysprosium–iron phase diagram Ref. (4) DyFe$_2$ forms peritectically from the liquid and solid DyFe$_3$ region at 1270°C.

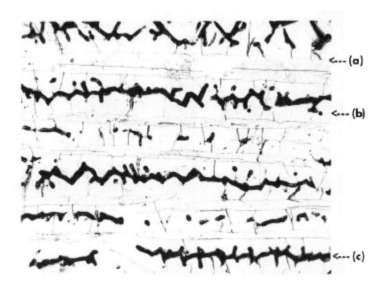

**FIGURE 1.30.** TbFe$_{1.99}$ - microstructure typical to peritectic formation. TbFe$_3$ (a) dendrites surrounded by TbFe$_2$ (b) in a eutectic matrix (c) dark phase. Magnification, × 100.

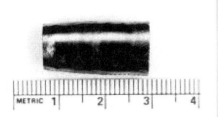

FIGURE 1.31. Single-crystal $TbFe_2$ grown in a horizontal cold crucible. The crystal is between 1 and 2.5 cm; growth rate, ~ 1 mm/hr.

experienced for $DyFe_2$ and reasonable production rates (>250 mm/hr) are expected.

It was recognized early that stoichiometric TERFENOL-D, $RFe_2$, is quite brittle and it often failed during machining, testing, or incorporation into devices. The strength/toughness can be improved by shifting the $RFe_2$ stoichiometry to lower iron contents, such as $RFe_{1.8}$, as reported by Peterson et al. (34). As shown is Fig. 1.32, the more ductile rare earth rich eutectic phase (dark) forms between the TERFENOL-D dendrites (light), and this matrix serves to markedly increase the toughness of the material to the user-friendly level. The sacrifice in strain due the stoichiometry shift is minimal and the strain as a function of Fe content in TERFENOL-D is reported by Verhoeven et al. (35). Table 1.9 shows how magnetostriction varies with iron content for $RFe_{1.9-1.95}$. The trade-off stoichiometry for TERFENOL-D transducer element development became $RFe_{1.9-1.95}$ and most of the material studies of the past decade are based on this stoichiometry range for iron.

**Table 1.9.** Magnetostriction (in ppm) for Different R:Fe Ratios of TERFENOL-D Measured at 13.5 MPa Prestress[a].

|  | Oe | | | |
| --- | --- | --- | --- | --- |
|  | 500 | 1000 | 1500 | 2000 |
| $Tb_{0.3}Dy_{0.7}Fe_{1.95}$ | 1084 | 1393 | 1544 | 1643 |
| $Tb_{0.3}Dy_{0.7}Fe_{1.92}$ | 1118 | 1358 | 1536 | 1678 |
| $Tb_{0.3}Dy_{0.7}Fe_{1.90}$ | 1054 | 1368 | 1552 | 1658 |

[a] Eight values at 500, 1000, 1500, and 2000 Oe applied filed, $H$.

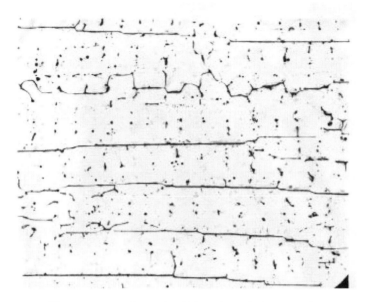

FIGURE 1.32. Transverse section of <112>-oriented free-standing zone melt "single crystal" $Tb_{0.3}Dy_{0.7}Fe_{1.95}$ sheet dentrites (light phase) separated by thin layers of eutectic (dark) phase. Magnification, × 80.

## Magnetostriction Anisotropy

A second huge anisotropy exists in the magnetostriction of TERFENOL-D. The strain is $\lambda_{111} \gg \lambda_{100}$ in these FCC Laves phase compounds and achieving the maximum strain in a transducer element is again a metallurgical problem. For TERFENOL-D, maximum strain is obtained when the <111> crystallographic direction is aligned along the drive or rod axis of the tranducer element. Although the <111> alignment has been achieved in some crystal growths by the Czochralski, Bridgman, and zone melt methods, such crystals were small, usually quite brittle, and the result of impractical, slow growth rates. Schematics depicting these types of crystal growth are shown in Fig. 1.33. The free-standing zone melt (FSZM) method was also used to produce grain-aligned rods of TERFENOL-D, $Tb_x Dy_{1-x} Fe_{1.9-1.95}$, where $x = 0.27$-0.3. This container-less method minimizes the process contamination of the material by avoiding any reaction between the materials and a container.

During the early attempts to obtain the preferred <111> direction along the rod axis it was realized that the natural growth direction of these materials is a <112> crystallographic direction. Using <112> seeds at the

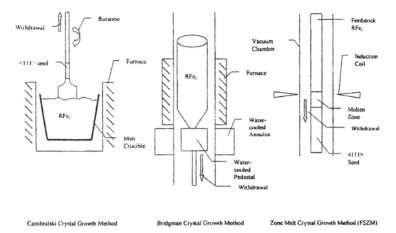

FIGURE 1.33. Schematics of Czochralski, Bridgman, and free-standing zone melt (FSZM) crystal growth methods. The FSZM method is containerless, and molten metal is not in contact with a container.

start of the zone melt operation it became fairly easy to produce single-crystal TERFENOL-D rods with a <112> along the growth axis at growth rates up to 500 mm/hr. Such growth rates (500 mm/hr) led to the development of the FSZM to produce large quantities of high-performance TERFENOL-D transducer elements with diameters up to 8 mm and lengths up to 300 mm. The crystal growth is the dendritic solidification of TERFENOL-D in the form of sheet dendrites which have a <111> crystallographic direction perpendicular to the plane of the sheets and orthogonal to a <112> growth axis. Figure 1.34 shows these <112>-oriented sheet dentrites in the cross section of a FSZM $Tb_{0.3} Dy_{0.7} Fe_{1.95}$ rod. Unfortunately, each sheet dendrite contains two or more parallel {111} twin planes which form during the solidification. Figure 1.35 shows these twin boundaries that constitute major crystalline defects which inhibit domain wall movement and thereby result in lower magnetostrictions in these twined single-crystal transducer elements at low applied fields. The strain losses are discussed on the basis of the crystallographic directions in the parent and twin portions of the crystal on either side of the twin boundary (36, 37).

Although some low-field strain losses are inherent in these <112>-twinned single-crystal TERFENOL-D drive elements, the FSZM method used to produce them proved to be a powerful research tool in further developments of these giant magnetostrictive materials. The consistency, repeatability, and relative ease of growing the "crystals" facilitated many

**FIGURE 1.34.** Cross section of <112>-oriented FSZM single-crystal TERFENOL-D. Metric scale, ~ 7 mm diameter; <111> perpendicular to sheet dendrites and <110> orthogonal.

studies in the 1980s and 1990s aimed at optimization of the magnetostrictive properties to meet the requirements of the seemingly unlimited number of applications and conditions. Clark *et al.* (*38*) provide an example in which the anisotropy compensation temperatures of $Tb_x \, Dy_{1-x}(Fe_{1-y}T_y)_{1.9}$ [$T = $ Co, Mn$(0.3 < x < 0.5)(0 < y < 0.3)$] were determined from magnetization and magnetostrictive measurements as a function of applied field, prestress, and temperature. The samples were

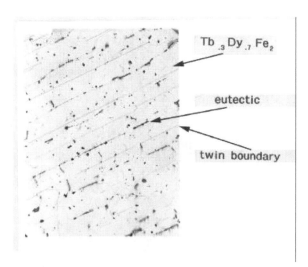

**FIGURE 1.35.** {111} twin boundaries formed at the center of each {111} sheet dendrite during the <112> dendritic solidification of (FSZM) of TERFENOL-D.

prepared by the FSZM method and in most cases were <112>-twinned single crystals. The metallurgical similarity of these rods lends credence in using these results to select values of $x$, $y$, and $T$ that yield maximum magnetostrictive performance over a wide temperature range.

Figures. 1.36 and 1.37 show typical strain curves for these high-performance TERFENOL-D transducer elements prepared by the FSZM method. The effect of prestress on strain shown in Fig. 1.36 depicts the load needed to orient the domains perpendicular to the drive axis before the field is applied. The maximum low-field strain is realized with prestress between 7 and 14 MPa when the applied field causes the perpendicular magnetization to rotate 90° toward the rod axis. For this reason, TERFENOL-D drivers are by design in compression when used in transducer applications. Figure 1.37 shows the effect of this mechanical bias on the strain at higher prestresses. From such a family of curves the portion of magnetization energy converted to work can be determined (39). The design engineer can match the requirements of the application to the field and load conditions used to generate these performance curves. These curves also provide the magnetic bias field as a function of prestress relationship which is conditional to the optimization of the energy conversion in many applications. The magnetization curves corre-

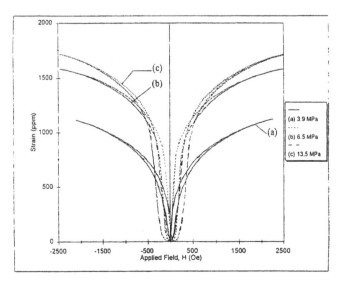

**FIGURE 1.36.** Magnetostriction of $Tb_{0.3}Dy_{0.7}Fe_{1.95}$ at 3.9, 6.5, and 13.5 MPa prestress.

sponding to the strain curves in Fig. 1.37 are shown in Fig. 1.38. These curves provide the hysteresis and permeability characteristics required by the applications engineer.

As shown in Figs. 1.36 and 1.38, TERFENOL-D drivers exhibit large magnetostriction jumps at low prestress and low fields. The model developed by Clark (11) predicts that this magnetization "jump" from the <111> perpendicular to the drive axis to the <111> oriented 19.5° from the <112> rod axis will result in a strain of 1067 ppm. In the absence of twins the strain would nearly double. In Fig. 1.39 the strain curve for FSZM $Tb_{0.3}Dy_{0.7}Fe_{1.92}$ at 6.5 MPa prestress is plotted and the projected curve based on the modelm (39) for untwinned <112>-oriented material is shown. The difference in strain due to the presence of this major crystal defect serves as incentive to continue the search to develop processes that yield the untwinned crystals for use in transducer applications. The projected strain curve for twin-free <111>-oriented TERFENOL-D is also shown in Fig. 1.39.

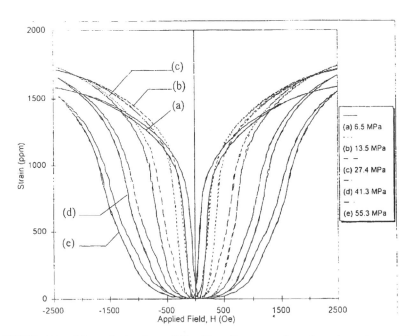

**FIGURE 1.37.**   Magnetostriction of $Tb_{0.3}Dy_{0.7}Fe_{1.95}$ as a function of prestress; 6.5–55.3 MPa.

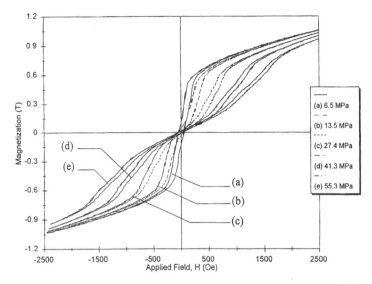

**FIGURE 1.38.** Magnetization curves corresponding to strain curves in Fig. 1.35 for $Tb_{0.3}Dy_{0.7}Fe_{1.95}$.

In Fig. 1.40 performance curves for FSZM twinned $Tb_{0.27}$ $Dy_{0.73}$ $Fe_{1.92}$ (fully compensated TERFENOL-D) are shown as a function of prestress. Magnetostriction jumps do not occur in these alloys and less hysteresis is present since the magnetization anisotropy is near zero for this stoichiometry. The result is a more linear approach to saturation. The corresponding magnetization versus field curves are shown in Fig. 1.41. Another crystalline defect that has a profound effect on the strain performance is shown in Fig. 1.42. This defect is a precipitate of $RFe_3$ in the $RFe_2$ grains referred to as Widmanstatten after the researcher who first observed this feature. At high temperatures, Widmanstatten precipitates in distinct crystallographic orientations. $RFe_2$ exhibits a solid solution range at elevated temperatures shown by the dotted line in Fig. 1.43 for $TbFe_2$. Solid solubility of $RFe_3$ in $RFe_2$ was found on the rare earth-rich side of $Tb_{0.27}$ $Dy_{0.73}$ $Fe_2$ (40). Widmanstatten precipitate has been observed in the $RFe_2$ phase of the alloys with compositions ranging from 65 to 70 at.% Fe in numerous binary, quasi-binary, and other specialty alloy systems of magnetostrictive importance. These crystalline defects inhibit domain wall motion and the magnetostrictive performance is greatly diminished as reflected by reduced strain, greater hysteresis, and lower $d_{33}$.

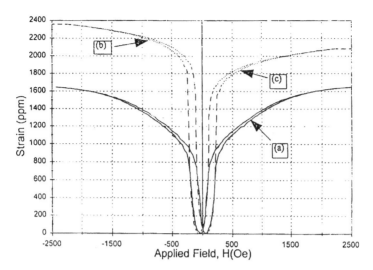

**FIGURE 1.39.** Magnetostriction for $Tb_{0.3}Dy_{0.7}Fe_{1.95}$ at 6.5 MPa prestress. a, experimentally determined; b, projected for <111> twin free; c, projected for <112> twin free.

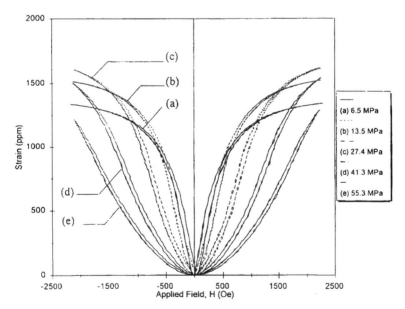

**FIGURE 1.40.** Magnetostriction of $Tb_{0.27}Dy_{0.73}Fe_{1.92}$ as a function of prestress 6.5–55.3 MPa.

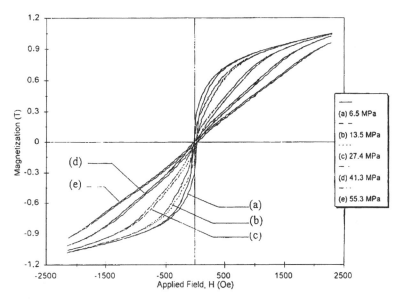

**FIGURE 1.41.** Magnetization curves corresponding to strain curves in Fig. 1.40 for $Tb_{0.27}Dy_{0.73}Fe_{1.92}$.

Upon casting, if the alloy surpasses $Fe_{1.96}$, then Widmanstatten precipitate can form either during crystal growth or during subsequent annealing. Obvious solutions are (i) start with excesses of rare earths to offset the losses incurred during processing [as is done in the commercial preparation of NdFeB alloys], (ii) to maintain high-vacuum integrity on all process equipment, and (iii) to limit the amount of time the molten alloy is in contact with container materials during the process. By use of process control techniques, this RFe3 precipitate defect does not appear in the high-performance transducer elements currently being produced by the FSZM and modified Bridgman (MB) crystal growth methods. The MB crystal growth method minimizes molten alloy contact with container materials by eliminating intermediate steps normally associated with the Bridgman method of crystal growth (i.e., melting, rod formation, and subsequent crystal growth). These three process steps occur in one step in the MB method.

As a result of the numerous investigations from 1980 to the present, it is possible to produce tailor-made TERFENOL-D-based giant magnetostrictive transducer elements with optimum performance required by devices for operation over a wide temperature and frequency range and under variable-load, input/output conditions.

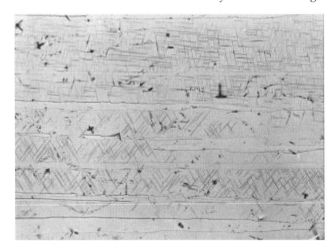

**FIGURE 1.42.** Transverse section of FSZM $Tb_{0.27}Dy_{0.73}Fe_{1.95}$. Needle shapes Widmanstatten precipitate in TERFENOL-D dendrites. Divorced eutectic phase separates the $RFe_2$ dendrites. Magnification, $\times$ 250 (II).

## SAMFENOL ($SmFe_2$)-Based Negative Giant Magnetostriction

The same metallurgical challenges exist in the development of high-performance negative-magnetostrictive $SmFe_2$ (SAMFENOL; $-\lambda >$ 2000 ppm) transducer elements. The anisotropy in the magnetization of $SmFe_2$ can be compensated by substitution of either Dy ($Sm_xDy_{1-x}Fe_2$; SAMFENOL-D) or Er ($Sm_xEr_{1-x}Fe_2$; SAMFENOL-E) for Sm. The high magnetostriction to anisotropy ratio can be achieved in these Laves phase compounds (see Table 1.5) in a manner similar to that accomplished in TERFENOL-D ($Tb_xDy_{1-x}Fe_2$) which compensates near room temperature for $x = 0.27$-0.3.

The huge anisotropy in the magnetostriction, $\|\lambda_{111}\| \gg \|\lambda_{100}\|$, also exists in the SAMFENOL systems. Thus, the <111> crystallographic direction alignment of these $RFe_2$ materials in single-crystal form along the drive axis is required to achieve the maximum magnetostrictive performance.

Single-phase $SmFe_2$ is more difficult to prepare than TERFENOL and TERFENOL-D. The added requirement of generating a controlled <111>, <112>, or <110> crystallographic direction along the rod axis is understandably even more difficult. Stoichiometric $SmFe_2$ forms peritectically, but only after the formation of $SmFe_3$ and $Sm_2Fe_{17}$ (and also peritectic reactions as shown in Fig. 1.43). A long-period (estimated to be 3 or 4 weeks) heat treatment would be required to convert this multiphase cast material to single-phase $SmFe_2$. Such single-phase material would probably not be in the single-crystal form with preferred crystallographic orientation that is necessary to achieve the high strains at practical applied fields for use in devices.

The preparation of transducer elements in these SAMFENOL-based systems is further complicated by: (i) the high vapor pressure of Sm which leads to process losses and shifts in stoichiometry and (ii) the high reactivity of both Sm (Sm oxidizes at room temperature when exposed to the atmosphere) and Sm alloys with the environment and process container materials at elevated temperatures. For the first complication, a process-dependent excess of Sm is used to maintain the stoichiometric alloy. For the second complication, several precautionary steps are required for successful alloy preparation. Sm raw materials should be stored and handled in an inert gas chamber, including weighing of components prior to alloying. The multiphase cast alloys should also be protected from the atmosphere prior to heat treatments that yield single-phase SAMFENOL-based alloys. The use of containers to form alloys from the melt is nearly prohibitive. Long-term heat treatments should be carried out with the samples suspended in sealed (e.g., Ta) crucibles at temperatures below the 720°C eutectic or below the 900°C peritectic temperature shown in Fig. 1.43.

Although the $SmFe_2$-based alloys are not amenable to processing techniques that would yield them as high-performance magnetostrictive transducer elements, a large portion of this strain could be realized with the single-phase alloy in the form of liquid-phase sintered compacts, particle aligned composites, or thin films. Some strain is sacrificed and higher applied fields may be necessary to achieve the strains. Both forms have the advantage of being producible in near-net shapes and potentially high-volume production quantities. Composites would be useful in some high-frequency applications and the huge strains possessed by these alloys make them candidates for use in sensor and micromechanical applications in which single crystals with preferred crystallographic alignment are not a requirement for use in such devices.

## Other Formulations

The light rare earths, Ce, Pr, and Nd, do not easily form $RFe_2$ when alloyed with Fe as encountered by Uchida and Mori (41). Melt-spinning and hot-press methods were necessary to synthesize polycrystalline samples of

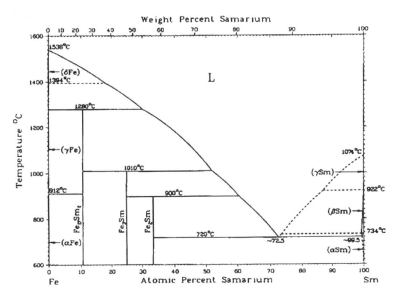

**FIGURE 1.43.** Samarium–iron phase diagram (reproduced from Kubaschewski with permission from ASM International). *ASM Handbook, vol 3: Alloy Phase Diagrams,* 1992.

$(Ce_xR_{1-x})Fe_2$, where R = Pr and Nd. High fields ($H = 0.4$-$10$ kOe) were required to produce strains of 60–600 ppm, respectively, in the best performance material studied $(Ce_{0.2}Pr_{0.8}Fe_{1.8})$. As noted previously, this reluctance persists for Sm and holds for the heavy rare earth Gd. Although the $RFe_2$ [where R = Tb, Dy, and Ho (holmium)] compounds melt noncongruently, psuedobinary combinations in slightly iron-deficient alloys behave as congruent melting compounds. $ErFe_2$ and $TmFe_2$ melt congruently above 1300°C. Similar information on the stability of the $RM_2$ Laves phases (where $M$ = Mn, Co, Ni, and Al) that form in these rare earth–metal (R–M) systems is available (42). Such information is useful in the predictions of which and how much of these R or M metals can be expected to substitute for Tb, Dy, or Fe in TERFENOL-D in the development of specialty alloys aimed at Improving a specific material and/or magnetic property. For example, Ho is known to reduce the hysteresis in TERFENOL-D. Ho substituted for both Tb and Dy in TERFENOL-D, $(Tb_xDy_yHo_z)Fe_{1.9-1.95}$, where $x + y + z = 1$. The values of $x$, $y$, and $z$ are application dependent and constitute a trade-off study between strain-sacrificed and hysteresis reduction.

# MAGNETIC DOMAIN AND GRAIN-RELATED PROCESSES

The preceding sections have established that TERFENOL-D possesses an appropriate strain capability suitable for use as an active transducer material. The characteristic strain response to an applied magnetic field, for compositions around $Tb_{0.3}Dy_{0.7}Fe_2$ under a suitable compressive stress, is composed of a rapid field-induced strain increase, which occurs at a particular jump or "burst" field, followed by a somewhat slower field response of strain increase. This section addresses some of the factors, associated principally with the magnetization processes, which have a direct bearing on the strain response to an applied field. Of particular note with regard to actuator applications are the seeming inability of the material to develop the total potential strain in a single rapid jump and the hysteretic strain response in both quasi-static and dynamic alternating applied fields.

The magnetization processes can, in principle, be inferred from knowledge of previously described strain and magnetization character-istics, from modeling investigations, and from observations of magnetic domain structures. Such observations have usually been made on materials in magnetic remanent states and under both applied magnetic fields and compressive stresses. The role of the microstructure is obviously extremely important in such considerations for any magnetic material. In highly magnetostrictive materials, this role is even more crucial due to the fact that most crystalline defects, such as inclusions, microcracks, second phases, grain boundaries, and dislocations, all have associated local stress distributions which can significantly affect the local magnetization distribution.

A brief summary of the crystallographic information pertinent to interpretations of the observed microstructural and magnetic domain structures is given, followed by a description of the particular defects which are considered to have a primary influence on the magnetos-trictive response. Some relevant domain observations are then described in the light of what magnetization distributions may be expected to exist. It is not intended to provide a detailed description of the physics of domains and domain walls but mainly to illustrate their significant microscopic influence on the magnetization and hence magnetostriction evolution under field and stress in a typical actuator material. There are numerous excellent contributions to the literature on magnetic domain phenomena from which further insight can be obtained—for example, the review article by Kittel and Galt (43), the texts of Craik and Tebble (44) and Chikazumi (45), and the much anticipated book by Hubert (46).

In the following section discussion will be centered on the $Tb_{0.3}Dy_{0.7}Fe_2$ composition, and all data and observations presented pertain to ambient temperatures such that the <111> crystallographic axes are axes along which the anisotropy energy is a minimum.

## Crystallographic Aspects

The production methods employed for TERFENOL-D are described in detail elsewhere in this chapter, and it is material formed by the free-standing zone process (37) which will be discussed here. Such material is known to solidify via a <112> dendritic growth front and the dendrites have a sheet morphology with (111) planes parallel to the sheet plane as illustrated in Fig. 1.44. The relations between the orthogonal crystallographic directions [111], $[11\bar{2}]$, and $[1\bar{1}0]$ are detailed in Fig. 1.44. The dendritic grain structure, the twinned nature of the dendrites, and the interdendritic rare earth phase material, indicated in Fig. 1.44, will be discussed in more detail.

### C15 Laves Phase

TERFENOL-D condenses in the cubic C15 Laves phase structure, the unit cell of which is shown in Fig. 1.45 along with the relationship of the three crystallographic directions shown in Fig. 1.44. It is these particular directions, and their associated planes, which will be referred to throughout this section.

### Stereographic Projections

Stereographic projections provide an extremely useful aid to visualizing the interrelationship between different crystal directions and planes. Two of the more frequently referenced planes to be used in discussion of domain structure-related images in this section are the $(1\bar{1}0)$ and (111) (Fig. 1.46). It can be seen from the $(1\bar{1}0)$ projection, in which the low index directions from both twin and so-called parent crystal orientations are included for the case of twinning about the (111) plane, that the twinned geometry exhibits high symmetry about the $[11\bar{2}]$ growth direction.

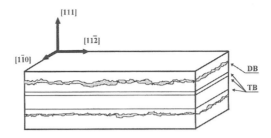

**FIGURE 1.44.** Schematic of a bar of TERFENOL-D showing the twin (TB) and dendrite bound (DB) and the relation between the $[11\bar{2}]$ growth direction and the orthogonal $[1\bar{1}0]$ and [111] directions. The [111] is normal to both the principal twinning plane and the sheet plane of the dendrite lamella. The interdendrite rare earth phase is shown shaded.

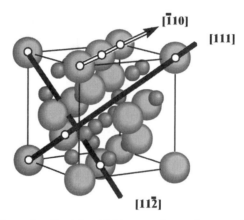

**FIGURE 1.45.** The unit cell of the C15 Laves phase showing the atom positions (large spheres; Tb and/or Dy; small spheres; Fe) and three useful orthogonal crystallographic directions.

## Crystal Defects

The microstructural defects present in any bulk ferromagnetic material can have a profound effect on the magnetization redistribution following the application of a magnetic field or stress. The principal effect is usually considered to be concerned with the interaction between domain walls and the disruption of the crystal lattice periodicity associated with a defected region. Such an interaction can lead to the pinning of a wall such

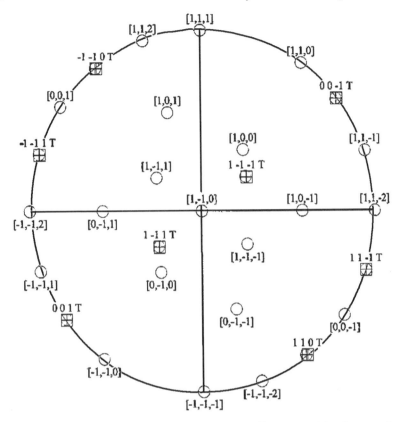

**FIGURE 1.46.** Stereographic projections of the (1$\bar{1}$0) plane (a), showing both parent and twin directions, and of the (111) plane (b).

that its motion through the material is impeded. In materials possessing high magnetostriction, any local or extended change in the periodic structure will, in most geometries, lead to a change in strain and, hence, to a change in the local magnetic anisotropy energy in the material.

The main types of defect found in TERFENOL-D material considered here are those associated with planar geometry and (compared to dislocations, for example) of macroscopic character. The dendrite, or grain boundaries, twin boundaries, and second-phase rare earth material will be described in more detail later. This will be done with respect to attempting to elucidate their influence on the magnetostrictive response through their effects on domain wall movement and magnetization configurations.

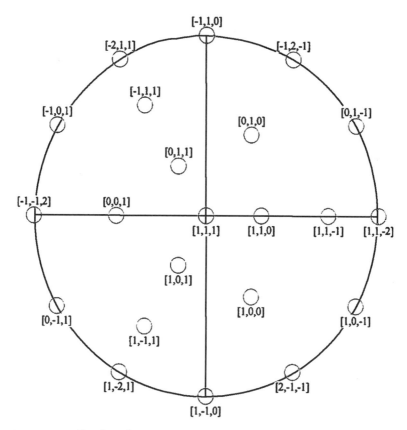

FIGURE 1.46. Continued

## Grain Boundaries

The dendritic growth front resulting from the zone rates employed in the production of free-standing zone TERFENOL-D leads to the dendrite sheet morphology shown in Fig. 1.44. The crystallographic misalignment between such sheets is usually less than a few degrees about the [111] orthogonal to the growth direction (5).

In material formed with a starting composition of approximately $RFe_{1.95}$, the interdendrite regions are usually found to contain deposits of a rare earth-rich phase, rejected from the dendrite solidification during growth (47–49). These deposits are not always continuous and vary both from sample to sample and in their local volume along the <112> growth direction and, orthogonally, in the (111) sheet plane. Such rare earth-rich material is also found to exist in a skeleton-like formation within the dendrite REFe$_2$ matrix phase.

## *Twin Boundaries*

An additional consequence of the zone rates used in the free-standing float zone process is that the dendrite sheets are often found to contain growth twins, the principal twinning plane being the (111) dendrite sheet plane (5). Such twin boundaries occur either singly or in even multiples within the dendrites, often running the entire length of the sheets, as illustrated in Fig. 1.44, or terminating within a sheet.

The atomic arrangement across such a growth twin, and associated parent and twin boundary directions, is depicted in Fig. 1.47 for the particular case of a (1$\bar{1}$0) plane. The [11$\bar{2}$] growth direction, which lies in the twin plane, is horizontal in this figure. Such boundaries have been the subject of considerable study using optical and transmission electron microscopy (50–53) and their presence was noted previously as being one of the major influences on the magnetostrictive characteristics observed in such materials. It is of interest to note that the twin structure, and its effect on the magnetic domain structure, has had a strong influence on the

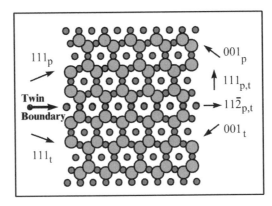

**FIGURE 1.47.** Atomic view of the (1$\bar{1}$0) plane showing a (111) plane twin boundary and associated in-plane parent (*p*) and twin (*t*) directions.

theory and modeling of magnetostriction particularly in the work of James and Kinderlehrer (54) and Zhao and Lord (55).

## RFe₃ *Precipitates*

The presence of second-phase material within the matrix phase is a distinct possibility for material grown with a starting composition more rich in iron than $RFe_{1.95}$. The occurrence of Widmanstatten precipitates, which form by a solid-state reaction in the $RFe_2$ matrix phase, have been observed by optical and electron microscopy (5,7,8,11) and have been shown to be of the composition $RFe_3$ (56, 57). Such precipitates form as narrow plates in the {111} planes of the matrix and would therefore be considered as a potential impedance problem for mobile domain structures in the material.

## Magnetic Domains

Magnetic domains within a crystalline ferro- or ferrimagnetic material are regions of very gradual or zero magnetization rotation. To a first approximation, which will be assumed here, they can be considered as volumes which are uniformly magnetized along particular directions in the crystal which are known as easy axes. The arrangement of the interconnections between domains which is established in such a material, the domain configuration, is that which will yield a minimum in the total energy for the body as a whole.

The competing energies in the case of TERFENOL-D are the magnetocrystalline anisotropy, elastic, magnetoelastic, magnetic field, demagnetization field, and stress energies. The particular elemental composition of this material results in the anisotropy energies having minima, or easy axes, directed along the <111> giving rise to the situation in which there are domains with magnetization in eight possible directions within the material. The resulting complexities in attempting to postulate what domain configurations can arise, and how they might change under magnetic field and stress application, are considerable (1–4, 13, 14).

Two additional factors, consequential on the previous energy considerations, are of importance in materials, such as TERFENOL-D, which exhibit large magnetostriction. First, the crystal within each domain volume will be subjected to a strain along the easy direction of magnetization of the domain resulting in a rhombohedral distortion of the cubic lattice. Second, when the domain configuration is terminated by a surface of the material, the demagnetizing, or magnetostatic, energy will in general attempt to reduce the volume of any domain having a magnetization component normal to the surface. This latter effect is responsible for the generation of so-called surface-closure domain configurations (1). Closure domain volumes decrease, often significantly, compared to those of internal, or bulk, domains.

## Domain Walls

The region separating neighboring domains is a domain wall in which rapid rotation of the magnetization from one easy direction to another occurs as a function of position in the crystal. The width of such walls is governed primarily by a competition between the anisotropy and exchange energies of the material. Examples of two such wall orientations commonly found in <111> easy axis materials are shown in Fig. 1.48 in which a wall is described in terms of the indices describing the plane of the wall and the rotation angle through which the magnetization vector rotates in moving from one domain easy direction to the next. Clark *et al.* (58) calculated the wall width and energy associated with the wall for the major expected wall configurations in TERFENOL-D. The two walls shown in Fig. 1.48 have wall energies on the order of $2 \times 10^{-3}$ Jm$^{-2}$, at least a factor of two less than wall energies for 180° rotation walls.

## Wall Projections on Particular (hkl) Planes

By assuming the easy directions of magnetization are at all times along <111>, it is possible to predict the wall planes associated with all combinations of pairs of such directed magnetizations. In terms of any domain observations on specifically orientated surfaces, it is therefore useful to establish the projections, or traces, of such wall planes on particular (hkl) surface planes. Such projections/traces for non-180° walls are given in Table 1.10 ( page 79) for the three orthogonal surface planes, $(11\bar{2})$, (111), and $(1\bar{1}0)$. It should be noted that a 180° wall plane can, in principle, adopt any orientation containing the easy axes associated with the wall.

Also detailed in Table 1.10 for two of the surface planes are the magnetic contrast and the topography (topo) contrast expected from such

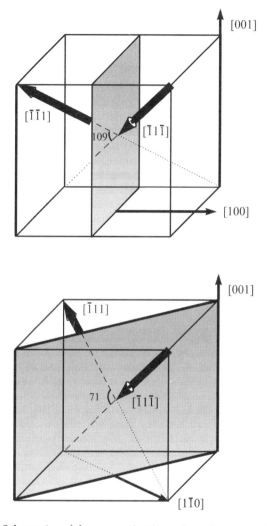

**FIGURE 1.48.** Schematics of the magnetization orientations associated with (top) a (100)109° domain wall and (bottom) a $(1\bar{1}0)$ 71° domain wall. The wall planes are shaded.

wall intersections with the planes. The magnetic contrast is given as a ratio of the components of magnetization normal to the particular surface plane arising from the two easy directions giving rise to the wall. Such magnetization components will give rise to the accumulation of surface

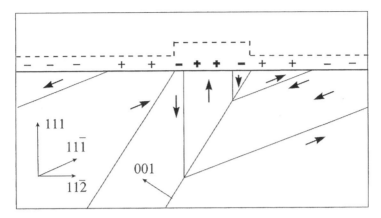

**FIGURE 1.49.** The topographical (dashed line) and magnetic charge accumulation ( $\pm$ ) resulting from the interaction of a possible domain configuration with a (111) plane surface.

magnetic charge. An illustration of this possibility is shown in Fig. 1.49 for the case of a possible surface closure domain configuration on a (111) surface. Also shown in Fig. 1.49 is the potential topographical surface distortion which could occur due to the emergence of a given domain system in a particular surface. The contrast information from Table 1.10 will be used in discussing the domain configurations imaged by magnetic force microscopy.

## Effects of Applied Stress and Magnetic Field

The application of either stress or magnetic field to a material such as TERFENOL-D results in a change to the anisotropy energy in the material. This in turn will cause the magnetization in the material as a whole to redistribute in order to reduce the new energy. In terms of the domain structure, this redistribution is accommodated by either inducing movement of the domain walls to reduce the volume of domains magnetized unfavorably with respect to the field or the stress or by appropriate rotation of the magnetization within the domains.

**Table 1.10.** The Non-180° Domain Walls Expected in a <111> Easy Direction Crystal Giving the Wall Traces, Anticipated Magnetic Charge, and Topographical Contrast on Particular Crystal Planes

| Easy directions | Wall plane | Wall type | Trace in (1̄10) | Magnetic contrast | Topocontrast | Trace in (1̄11) | Magnetic contrast | Topocontrast | Trace in (112̄) |
|---|---|---|---|---|---|---|---|---|---|
| 1̄11, 11̄1̄ | 110 | 109 | 001 | 0:0 | None | 1̄10 | 3:2 | Weak | 1̄10 |
| 111, 1̄1̄1 | 101 | 109 | 11̄1̄ | 0:2 | Weak | 101̄ | 3:2 | Weak | 1̄31 |
| 111, 1̄11 | 011 | 71 | 11̄1̄ | 0:−2 | Weak | 011̄ | 3:2 | Weak | 31̄1 |
| 111, 11̄1̄ | 100 | 71 | 001 | 0:2 | Weak | 01̄1 | 3:−1 | Strong | 021 |
| 111, 1̄1̄1 | 010 | 71 | 001 | 0:−2 | Weak | 101̄ | 3:−1 | Strong | 201 |
| 111, 1̄11 | 001 | 71 | 110 | 0:0 | None | 1̄10 | 3:−1 | Strong | 1̄10 |
| 11̄1, 1̄11 | 100 | 71 | 001 | 0:2 | Weak | 011̄ | 1:1 | None | 021 |
| 11̄, 1̄11 | 010 | 71 | 001 | 0:−2 | Weak | 101̄ | 1:1 | None | 201 |
| 1̄1̄, 1̄11 | 001 | 71 | 110 | 2:−2 | Strong | 110 | 1:1 | None | 1̄10 |
| 11̄, 11̄1̄ | 101 | 109 | 111 | 0:−2 | Weak | 1̄11 | 1:−1 | Strong | 111 |
| 111, 1̄1̄1̄ | 011 | 109 | 111 | 0:−2 | Weak | 2̄11 | 1:−1 | Strong | 111 |
| 1̄11, 11̄1̄ | 110 | 109 | None | 2:2 | None | 112̄ | 1:−1 | Strong | 111 |

In considering an ideally demagnetized state of the material, in which the overall magnetization is zero, it is useful to imagine the magnetization to be equally distributed in the eight easy directions. The application of field or stress then results in a change in occupancy of these eight directions. The domain configurational changes are not as easily imagined because all the domain magnetizations are coupled together via the domain wall configuration. The walls are restrained from movement in some cases by the defect nature of the material and in all cases by their wall energy interactions with the total energy of the material. It is of interest that, under the action of stress and field, an indication of the energy barriers to wall movement can be obtained. If a material is subjected to the application and then removal of a 40 MPa stress along the [11$\bar{2}$] direction, a residual strain is found to exist in the material. This indicates that the magnetization redistribution into the [111] upon stress application does not reform identically when the stress is removed. The change in strain and magnetization variations with applied field ($d_{33}$ and $\chi$, respectively) following such stress cycling can yield values for the energy barriers to domain movement, or coercivities, which can be attributed to different wall configurations.

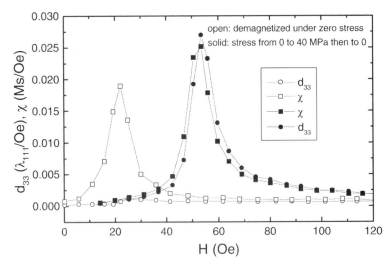

FIGURE 1.50. The variation of the strain response, $d_{33}$, and magnetization response, $\chi$, to a magnetic field applied along the [11$\bar{2}$] direction of a TERFENOL-D sample subjected to different stress histories.

The field dependencies of $d_{33}$ and $\chi$, measured under zero stress but with different stress histories, are shown in Fig. 1.50. In the case with no stress cycling, a peak in $\chi$ occurs at 21 Oe; however, the corresponding $d_{33}$ is very small. This implies that the magnetization redistribution by domain wall movement has made only a small contribution to the strain change. It is therefore obvious that this peak in susceptibility must be the result of 180° domain wall motion, mostly from magnetization redistributing from the $[\bar{1}\bar{1}1]$ to the $[11\bar{1}]$ direction (and the equivalent directions in any twin material) following field application along the $[11\bar{2}]$. Hence, the coercivity for such 180° domain wall motion can be determined as being 21 Oe. In the case of magnetization following the application and removal of a 40 MPa compressive stress along $[11\bar{2}]$, peaks in both $d_{33}$ and $\chi$ can be seen at an applied field of 53 Oe. These are believed to arise from non-180° domain wall motion, mostly from the perpendicular easy axis to the $[11\bar{1}]$, with a coercivity, or activation field, of 53 Oe. This value is much higher than that from 180° domain wall motion and is considered to be due to the effect of impedance of inhomogeneous stresses due to the defect nature of the material on the motion of the domain walls.

## Domain Observations

The first reported domain studies, on a single crystal (110) disc of $Tb_{0.27}Dy_{0.73}Fe_2$, were carried out by Clark et al. (58) using synchrotron radiation X-ray topography. In this technique the magnetic domain structure is revealed due to the differing lattice strains across non-180° domain walls. The domain configurations were interpreted in terms of low-energy 71 and 109° walls, and the slow approach to magnetic saturation was attributed to the formation of immobile <111> axis domains inclined to the surface of the crystal. Additional initial X-ray topography studies on similar material by Lord et al. (59) revealed that $RFe_3$ Widmanstatten precipitates severely disrupted the domain structure symmetries.

The surface distortion caused by domain configurations having magnetization components normal to the surface, indicated in Fig. 1.49, was initially observed using optical microscopy by Parpia et al. (60). The well-established technique of revealing magnetic domains using magnetic colloid, the Bitter technique (2), was first used by Janio et al. (61) to reveal extremely complex, closure-type structures on various surface orientations. Such structures have since been the subject of detailed investigation using differential interference contrast microscopy (62, 63);

transmission electron Lorentz microscopy (*10–12*), and scanning probe microscopy (*64*).

The use of scanning probe atomic and magnetic force microscopy (*65*) is highly appropriate, in principle, for investigations concerning TERFENOL-D samples. The surface distortion, should it be present, can be quantified directly using atomic force microscopy (AFM) techniques, and magnetic force microscopy (MFM) has the capability of providing information on the magnetization distribution from the specimen. The interpretation of MFM images is still open to question for certain types of materials and probes, but the model proposed by Hubert *et al.* (*66*) is adopted here such that the contrast observed in a magnetic image can be directly associated with the density of magnetic surface charge on the specimen.

The AFM and MFM images presented in this section have all been produced using a Digital Instruments Dimension 3000 scanning probe microscope. A silicon cantilever, with a tip coated with 35 nm of CoCr alloy, was used in the resonant mode of operation. The magnetic images, taken at a lift height above the specimen surface of between 30 and 50 nm , were formed by monitoring the difference in phase between the cantilever resonance and the driving potential. Such a phase image is induced by the magnetic force gradient normal to the specimen surface arising from the magnetic domain configuration. Observations discussed in the following sections were obtained from particular surface planes, aligned via the use of X-ray diffraction, and prepared in ideally strain-free conditions by mechanical polish and chemical lapping.

### Images from {110} Planes: Effects from Twins

Observations on $(1\bar{1}0)$ surfaces provide the opportunity to see the effects of the emergence of both the dendrite sheet and the twin boundaries as can be seen in Fig. 1.44. This surface plane also offers a potential advantage to understanding domain evolution under field and stress because it contains two of the <111> easy axes and the $[11\bar{2}]$ growth direction (see Fig. 1.46). An optical differential interference contrast (DIC) micrograph from such a plane of $Tb_{0.3}Dy_{0.7}Fe_{1.95}$, the material used in all subsequent images, is shown in Fig. 1.51. The horizontal contrast in this image occurs as a result of dendrite grain and growth twin boundaries running parallel to the horizontal growth direction. The zig-zag, or chevron, contrast bands seen throughout the figure occur as a result of distortions of the surface topography caused by the magnetic domain configuration of the material. Such contrast, running in bands along the [110] direction in both parent and twin regions of the specimen, clearly demonstrates the apparently complete compatibility of such structures as they traverse the microstructure of the sample (*13, 21, 22*).

An interpretation in terms of the possible domain structures associated with the contrasts shown in Fig. 1.51 is given in Fig. 1.52, in which the internal structure is composed of (001)109° walls having wall traces along the [110] and possessing out-of-plane magnetization components. A possible closure structure of 180° walls is indicated in Fig. 1.52. Such a configurational interpretation is partially confirmed in the AFM and MFM images shown in Fig. 1.53. The topographic contrast shown in the height image is composed of broad, 25-$\mu$m-wide bands which correlate well in dimension and direction to the chevron bands in Fig. 1.51. Detailed scans of the height variation in such bands by Holden *et al.* (*67*) confirmed the interpretation of the model in Fig. 1.52 and also allowed an estimate of the magnitude of the magnetostrictive distortion in such domains to be obtained.

The MFM image in Fig. 1.53 shows excellent geometric correlation between the topography bands and magnetic contrast, running from

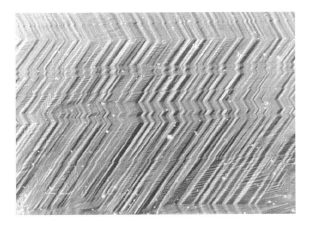

**FIGURE 1.51.** Optical DIC micrograph of a (1$\bar{1}$0) plane surface showing topography contrast associated with surface distortions across dendrite and twin boundaries. The in-plane [11$\bar{2}$] direction is horizontal and the image length is 1 mm.

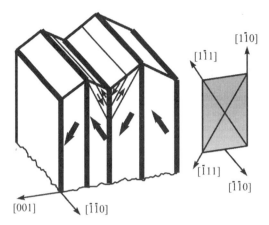

**FIGURE 1.52.** Schematic of the surface distortions produced in a $(1\bar{1}0)$ plane surface due to magnetostrictive strains associated with (001)109° wall configurations. Indicated are possible {112} 180° wall surface closures structures.

bottom left to top right in the [110] direction. The contrast occurs mainly as a result of the charge density associated with the components of magnetization normal to the surface. Finer magnetic contrast is seen to run parallel to these features which could be associated with the presence of 180° wall closure structures, although closer inspection

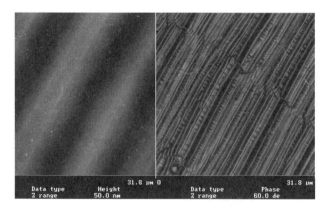

**FIGURE 1.53.** AFM (left) and MFM (right) images from a $(1\bar{1}0)$ surface. The sample is in a remanent magnetic state following field application along the horizontal $[11\bar{2}]$ axis.

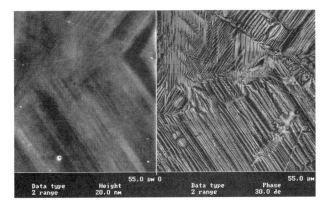

**FIGURE 1.54.** AFM and MFM images from a $(1\bar{1}0)$ surface showing topographic and magnetic contrast across twin boundaries. The sample is in a remanent magnetic state following field application along the horizontal $[11\bar{2}]$ axis.

reveals a highly detailed set of contrasts of ever-decreasing sub-micrometer dimensions.

In areas such as imaged in Fig. 1.54, in which $(001)109°$ wall configurations are shown interacting with twin boundaries, it can be seen that exact compatibility of the magnetic contrast is not maintained across

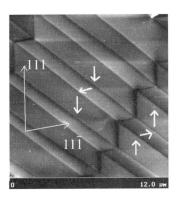

**FIGURE 1.55.** Magnetic contrast from a twin-oriented region of a $(1\bar{1}0)$ surface in which all the magnetization is confined to the magnetically easy [111] and $[11\bar{1}]$ axes in the surface plane as indicated.

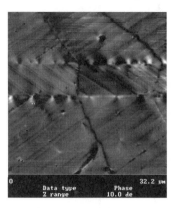

FIGURE 1.56.   Magnetic contrast arising from interactions across a double twin boundary in a (1$\bar{1}$0) surface in which all the magnetization is confined to the surface plane. The twin boundaries are parallel to the horizontal [11$\bar{2}$] axis.

each defect. The topographic AFM contrast clearly defines the changes in surface distortion which occur in these interactions with configurations composed of domains with out-of-plane components.

Most {110} surfaces reveal the complex closure structures discussed previously, but if the sample is carefully field demagnetized then it is possible to image in-plane domain configurations as shown in Fig. 1.55. Here, the contrast occurs from the domain walls and the configuration

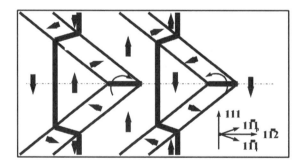

FIGURE 1.57.   A possible domain configuration to account for the magnetic contrast displayed in Fig. 1.56 for in-plane magnetization interactions at a horizontal twin boundary on a (1$\bar{1}$0) surface.

**FIGURE 1.58.** Magnetic contrast arising from interactions across a double twin boundary in a (11$\bar{2}$) surface. The twin boundaries are parallel to the horizontal [1$\bar{1}$0] axis.

comprises (110)71° and both (112) and (11$\bar{2}$)180° walls. Note the stronger phase image contrast from the 180° walls and the domain widths which are on the order of 2 $\mu$m. When such in-plane wall configurations interact with twin boundaries, an example of which is shown in Fig. 1.56, unusual contrast features are imaged in the vicinity of the twin at the places in which the wall structure crosses the boundary. Almost exact compatibility of the wall structure across the boundaries is observed and a possible interpretation of the domain configuration is illustrated in Fig. 1.57. This shows {110}71$^0$ and {112}180° walls such that it may be possible for magnetization in the [111] direction, which is a common direction for both parent and twin-oriented material, to possess continuity across the twin boundary. The unusual contrast is considered to arise from (111)140° wall segments lying in the twin boundary which would possess alternating magnetic charge distributions on the walls associated with neighboring segments. It should be noted that all magnetization directions incident at nonnormal inclinations to a twin boundary will of necessity form a domain wall which lies within the twin boundary. It has been suggested *(12)* from transmission electron microscope investigations that such twin boundary walls, as indicated in Fig. 1.57, once formed are likely to be strongly pinned to the boundary. It is thus possible that such a pinning mechanism may be responsible for the twin influence on the magnetostrictive response of this material.

The image shown in Fig. 1.58 is from a (11$\bar{2}$) fractured surface *(67)* and

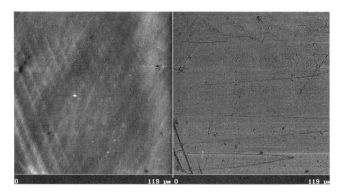

FIGURE 1.59.   AFM and MFM images from a large, twin-free area of a $(1\bar{1}0)$ surface showing pronounced distortion in the topography and magnetic domain wall contrast from an in-plane magnetization distribution.

shows that domain structures can form across, in this case, a double twin boundary which reveal little obvious signs of compatibility. It is considered that such configurations would be more difficult to move with field or stress application.

In areas of {110} plane surfaces well away from twin or dendrite boundaries, careful demagnetization reveals large-volume, in-plane domains with widths of approximately 20 $\mu$m. Such an area is shown in Fig. 1.59, in which the magnetic contrast arises from the domain walls. However, of some significance is the topographic contrast associated with this area from which could be surmised that the domain configuration would have normal components to the surface because the orientations of the surface distortions are along the same <110> directions as seen in Figs. 1.53 and 1.54.

## Images from {110} Planes: Effects from Defects, Fields, and Stress

Observations of the topography and magnetic contrast from (111) plane surfaces can reveal interesting information regarding configurations because this plane does not contain any easy magnetization directions or twin boundaries. Figures 1.60 and 1.61 show the changes which occur from field saturation and removal in different crystallographic directions. For the case of in-plane remanence, the magnetic contrast mainly occurs from magnetizations in the nonpolar easy axes. The image in Fig. 1.61, formed following polar field application, contains contrast mainly occurring from polar-oriented magnetization.

The magnetic structures revealed from specimens which have been field demagnetized with in-plane fields show relatively simple configurations, still showing <111> easy axis magnetizations, as shown in Fig. 1.62. The petal-like structures are formed from 180° surface closure domains.

The effect of fine Widmanstatten precipitates on domain configurations is shown in Fig. 1.63 from the (111) surface of a nominal single-crystal sample of $Tb_{0.3}Dy_{0.7}Fe_2$. Here, the severe influence of the precipitates on the magnetic structure can clearly be seen. The AFM image shows the threefold symmetry of the precipitates which condense in the {111} planes of the TERFENOL-D matrix. It is not difficult to imagine that such magnetic–defect interactions will give rise to pinning under field or stress application.

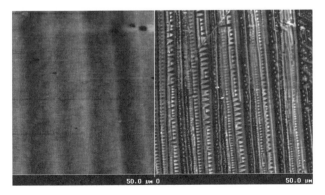

**FIGURE 1.60.**   AFM and MFM remanent state images from a (111) surface following field application along the (horizontal) [11$\overline{2}$] axis.

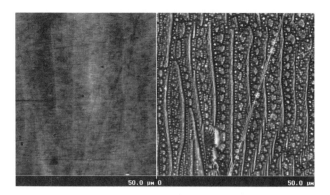

**FIGURE 1.61.** AFM and MFM remanent state images from a (111) surface following field application along the [111] direction normal to the surface. The [11$\bar{2}$] axis is horizontal.

When the starting composition of the material is even more Fe rich than in the previously imaged specimen, then large plate-like laths of $RFe_3$ material condense in the $RFe_2$ matrix. The effect of such second-phase material on the magnetic configuration, shown in Fig. 1.64, is to completely remove any magnetization continuity. The MFM image also illustrates the magnetic structure within the $RFe_3$ plate which is typical of that displayed from the basal plane of a hexagonal material possessing a polar, uniaxial anisotropy.

Regarding images from float-zoned TERFENOL-D, the effect of the interdendritic rare earth deposits (see Fig. 1.44) would be expected to

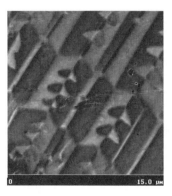

**FIGURE 1.62.** Magnetic contrast from the (111) surface of a field-demagnetized sample showing the "petal-like" contrast from 180° closure domains.

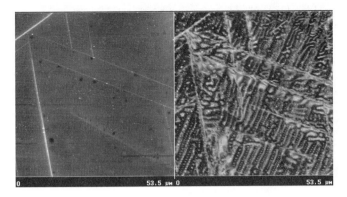

**FIGURE 1.63.** AFM and MFM remanent state images from the (111) surface of a sample containing fine Widmanstatten precipitates. The influence of the precipitates on the magnetic structure can be clearly seen (sample courtesy of Xeugen Zhao).

have a similar influence on domain configurations as that demonstrated in Fig. 1.64. The skeleton network character of such deposits becomes of interest for (111) plane surfaces in which isolated finger-like rare earth second-phase material, of $\mu$micrometer dimensions, can emerge from the surface. Such material can often fracture during surface preparation, leaving surface cavities and microcracks in the surface region, and the

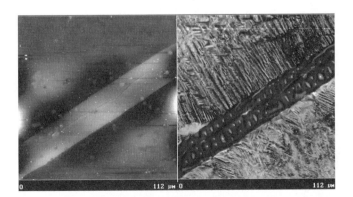

**FIGURE 1.64.** AFM and MFM remanent state images from the (111) surface of a sample containing large $RFe_3$ plate-like precipitates. The magnetic structure is imaged in both the TERFENOL-D and the $RFe_3$ phases.

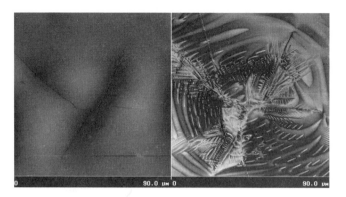

**FIGURE 1.65.** AFM and MFM remanent state images from the (111) surface of a sample showing the pronounced influence on the magnetic contrast from a connected series of surface microcracks.

remaining rare earth material is prone to oxidation. Such mechanical damage will always be present in the surface region of any sample. The influence of such coarse defected areas on the domain configuration can be extensive, as shown in Figs. 1.65 and 1.66. The long-range stress and microcrack distribution in Fig. 1.65 is thought to occur as a result of the volume change associated with second-phase, oxidized rare earth. The MFM image in Fig. 1.66 shows that the magnetic structure is still disturbed from regularity at distances $>70\,\mu$m from a crack termination.

**FIGURE 1.66.** AFM and MFM images from the (111) surface of a sample showing the long-range effect on the magnetic structure of the stress distribution from termination of a surface crack.

Of significant interest to the use of TERFENOL-D in actuation is the response of the magnetization, and hence magnetostrictive strain, to any applied field and stress. MFM images from a (111) surface of a sample subjected to various applied compressive stress along the $[11\bar{2}]$ axis are shown in Fig. 1.67. The sample was initially in a remanent magnetic state following field saturation and removal along the $[11\bar{2}]$ growth direction. The magnetic structure changes which occur, as the stress is cycled from zero to 10 MPa, 20 MPa, and back to zero, indicate that the magnetization redistributes into the <111> directions orthogonally inclined to the stress application direction as expected (see Fig. 1.46 and Table 1.10). The redistribution is far from complete, although there is no doubt that strain hysteresis is present because the final configuration displays differences from the initial configuration.

The effects on magnetic structure of an applied magnetic field are shown in Fig. 1.68 for a sample initially in a remanent state resulting from

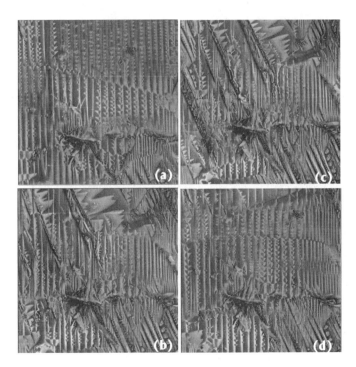

**FIGURE 1.67.** Magnetic contrast images from the (111) surface of a sample subjected to compressive stresses, applied along the $[11\bar{2}]$ axis (vertical in the figure), of (a) 0, (b) 10 MPa, (c) 20 MPa, and (d) 0.

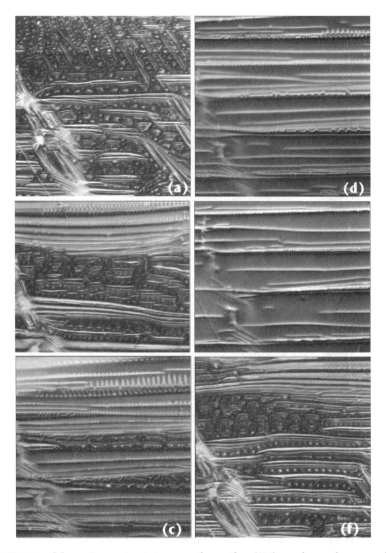

**FIGURE 1.68.** Magnetic contrast images from the (111) surface of a sample subjected to applied magnetic fields along the $[11\overline{2}]$ axis (vertical in the figure) of (a) 0, (b) $\sim$ 200 Oe, (c) $\sim$ 300 Oe, (d) $\sim$ 450 Oe, (e) $\sim$ 800 Oe, and (f) 0. The sample was initially in a remanent state following field application along the [111] surface normal.

field application and removal along the <111> polar axis. The field was applied along the $[11\bar{2}]$ growth axis and significant changes to the configuration can be observed as the field effectively pulls the magnetization from the polar axis and principally into the $[11\bar{1}]$ direction (see Fig. 1.46). It is interesting to note that, although the sample is under zero applied stress, using the same maximum applied field the magnetostrictive strain developed by the sample would be near saturation. Evidence of hysteretic behavior is also apparent from these data because the initial and final zero field configurations are not the same.

The MFM data in Fig. 1.68 show a common feature with all previous domain observations associated with field applications (*11, 17, 21, 22*) in that it has always been difficult to entirely remove the surface closure domains from a sample. The consequence of this general result is that it will require extremely large values of applied field to reach total magnetic saturation and, hence, maximum strain capability of such materials. The fraction of the specimen volume that is occupied by such closure domain structures remains unknown and requires further investigation.

## Summary

Consideration has been given in this section to the effects of particular crystal defects on observed magnetic domain configurations. Such observations have enabled much insight to be gained with respect to our understanding of the magnetization processes giving rise to the development of magnetostrictive strain. TERFENOL-D is a fascinating material in that it possesses extremely attractive characteristics with regard to actuation devices and also has yielded images showing the complexities of magnetic domain configurations. The major problem of any domain observation technique is the difficulty of determining the relations between observed surface closure and interior or bulk magnetic structure configurations. Perhaps TERFENOL-D, with the benefit of the significant magnetoelastic strain capability, may help yield some answers to this problem in the future.

## MANUFACTURE OF TERFENOL-D

Giant magnetostrictive TERFENOL-D has the highest room-temperature strain achievable at practical drive fields currently known. The most technologically advanced of these alloys is $Tb_{0.3}Dy_{0.7}Fe_{1.9-1.95}$. The advantages of TERFENOL-D compared to existing transducer materials have driven the demand for high-performance TERFENOL-D transducer elements to large-volume production. The following methods of manufacture have been developed based primarily on the applications:

1. Sintered powder compacts similar to permanent magnetic manufacturing: Most suitable for mass production of small irregular shapes.

2. Particle-aligned polymer matrix composites: Near-net shapes using high-resistivity materials for high-frequency applications.

3. Produce a "near" single crystal with the highest strain performance of any method limited to rods up to 8 mm in diameter: Directional solidification by the containerless FSZM method.

4. Crystal growth <112> directional solidification using the MB method: This method is the largest volume process practiced today and capable of rod sizes up to 70 mm in diameter and yields strain performance near that of FSZM material.

Manufacture of high-performance TERFENOL-D transducer elements using the MB method of crystal growth is described in this section. Manufacturing involves more than producing these giant magnetostrictive materials in their optimum form, i.e., crystal growth with preferred crystallographic direction. The process also includes precision machining of laminations, final diameters, and parallel ends of the cut-to-length drivers. Accurate testing/performance evaluation of the finished transducer elements is also a critical step in the manufacturing process. An overlying factor throughout the process is cost. The process must yield affordable drivers for any given application. To accomplish all these goals the manufacturing process must be flexible enough to accommodate minor variations dictated by cost/performance trade-offs but stringent enough to produce consistent, high-performance transducer elements in large quantities with high yields.

## History of Manufacturing

The ideal room-temperature giant magnetostrictive transducer drive element was defined in the 1970s as a defect-free single crystal of TERFENOL-D ($Tb_{0.3}Dy_{0.7}Fe_2$) with the <111> crystallographic direction aligned along the drive axis of the transducer element. The compromises/trade-offs necessary to the development of the large-scale, cost-effective production of these drive elements were determined from the extensive research of the 1970s.

## Research Quantities: 1970s and 1980s

The FSZM method of crystal growth was developed primarily to provide high-performance drive elements for R&D purposes. Near single crystals of $Tb_{0.3}Dy_{0.7}Fe_{1.9-1.95}$ with the <112> crystallographic direction along the drive axis were produced with diameters up to 8 mm and at growth rates up to 500 mm/hr (37, 68). Substitution for Tb, Dy, and/or Fe as well as different Tb to Dy ratios were explored. The FSZM method is currently used in the manufacture of high-performance magnetostrictive transducer elements (35).

For larger diameter rods (10–50 mm) the Bridgman (BRDG) or modified Bridgman crystal growth method was used during the 1970s and 1980s. The <112> crystallographic directional solidification along the drive axis is attained but multiple crystals exist in these Bridgman-grown drivers. This results in slightly lower strains and a moderate increase in hysteretic behavior. However, the production of these larger diameter rods is more cost-effective because of substantial reductions in both labor and material losses. Applications that require large projection areas (e.g., sonar and vibration control) would require extensive machining of small-diameter FSZM rods to form bundles to obtain a large cross-sectional area. With the large-diameter BRDG rods, a single rod could be produced to meet the size requirement at a much lower cost while providing 80–90% of FSZM performance.

## Pilot Plant Production in 1988

In order for this new technology to advance beyond the research stage, a reliable source of these giant magnetostrictive transducer elements was required. This was accomplished at ETREMA Products, Inc., in 1987 with the technology transfer of the processes from the laboratory to a commercial setting. A continuous effort to produce higher performance/lower cost transducer elements for use in prototype device developments led to the next level of operation for this technology. During the early 1990s the worldwide transducer industry became aware of the extraordinary transduction capabilities of these materials. This increasing interest required an increased availability in sizes, shapes, and stoichimetries and of large production quantities of devices using these drivers.

## Full-Scale Production in 1995

Full-scale production was realized with the development and installation of the next-generation MB system referred to as the ETREMA Crystal Growth (ECG) system. This system is fully automated, computer controlled, and capable of producing drivers as large as 65 mm in diameter in 175-mm lengths or multiple smaller diameter rods in 300-mm lengths. The appropriate machining and testing capabilities were developed in parallel with the ECG system. The current manufacturing system can process more than 2.5 metric tons of TERFENOL-D alloy per year on a one-shift basis.

## TERFENOL-D Manufacturing Process

The manufacturing process used to produce high-performance, affordable TERFENOL-D transducer elements is described here. The various sizes, shapes, stoichiometries, etc. required by the transducer industry are incorporated in the process, which is the only known large-volume process in operation today. The FSZM method to produce near single-crystal drivers with diameter up to 8 mm was previously described. This section describes the MB method of crystal growth used to produce high-performance transducer elements with diameters between 10 and 65 mm in large quantities. The associated machining and testing techniques used to meet the tolerances and performance specifications of the finished drivers are also described.

## Manufacturing Process Flow Sheet

The materials flow sheet for the manufacture of TERFENOL-D transducer elements is shown in Fig. 1.69. Each of these steps is discussed in detail.

## Alloy Preparation

Homogeneity of the stoichiometric alloy being processed is critical to the performance and therefore the yield of the finished transducer element. Alloy formation is accomplished in a water-cooled, copper-hearth, nonconsumable electrode arcmelter. Rare earths, Tb and Dy, in appropriate weights are melted and homogenized by repeated melting before alloying with iron to form the $RFe_{1.92}$ Laves phase TERFENOL-D. $Tb_xDy_{1-x}$ is a solid solution for all values of $x$, and an example of performance as a function of $x$ is shown in Fig. 1.70. $Tb_{0.27}Dy_{0.73}Fe_{1.92}$ is near the totally compensated stoichiometry and exhibits a more linear approach to saturation and low hysteresis. The changes in magnetostrictive response in the $Tb_{0.3}Dy_{0.7}Fe_{1.92}$ alloy show the effects of the temperature-dependent magnetization anisotropy constants of Tb and Dy on the strain performance. This alloy exhibits the low field jump in the strain and slightly higher hysteresis. This comparison (Fig. 1.70) illustrates the importance of initial preparation of a homogeneous alloy to avoid microsegregation and inconsistent magnetostrictive performance in the finished drivers. The alloy, $Tb_{0.3}Dy_{0.7}Fe_{1.92}$, is by far the most common stoichiometry produced.

## Crystal Growth-Production Scale

The current production-scale crystal growth system is based on the patented MB process (69). This ECG system is shown in Fig. 1.71. This

## Process Flow

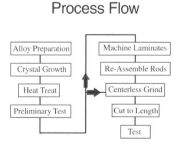

**FIGURE 1.69.** Manufacturing process flow sheet for production of TERFENOL-D transducer elements.

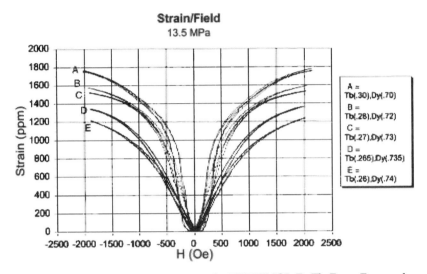

**FIGURE 1.70.** Magnetostriction curves for TERFENOL-D, $Tb_xDy_{1-x}Fe_{1.92}$, where $x = 0.3, 0.28, 0.27, 0.265$, and $0.26$. FSZM material at 13.5 MPa (1.95 ksi) prestress.

system is capable of producing high-performance TERFENOL-D drivers as large as 65 mm in diameter in 175-mm lengths or several smaller diameter rods in 300-mm lengths in a given heat. The system is fully automated and computer controlled which results in direct labor cost reductions. Additional cost reductions are realized in the higher yields and larger volumes of transducer elements produced in this system. Crystal growth is directional solidification of TERFENOL-D sheet dendrites along the <112> crystallographic direction and obtaining that orientation along the rod axis.

Unlike the FSZM-seeded near-single-crystal drivers (limited to 8-mm diameter), these larger diameter (10–65 mm) MB rods contain several <112>-oriented crystals within each cross section. Some strain is lost and the hysteresis is slightly increased due to the misorientation of the multiple crystals with respect to one another in a given cross section. Figure 1.72 shows a comparison of the strain performance of near-single-crystal FSZM driver and <112> directionally solidified MB driver growth in the ECG system. Some strain loss has been experienced as the diameter is increased. This loss is primarily due to excessive radial heat transfer

**FIGURE 1.71.** The ECG system.

during solidification which reduces the level of <112> alignment along the drive axis. Recent refinements in the crystal growth parameters have resulted in higher strains (less loss) in the larger diameter transducer elements.

End-to-end and lot-to-lot consistency is important to the overall yield which impacts the price of the production drivers. The bottom-to-top performance (Fig. 1.73) shows the ECG production material to be consistent in strain performance, which is a reflection of the low level of macrosegration during the crystal growth. The lot-to-lot consistency is a matter of controlling the growth parameters in ECG operation to generate the conditions necessary for high yield of high-performance TERFENOL-D transducer elements. Figure 1.74 shows the consistency of the drivers produced as a function of time.

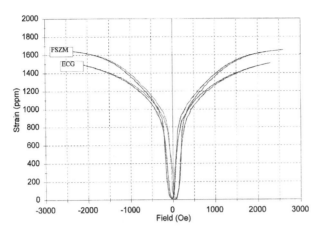

**FIGURE 1.72.** Magnetostrictive strain plots for FSZM and MB TERFENOL-D.

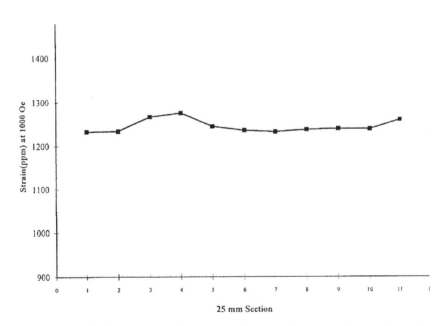

**FIGURE 1.73.** End-to-end performance of production transducer elements. Average strain at 1000 Oe for 10 MPa prestress; 25-mm-long sections from 30-mm-diameter rods with lengths up to 350 mm.

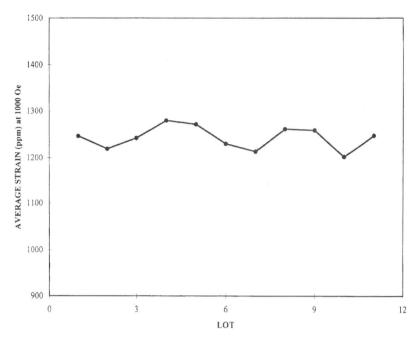

**FIGURE 1.74.**    Lot-to-lot average strain at 1000 Oe and 10 MPa prestress.

## Heat Treatment

Proper heat treatment of the <112>-oriented as-grown TERFENOL-D drive rods has a pronounced positive effect on the strain performance (*35, 70, 71*). The low-field burst effect or magnetostriction jumps (*39*) only occur in rods that have been heat treated above 900°C (Fig. 1.75). A significant increase in the saturation strain, e.g., $\lambda$ at 2.5 kOe, is also achieved by heat treatment. Most of this strain enhancement can be accomplished by heat treatments at 900–1000°C for periods of 20 min to 6 hr. Similarly, a significant improvement in the magnetomechanical coupling coefficient, $k_{33}$, was measured on TERFENOL-D samples treated at 1000°C for 24 hr (*72*). This study showed that some enhancement of $k_{33}$ could be obtained by increasing the time at temperature to 7–10 days.

Based on these and other investigations, it is concluded that a cost-effective heat treatment of production-scale TERFENOL-D transducer

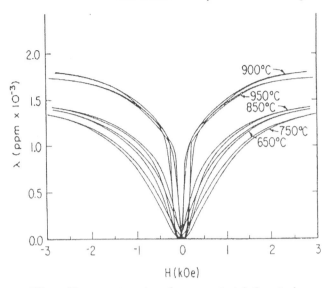

**FIGURE 1.75.**   Effect of heat treatment on the magnetostrictive strain properties of TERFENOL-D ($Tb_{0.31}Dy_{0.69}Fe_{1.96}$) at 6.9 MPa prestress (70).

elements can be accomplished at 950–1000 °C for a time period of less than 24 hr. The temperature and time periods vary as a function of composition and are application dependent.

## Machining

TERFENOL-D is a brittle intermetallic compound and is difficult to machine. It is not conducive to normal machining operations such as turning in a lathe, milling of flat surfaces, and drilling of holes. Abrasive methods, such as belt grinding, centerless grinding, and high-speed diamond saws, must be used to meet the surface finish and geometric tolerances required by the end user. Coolants are necessary in each of these operations to avoid localized heating and oxidation of the surface and powder filings.

Precise machining to the final dimensions and tolerances is essential to the optimum performance of the devices. Parallelism and perpendicularity of the cut-to-length ends of the drive rods is critical to accomplishing a uniform prestress with no bending of the transducer.

These precise cuts are made using a high-speed abrasive diamond saw which is automated and controlled using CNC programs. Surface finish of the ends (roughness) and machining a radius on the corners to prevent chipping are often factors in the final specifications. Figure 1.76 provides examples of some sizes and shapes of solid and laminated TERFENOL-D transducer drive elements.

Diamond saws are also used to produce laminations as thin as 0.75 mm and in lengths of 200 mm. Laminated drivers reconstructed with magnetic isolation epoxy joints between each slab are required to reduce eddy current losses in high-frequency applications.

Conventional drilling and turning techniques cannot be used to produce complex shapes and axial holes in TERFENOL-D. Axial holes in the TERFENOL-D drive elements are a common requirement to accommodate prestress assemblies or serve as cooling channels in some transducers. Proprietary processes have been developed to accomplish a variety of shapes.

Small square elements (1–3 mm$^2$) can be cut from large-diameter rods using a diamond saw. This process is cost-effective, with most costs

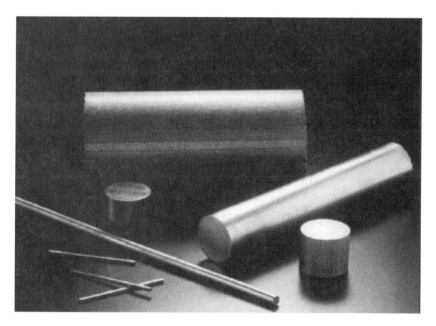

**FIGURE 1.76.** Examples of some sizes and shapes of solid and laminated TERFENOL-D transducer drive elements.

attributed to the materials loss due to the saw kerf. Techniques to reduce these costs are being developed which will make the manufacture of these small square elements even more effective.

Some of the standard machining tolerances for TERFENOL-D drivers are listed in Table 1.11. These tolerances meet the critical dimensional and

**Table 1.11.** Standard Machining Tolerances for TERFENOL-D

|  | *mm* | *Tolerance* |
|---|---|---|
| Diameter | 1.91–6.347 | ± 0.08 |
|  | 6.35–25.37 | ± 0.05 |
|  | 25.40–38.07 | ± 0.08 |
|  | 38.10–68.00 | ± 0.13 |
| Length | 1.00–50.77 | ± 0.05 |
|  | 50.80–101.57 | ± 0.08 |
|  | 101.60–152.37 | ± 0.10 |
|  | 152.40–203.17 | ± 0.15 |
|  | 203.20–304.77 | ± 0.25 |
| Parallelism | 0.05 | |
| Perpendicularity | 0.04 | |
| Surface finish | | |
|   On rod end faces | $R_a \leq 0.5\,mm$ | |
|   On rod outer diameter | $R_a \leq 1.2\,mm$ | |
| Radiusing of rod ends | | |
| (To avoid chipping, sharp edges of rods are either "broken" by hand or a radius is machined) | | |
|  | Nominal diameter | Maximum edge break |
|  | 19.10–max | 1.52 mm |
|  | 10.10–19.09 | 1.27 mm |
|  | 5.00–10.09 | 1.02 mm |
|  | 1.91–4.99 | 0.76 mm |

surface finish specifications for TERFENOL-D transducer drive elements used in devices. These tolerances are critical to the final application for the following reasons:

1. Material used under prestress requires even loading.

2. Material is brittle—uneven loading will lead to chipping or cracking.

3. Uneven loading will lead to distortions in transducer output.

## GIANT MAGNETOSTRICTIVE COMPOSITES

Ordinary bulk TERFENOL-D rods of some millimeter diameter work reasonably well in applications up to approximately some kiloHertz. For higher frequencies, eddy currents appear which heat up the material and also prevent the applied magnetic field from penetrating the material. To increase the bandwidth, it may be necessary to laminate the rods, an operation which is both costly and time-consuming.

Giant magnetostrictive composites based on TERFENOL-D particles have been known for several years (73, 74). These new types of magnetostrictive composites are formed together with a binder and an optional, additional electrical insulator layer. The binder creates an electrical insulating layer between the particles, which increase the resistivity, and makes magnetostrictive materials suitable for high-frequency applications. Therefore, the basic purpose of these giant magnetostrictive composites is to create a complement to the bulk TERFENOL-D and consequently broaden the useful range of the giant magnetostrictive materials into the ultrasonic regime, in which competes with piezoelectric ceramics when a very high power density is needed.

The composite materials have been shown to be easier to manufacture in different shapes and easier to machine depending on a less brittle nature in comparison with bulk Terfenol. In addition, they have a high tensile strength which makes them suitable for applications in which the tensile strengths are higher than those required of bulk TERFENOL-D, for example, in resonant applications. The use of magnetostrictive powders also provides the possibility of aligning the particles to the desired <111> direction parallel to the central axis, which has to date proven impossible with ordinary metallic Terfenol-D. The giant magnetostrictive composites are under development. Here, physical aspects of the magnetostrictive composites will be considered.

## The Giant Magnetostrictive Powder Composite

### Magnetostrictive and Physical Properties

Since the giant magnetostrictive powder composite (GMPC) consists of considerable amounts of polymer binder, the density and bulk magnetization will be proportionally lower compared to those of metallic TERFENOL-D. Other properties will also be affected (see Table 1.12 for a comparison).

### Fabrication

There are four principally different production methods:

- Uniaxial pressing with randomly oriented particles
- Uniaxial pressing with magnetically aligned particles

**Table 1.12.** Comparison of Typical Values of Material Properties for GMPC and Bridgman-Grown TERFENOL-D with Composition $Tb_{0.3}Dy_{0.7}Fe_{1.95}$

| Property | Powder composite (GMPC) | TERFENOL-D (Bridgman) |
|---|---|---|
| Density (kg/m$^3$) | 6750–8000 | 9250 |
| Young's modulus (GPa) | 16–25 | 25–35 |
| Sound propagation speed (m/s) | 1400–1800 | 1650–1950 |
| Tensile strength (MPa) | 55 | 28 |
| Compressive strength (MPa) | 300 | 700 |
| Material resistivity ($\mu\Omega$ m) | 300–600 | 0.6 |
| Saturation magnetization (T) | 0.7–0.85 | 1.0 |
| Relative permeability (–) | 2–6 | 3–20 |
| Coupling coefficient (–) | 0.24–0.4 | 0.75 |
| Piezomagnetic constant (nm/A) | 3–9 | 5–15 |

- Isostatic pressing with randomly oriented particles
- Isostatic pressing with magnetically aligned particles

The manufacturing process consists of the steps shown in Fig. 1.77. The starting material should be a casted alloy with a suitable composition (e.g., $Tb_{0.27}Dy_{0.73}Fe_{1.95}$). Since terbium and dysprosium have very stable oxides, the crushing must be made in inert atmosphere to prevent TERFENOL-D from oxidizing. After the grinding process the powder must be sifted and optimal size distributions blended. The size distribution(s) should be chosen both so that the eddy-current losses are reduced at the intended working frequency of the material and so that the compacted bulk density of the material is as high as possible. Theoretically, this can be achieved with a bimodal distribution.

The optional surface coating of the particles increases the bulk resistivity of the composite. It also decreases the bulk density, which decreases the energy density. The preferred surface coating is a very thin oxide layer. After the layer is applied, the magnetostrictive powder should be mixed with the binder.

The optional magnetic aligning aims to align the particles with the highly magnetostrictive, easy <111> crystal direction parallel to the working direction of the GMPC rod. The aligning must be carried out in a static magnetizing field prior to any compaction so that the rotation of the particles is not hindered.

The compacting step can be carried out using either a uniaxial pressing tool or an isostatic press. Since the particles are very brittle, caution must be taken not to apply too high pressure so that the particles may crack. On the other hand, a high-performance powder composite

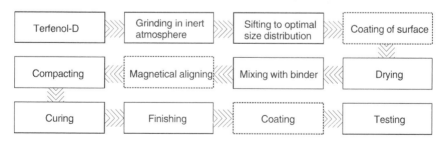

**FIGURE 1.77.** The manufacturing process for GMPC. The dashed line indicates optional process steps.

must have a high density, which is very much dependent on the compacting pressure.

If the binder is a thermosetting resin or a thermoplastic resin, a thermal curing in a controlled atmosphere must be carried out. The finishing operation can be carried out with conventional cutting methods (e.g., drilling, turning, and cutting). The optional coating operation of the finished detail is performed for appearance and corrosion resistance.

A frequency response of the displacement for an applied magnetizing field with constant amplitude is usually sufficient to categorize the rod. From the frequency response, $Q$-value, damping, modulus of elasticity, coupling coefficient, and resistivity can be estimated.

## *Uniaxial Pressed Composites*

Uniaxial pressing results in a composite that can be highly compacted, thus resulting in a high-density isotropic composite, given that the geometry is suited for uniaxial pressing. As a rule of thumb, the (axial) length of the rod should not exceed two diameters. Also, the resistivity tends to be lower axially compared to the perpendicular directions; this results in an improved mechanical performance and bulk coupling coefficient since the highly ductile binder phase will be thinner in the axial direction.

Magnetic aligning of the particles is possible by letting the particles fall through a homogeneous, DC-magnetizing field of appropriate strength. Since both the compressibility and the ductility of the powder are insignificant, the optimal compacting pressure will be moderate, thus making it possible to use nonferromagnetic material for the pressing tool. The effect of aligning can be improved if a sinusoidal magnetizing field is superimposed on the DC field, thus using the magnetostriction to vibrate the particles and thereby making it easier for the particles to rotate in the pressing tool. If the aligned powder is sufficiently compacted, the particles will not rotate away from the preferred direction at the final compacting stage.

## Isostatically Pressed Composites

Isostatic pressing will, by definition, result in a mechanically isotropic composite. In this process it is both possible to align the particles before pressing and to keep the particles in the preferred direction during the compacting process. If a thermoplastic binder is used, a hot isostatic press must be used to keep the powder together due to the poor compressibility of the powder.

## Magnetic Alignment of Magnetostrictive Powders

In a 1979 patent by Malekzadeh and Milton (75), a manufacturing process for sintered rare earth–iron Laves phase magnetostrictive alloy product, characterized by a grain-oriented morphology, was described. The grain-oriented morphology is obtained by magnetically aligning powder particles of the magnetostrictive alloy prior to sintering. The aligning procedure increases the performance so that the $|\lambda_{\parallel} - \lambda_{\perp}|$ magnetostriction is increased by approximately 20%. An explanation of how and why the performance is improved is provided in this section.

Assume that the magnetostrictive grains have a spherical geometry and that the binder does not hinder the alignment of the grains' <111> easy direction[7] with a direction collinear with the external magnetizing field. Thus, the shape anisotropy will influence the direction of orientation of the grains (Fig. 1.78). Shape anisotropy can be explained as the reduction in the magnetizing field in the magnetized material due to the appearance of magnetic poles. The magnetic poles cause a demagnetizing field, $H_d$, directed in the opposite direction to the magnetizing field. The size of the demagnetizing field depends on two parameters: the geometrical shape of the grains and the magnetization of the grains, $M$.

The demagnetizing field is a product of the magnetization of the material (magnetical pole strength) and the shape of the material (distance between the magnetic poles):

$$H_d = N \cdot M \tag{1.40}$$

---

[7]The following convention is followed for the indices of planes and directions. Planes of a form are planes related by symmetry, such as the six faces of a cube: (100), (010), (001), ($-100$), ($0-10$) and ($00-1$). The indices of any one enclosed in braces {100} stand for the whole set. The indices of particular directions are enclosed in square brackets, such as the six cube-edge directions: [100], [010], [001], [$-100$], [$0-10$], and [$00-1$]. These are directions of a form, and the whole set is designated by the indices of any one, enclosed in angular brackets <100>.

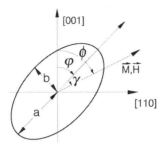

**FIGURE 1.78.** The direction of the magnetization vector, $\phi$, and the direction of the symmetry axis of the grain, $\varphi$, related to the principal axes of the crystal.

The demagnetizing factor, $N$, can only be determined exactly for shapes of second order, i.e., spheres and ellipsoids. Assuming that an oblong ellipsoid can approximate the shape of a TERFENOL-D grain, it is possible to determine the state of energy for different directions of the symmetry axes of the ellipsoid with respect to the direction of magnetization (Fig. 1.78):

$$\begin{cases} N_a = \frac{1}{(c^2-1)} \left( \frac{c}{\sqrt{c^2-1}} \pm \cdot \ln\left(c + \sqrt{c^2-1}\right) - 1\right) \\ N_b = \frac{1-N_a}{2} \\ N_a > N_b \\ c = \frac{a}{b} \end{cases} \tag{1.41}$$

where $N_b$ is the demagnetizing factor in the "long" direction of the ellipsoid and $N_a$ represents all directions perpendicular. Assuming that the grain is a monocrystal magnetized to saturation (i.e., a single domain), the total energy of that grain under an external magnetizing field, $H_{ext}$, can then be written (76):

$$E_{tot} = \underbrace{-\mu_0 \int \boldsymbol{H} \cdot d\boldsymbol{M}}_{E_{sh}} + E_a \tag{1.42}$$

where $E_a$ is the crystal anisotropy energy and $E_{sh}$ is the shape anisotropy energy. By expressing the total magnetizing field $H$ in terms of the external magnetizing field, $H_{ext}$, and using Eqs. (1.41) and (1.42), the shape energy can be written as

$$E_{sh} = -\mu_0 \int (\boldsymbol{H}_{ext} - N\boldsymbol{M})d\boldsymbol{M} = -\mu_0 \left( \boldsymbol{H}_{ext} \cdot \boldsymbol{M} - \frac{N\boldsymbol{M} \cdot \boldsymbol{M}}{2}\right) \tag{1.43}$$

Separation into directional energies yields

$$
\begin{aligned}
E_{\text{sh}} &= -\mu_0 H_{\text{ext}} M + \frac{\mu_0}{2} \left( N_a (\cos \gamma \cdot M)^2 + N_b (\sin \gamma \cdot M)^2 \right) \\
&= -\mu_0 H_{\text{ext}} M + \frac{\mu_0}{2} M^2 \left( N_a + \sin^2 \gamma \cdot (N_b - N_a) \right)
\end{aligned}
\tag{1.44}
$$

where $\gamma$ is the angle between the long axis of the ellipsoid and the magnetization vector. In order to simplify the understanding of the state of energy in the crystal, the motion of the magnetization vector is restricted to the $(1\bar{1}0)$ plane.

In 1929, the Russian physicist Akulov showed that the anisotropy energy of the crystal, $W_a$, for a cubic crystal can be expressed in terms of a series expansion of the direction cosines, $\alpha_1$, $\alpha_2$, $\alpha_3$, of the magnetizing vector, $M_s$, relative to the crystal axes (77) (see Fig. 1.79). The constants $k_0$, $k_1$, $k_2$, $k_3 (\text{J/m}^3)$ are specific for each crystalline ferromagnetic material. The values of the constants are often known with less than two significant data points and the accuracy decreases with higher constants, i.e., $k_2$ has less accuracy than $k_1$. There is seldom gain in utilizing constants higher than $k_2$. See Eq. (1.25), where $k_1$ corresponds to $K_4$ and $k_2$ to $K_6$. The intrinsic crystal binding energy $k_0$ is constant and can be set to 0.

$$
E_{\text{anis}} = k_0 + k_1 (\alpha_1^2 \alpha_2^2 + \alpha_2^2 \alpha_3^2 + \alpha_3^2 \alpha_1^2) + k_2 (\alpha_1^2 \alpha_2^2 \alpha_3^2) + \dots
\tag{1.45}
$$

Using definitions of direction cosines, the anisotropy energy can be written as

$$
E_{\text{anis}} = k_1 \left( \frac{\sin^4 \phi}{4} + \sin^2 \phi \cdot \cos^2 \phi \right) + k_2 \left( \frac{\sin^4 \phi \cdot \cos^2 \phi}{4} \right)
\tag{1.46}
$$

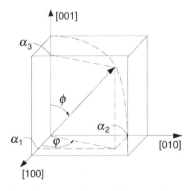

**FIGURE 1.79.** Directional cosines in a cubic crystal.

The expression for the total magnetic energy of the grain, $W_{tot}$, can be written as

$$E_{tot}(\varphi, \phi) = -\mu_0 H_{ext} M + \frac{\mu_0}{2} M^2 \left(N_a + \sin^2 \gamma \cdot (N_b - N_a)\right)$$
$$+ k_0 + k_1 \left(\frac{\sin^4 \phi}{4} + \sin^2 \phi \cdot \cos^2 \phi\right) + k_2 \left(\frac{\sin^4 \phi \cdot \cos^2 \phi}{4}\right) \tag{1.47}$$

Figure 1.80 shows the angle between the [001] axis and the magnetization vector, $\phi$, corresponding to the minimum energy for different ellipsoid shapes ($1 < c < 2$) and different angles between the [001] axis and the symmetry axes of the ellipsoid, $\varphi$. The values of the anisotropy energy constants, $k_1 = -80 \, \text{kJ/m}^3$ and $k_2 = -180 \, \text{kJ/m}^3$, represent the crystal anisotropy (78) of $Tb_{0.27}Dy_{0.73}Fe_{1.95}$ at 300 K. For instance; for $c = 1$ and $\phi = 54.7$, for all possible values of $\varphi$ (i.e., for a spherical grain), the state of energy is determined only by the crystal anisotropy; that is, the lowest energy will always coincide with the easy direction of magnetization, <111>. As $c$ increases the difference between $\phi$ and $\varphi$ will decrease, i.e., the magnetization vector will successively align itself with the symmetry axis of the ellipsoid.

## The Effect of Magnetic Aligning

Since ternary TERFENOL-D is a result of minimization of the anisotropy energy rather than maximization of the magnetostriction, the magneti-

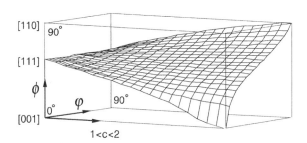

**FIGURE 1.80.**   The angle from the principal crystal axes corresponding to the least energy of an oblong ellipsoid for $1 < c < 2$.

cally easy direction at room temperature is undefined for the composition $Tb_{0.27}Dy_{0.73}Fe_2$. At temperatures higher than room temperature the easy axis is <111>, whereas for lower temperatures the easy axis is <100>. Therefore, magnetic alignment of the grains can have an effect opposite to the desired effect (i.e., aligning the <100> axis collinear with the central axis of the rod), thus reducing the saturation magnetostriction ($\lambda_{100} \approx 90 \cdot 10^{-6} < < \lambda_{111} \approx 1700 \cdot 10^{-6}$).

One way to ensure that the highly magnetostrictive <111> axis is collinear with the central axis is to use a terbium-rich composition, thus increasing the crystal anisotropy energy, i.e., making <111> a more dominant easy axis. A drawback associated with a high terbium content (besides having a high anisotropy energy) is that the magnetostriction exhibits an increased hysteresis as a function of the magnetizing field. Figure 1.81 shows the magnetostriction of a composite with the composition $Tb_{0.285}Dy_{0.715}Fe_2$ as a function of the magnetization. Note that, at a prestress >12 MPa, the curves are almost identical, suggesting that the domains at low-magnetizing fields mainly populate the directions which are nearly perpendicular to the easy <111> axis along which the stress is applied, thus resulting in an almost full magnetization rotation process. For full magnetization rotation of a TERFENOL-D crystal, the corresponding magnetostriction is $\lambda_s = 2400$ ppm. At a

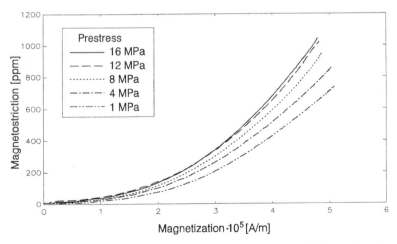

**FIGURE 1.81.** Magnetostriction of a composite TERFENOL-D rod with the composition $Tb_{0.285}Dy_{0.715}Fe_2$ as a function of the magnetization at different prestresses.

volume density of 71.5% for the aligned terbium-rich composite the saturation magnetostriction can be assumed (79) to be $0.715 \cdot \lambda_s$. Since the magnetization of a TERFENOL-D crystal is $M_s = 800 \, \text{kA/m}$, the saturation magnetization of the composite can be estimated to be $0.715 \cdot M_s$.

$$\lambda = \lambda \left( \frac{M}{M_s} \right)^2 \tag{1.48}$$

Figure 1.82 shows the magnetostriction as a function of magnetization at a prestress of 16 MPa for the terbium-rich aligned sample. The measured data (circles) are compared with the full magnetization rotation.

If the theoretical curve is shifted 42 kA/m to the right, the agreement between the curves becomes good, suggesting that the alignment of the terbium sample is almost perfect. The shifting of the magnetostriction verses magnetization curve is done to exclude the domain wall movements from the model, thus allowing only domain rotation. This difference can be explained by nonmagnetostrictive 180° domain wall motion.

In order to clarify the role of the anisotropy energy for the degree of alignment, two samples with the composition $Tb_{0.27}Dy_{0.73}Fe_2$ were prepared—one magnetic aligned and the other magnetically isotropic—

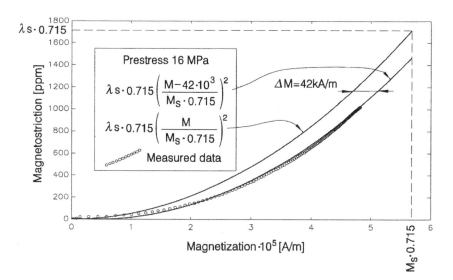

**FIGURE 1.82.**  Magnetostriction as a function of magnetization. The measured data (circles) are compared with the theoretical full magnetization rotation.

and one magnetically aligned sample with the composition $Tb_{0.285}Dy_{0.715}Fe_2$ was also prepared. Figure 1.83 shows the magnetostriction versus magnetization for the three samples. Clearly, the alignment is less efficient for the low-terbium aligned sample than for the $Tb_{0.285}Dy_{0.715}Fe_2$ sample. This is mostly due to the less dominant easy <111> direction of the low-terbium sample (i.e., the reduced anisotropy energy). The gray zone in Fig. 1.83 is limited by the magnetostrictive curve for a single-crystal, full-magnetization rotation, modified with respect to the density and the 180° domain wall motion:

$$\lambda = \lambda_s \cdot 0.715 \left( \frac{M - 42 \cdot 10^3}{M_s \cdot 0.715} \right)^2 \tag{1.49}$$

The lower boundary of the gray zone is determined by a theoretical expression for a randomly oriented isotropic magnetostrictive polycrystal (77):

$$\lambda = \frac{2\lambda_{100} + 3\lambda_{111}}{5} \cdot \left( \frac{M - 42 \cdot 10^3}{M_s \cdot 0.715} \right)^2 \tag{1.50}$$

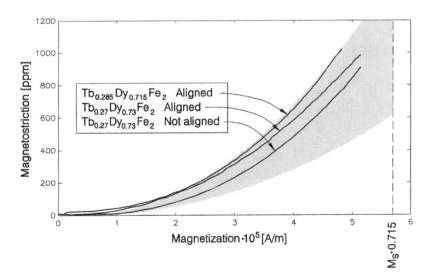

**FIGURE 1.83.** Comparison between the efficiency of magnetic alignment for different compositions of the magnetostrictive grains.

This equation represents an extreme assumption; that is, the strain is uniform throughout the sample, but the stress varies from grain to grain. This assumption is usually considered to be physically more realistic than assuming uniform stress and varying strain. A comparison with the curve for the magnetically isotropic composite with the composition $Tb_{0.285}Dy_{0.715}Fe_2$ shows that this assumption is not valid, but an assumption not as extreme as those of uniform stress and uniform strain might lead to an expression which better represents the magnetization process.

## Physical Properties

### The Tensile Strength of GMPC

The mechanical strength of GMPC is very important because giant magnetostrictive composites are usually used in resonance in which the amplitude of the stress can be extremely high. TERFENOL-D is very brittle and its ability to withstand mechanical tensile stress is poor, thus reducing its suitability for high-power generation of mechanical vibrations. However, Petterson et al. (80) showed that by reducing the iron content, the strength of TERFENOL-D could be dramatically increased.

TERFENOL-D, $Tb_{0.27}Dy_{0.73}Fe_x$, solidifies as very brittle dendrites. If $x < 2$, the material between the dendrites consists of a two-phase eutectic mixture of the rare earth metals plus TERFENOL-D or it is a pure rare earth metal via a divorced eutectic mechanism. The ductile rare earth metal behaves as an interconnected skeleton network throughout the TERFENOL-D primary phase. The presence of a ductile skeleton network is expected to enhance the strength by retarding crack propagation throughout the brittle primary phase.

For GMPC, there are no obvious advantages in keeping the Fe content, $x$, under 2 since $x = 2$ have the best magnetostrictive properties (Fig. 1.84). Furthermore, the binder will form a ductile matrix (in fact, it is very similar to the ductile rare earth phase in $Tb_{0.27}Dy_{0.73}Fe_x, x < 2$) surrounding the TERFENOL-D particles, thus enhancing the mechanical properties compared with ordinary TERFENOL-D rods. In order to examine the tensile strength of GMPC, some rods were prepared with a uniaxial pressing technique. The rods were broken in a three-point bending test machine. Since the rupture is brittle (i.e., there is no plastic deformation prior to fracture), the maximum stress, calculated from the

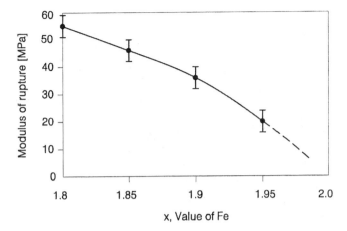

**FIGURE 1.84.** Variation in the modulus of rupture with Fe stoichiometry in $Tb_{0.27}Dy_{0.73}Fe_x$.

fracture loading, is equivalent to the maximum tensile strength of the sample.

Table 1.13 shows the modulus of rupture for seven GMPC rods. The large range in the results is probably due to imperfect binder coating of the TERFENOL-D particles. The fracture mechanism of the composite is not particle pull-out from the matrix but rather cracking throughout the

**Table 1.13.** Maximum Tensile Stress at Fracture Measured in Three-Point Bending Tests

| Sample No. | Rupture stress (MPa) |
| --- | --- |
| 1 | 52.7 |
| 2 | 35.3 |
| 3 | 50.7 |
| 4 | 55.8 |
| 5 | 38.0 |
| 6 | 67.5 |
| 7 | 60.0 |

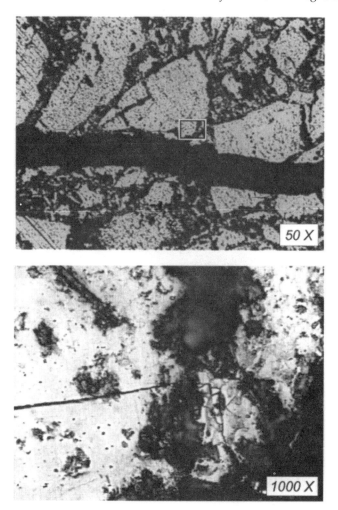

**FIGURE 1.85.** Fracture of TERFENOL-D particles in a GMPC rod. (Top) At 50 times enlargement, a crack dividing TERFENOL-D particles can be seen. (Bottom) Detail enlargement showing content of the square indicated in the top figure.

brittle TERFENOL-D particles. Figure 1.85 (top) shows a crack on the side of the GMPC rod subjected to tensile stress, photographed at 50 times enlargement. It is obvious that the particles have been split. Figure 1.85 (bottom) shows an additional enlargement of the crack in the area marked in the top of the figure.

# REFERENCES

*a.* E. du Tremolet de Lacheisserie, "Magnetostriction: Theory and Applications of Magnetoelasticity," CRC Press, Boca Raton, FL, 1993.

1. A. E. Clark, J. B. Restorff, M. Wun-Fogle, and J. F. Lindberg, Magnetoelastic coupling and $\Delta E$ effect in $Tb_x Dy_{1-x}$ single crystals, *J. Appl. Phys.* **73**, 6150, (1993). [See also A. E. Clark, J. B. Restorff, M. Wun-Fogle, and J. F. Lindberg, Piezomagnetic properties and $\Delta E$ effect in TbZn at 77K, *J. Magn. Magn. Mat.* **140–144**, 1151 (1995)].

2. See any standard quantum mechanics text, for example, R. Eisberg and R. Resnick, "Quantum Physics of Atoms, Molecules, Solids, Nuclei, and Particles," Chapters 8–10. Wiley, New York, 1974.

3. T. Thole, *in* R. Coehoorn, *in* "Supermagnets, Hard Magnets" (G. J. Long and F. Grandjean, Eds.), pp.133–170. Kluwer Academic, Dordrecht, 1991.

4. E. P. Wolfarth, *in* "Ferromagnetic Materials" (E. P. Wolfarth, Ed.), Vol. 1, pp. 1–70. North-Holland, Amsterdam, 1980.

5. E. Callen and H. Callen, Magnetostriction forced magnetostriction and anomalous thermal expansion in ferromagnets, *Phys. Rev.* **139**, A455–A471, (1965).

6. A. E. Clark, B. F. DeSavage, and E. R. Callen, Magnetostriction of single crystal dysprosium, gadolinium iron garnet and dysprosium iron garnet, *J. Appl. Phys.* **35**, 1028, (1964).

7. A. E. Clark, R. Abbundi, H. T. Savage, and O. D. McMasters, Magnetostriction of rare-earth-$Fe_2$ laves phase compounds, *Physica* **73**, 86–88, (1977).

8. S. Chikazumi, "The Physics of Magnetism," p. 57. Wiley, New York, 1964. [Most quantum mechanics texts contain a short discussion of spin-orbit coupling. See for example L. I. Schiff, "Quantum Mechanics," 3rd ed., p. 433. McGraw-Hill, New York, 1968.]

9. N. S. Akulov, *Z. Physik* **52**, 389, (1928).

10. R. Becker and W. D. Doring, "Ferromagnetismus." Springer, Berlin, 1939.

11. A. E. Clark, Magnetostriction rare-earth-$Fe_2$ compounds, *in* "Ferromagnetic Materials" (E. P. Wolfarth, Ed.), Vol. 1, pp. 531–589. North-Holland, Amsterdam, 1980.

12. R. Wu, L. J. Chen, A. Shick, and A. J. Freeman, First-principles determinations of magnetocrystalline anisotropy and magnetostriction in bulk and thin-film transition metals, *J. Magn. Magn. Mat.* **177–181**, 1216, (1998).

13. M. L. Spano, K .B. Hathaway, and H. T. Savage, Magnetostriction and magnetic anistropy of field annealed metglas 2605 alloys via dc M-H loop measurements under stress, *J. Appl. Phys.* **53**, 2667, (1982).

14. A. E. Clark and H. T. Savage, Magnetostriction of rare earth-$Fe_2$ compounds under compressive stress, *J. Magn. Magn. Mat.* **31–34**, 849 (1983).

15. J. R. Cullen, S. Rinaldi, and G. V. Blessing, Elastic versus magnetoelastic anisotropy in rare earth-iron alloys, *J. Appl. Phys.* **49**, 1960, (1978).

16. J. P. Joule, *Ann. Electr. Magn. Chem.* **8**, 219 (1842); *Phil. Mag.* [3] **30**, 76 (1847).

17. A. E. Clark, B. F. DeSavage, and R. M. Bozorth, Anomalous thermal expansion of single crystal dysprosium, *Phys. Rev.* **138**, A216, (1965).

18. Stevens, K. W. H., Matrix elements and operator equivalents connected with the magnetic properties of rare earth ions, *Proc. Phys. Soc.* **65**, 209, (1952).

19. M. L. Spano, A. E. Clark, and M. Wun-Fogle, Magnetostriction of Dy-rich $Tb_xDy_{1-x}$ single crystals, *IEEE Trans. Magn.* **25**, 3794, (1989).

20. A. E. Clark, M. Wun-Fogle, J. B. Restorff, and J. F. Lindberg, Magneto-mechanical properties of single crystal TbxDy1-x under compressive stress, *IEEE Trans. Magn.* **28**, 3156, (1992).

21. A. E. Clark and H. S. Belson, Giant room-temperature magnetostrictions in $TbF_2$ and $DyFe_2$, *Phys. Rev.* B5, 3642, (1972); "Magnetostriction of Terbium Iron and Erbium Iron Alloys," IEEE Trans. on Magnetics MAG-8, 477, (1972).

22. C. Williams and N. Koon, Anisotropy of single crystal $Ho_xDy_y$ $Tb_{1-x-y}Fe$ laves phase compounds, *Physica* **86-88B**, 14, (1977).

23. K. B. Hathaway and A. E. Clark, Magnetostrictive materials, *MRS Bull.* **18**, 24–41, (1993).

24. R. Abbundi and A. E. Clark, Anomalous thermal expansion and magnetostriction of single crystal $Tb_{.27}Dy_{.73}Fe_2$, *IEEE Trans. Magn* **MAG-13**, 1519, (1977).

25. A. E. Clark, and D. N. Crowder, High temperature magnetostriction of TbFe2 and $Tb_{.27}Dy_{.73}Fe_2$, *IEEE Trans. Magn.* **MAG-21**, 1945, (1985).

26. J. B. Restorff, M. Wun-Fogle, J. P. Teter, J. R. Cullen, and A. E. Clark, Magnetic and magnetoelastic properties of single crystal Tb.5Dy.5Zn, *IEEE Trans. Magn.* **32**, 4782, (1996).

27. A. E. Clark, J. B. Restorff, M. Wun-Fogle, and J. F. Lindberg, Piezomagnetic properties and $\Delta E$ effect in TbZn at 77 K, *J. Magn. Magn. Mat* **140–144**, 1151, (1995).

28. J. R. Cullen, A. E. Clark, and K. B. Hathaway, *in* "Materials Science and Technology" (K. J. H. Buschow, R. W. Kahn, P. Hassen, and E. J. Kramer, Eds.), Vol. 3B, pp. 529–565. VCH, Weinheim, 1994.

29. A. E. Clark, Magnetostrictive rare-earth-$Fe_2$ compounds, *in* "Ferromagnetic Materials" (E.P. Wohlforth, Ed.) Vol. 1 pp. 531–589. North-Holland, Amsterdam, 1980.

30. K. A. Gschneidner, Jr., Systematics of the intra-rare-earth binary alloy systems, *J. Less Common Metals* **114**, 29–42, (1985).

31. M. P. Dariel, J. T. Holthuis, and M. R. Pickus, The terbium-iron phase diagram, *J. Less Common Metals* **45**, 91–101, (1976).

32. A. S. Van Der Goot and K. H. J. Buschow, The dysprosium-iron system: Structural and magnetic properties of dysprosium-iron compounds, *J. Less Common Metals*, **21**, 151–157, (1970).

33.  O. D. McMasters, G. E. Holland, and K. A. Gschneidner, Jr. Single crystal growth by the horizontal levitation zone melting method, *J. Crystal Growth* **43**, 577–583, (1978).

34.  D. T. Peterson, J. D. Verhoeven, O. D. McMasters, and W. A. Spitzig, Strength of TERFENOL-D, *J. Appl. Phys.* **65**(9), 3712–3713 (1989, May 1).

35.  J. D. Verhoeven, J. E. Ostensen, E. D. Gibson, and O. D. McMasters, The effect of composition and magnetic heat treatment on the magnetostriction of $Tb_xDy_{1-x}Fe_y$ twinned single crystals, *J. Appl. Phys.* **66**(2), 772–779 (1989, July 15).

36.  A. E. Clark, J. D. Verhoeven, O. D. McMasters, and E. D. Gibson, Magnetostriction in twinned [112] crystals of $Tb_{.27}Dy_{.73}Fe_2$, *IEEE Trans.* **MAG-22**, 973–975, (1986).

37.  J. D. Verhoeven, E. D. Gibson, O. D. McMasters, H. H. Baker, The growth of single crystals, *Met. Trans. A* **18A**, 223–231, (1987).

38.  A. E. Clark, J. P. Teter, and M. Wun-Fogle, Anisotropy compensation and magnetostriction in $Tb_xDy_{1-x}(Fe_{1-y}T_y)_{1.9}$ (T = Co, Mn), *J. Appl. Phys.* **69**(8), 5771–5773 (1991, April 15).

39.  A. E. Clark, J. P. Teter, and O. D. McMasters, Magnetostriction "jumps" in twinned $Tb_{.3}Dy_{.7}Fe_{1.9}$, *J. Appl. Phys.* **63**(8), 3910–3912 (1988, April 15).

40.  P. Westwood, J. S. Abell, and K. C. Pitman, Phase relationships in the Tb-Dy-Fe ternary system, *J. Appl. Phys.* **67**(9), 4998–5000 (1990, May 1).

41.  T. Uchida and T. Mori, Magnestostrictive properties of laves phase $RFe_2$ (R = Ce, Pr, Nd), *in* "Proceedings of the International Symposium on Giant Magnetostrictive Materials and Their Applications," Tokyo, Japan, Nov. 5–6, 1992, pp. 169–174.

42.  "ASM Handbook," Vol. 3: Alloy Phase Diagrams, 1992.

43.  C. Kittel and J. J. Galt, *in* "Solid State Physics, Vol. 3" (F. Seitz and D. Turnbull, Eds.). Academic Press, New York, 1956.

44.  D. J. Craik and R. S. Tebble, "Ferromagnetism and Ferromagnetic Domains." North-Holland, Amsterdam, 1965.

45.  S. Chikazumi, "Physics of Magnetism." Wiley, New York, 1998.

46.  A. Hubert, "Magnetic Domains," in press.

47.  J. S. Abel and D. G. Lord, Microstructure observations of rare earth iron alloys, *J. Less Common Metals* **126**, 107, (1986).

48.  Y. J. Bi and J. S. Abell, Microstructure of rare earth intermetallic single crystals, *J. Alloys Comp.* **207/208**, 321, (1994).

49.  Y. J. Bi, J. S. Abell, and A. M. H. Hwang, Defects in TERFENOL-D crystals, *J Magn Magn Mat.* **99**, 159, (1991).

50.  A. G. I. Jenner, D. G. Lord, and C. A. Faunce, Microstructure and magnetic properties of TbDyFe, *IEEE Trans. Magn.* **24**, 1865, (1988).

51.  A. G. Jenner, D. G. Lord, and C. A. Faunce, Magnetoelastic properties of terbium-dysprosium-iron compounds, *J. Appl. Phys.* **69**, 5780, (1991).

52. M. M. Al-Jiboory, D. G. Lord, Y. Bi, J. S. Abell, and J. P. Teter, Magnetic domains and microstructure in TERFENOL crystals, *J. Appl. Phys.* **73**, 6168, (1993).

53. A. P. Holden, D. G. Lord, and P. J. Grundy, Transmission electron microscopy of TERFENOL-D crystals, *J.Appl. Phys.* **79**, 4650, (1996).

54. R. D. James and D. Kinderlehrer, Theory of magnetostriction with application to TERFENOL-D, *J.Appl. Phys.* **76**, 7012, (1994).

55. X. G. Zhao and D. G. Lord, Effect of twin boundaries on the magnetization processes in TERFENOL-D, *in* "Proceedings of the International Actuator Conference '98", Bremen, Germany, 1998, p. 73.

56. H. T. Savage, R. Abbundi, A. E. Clark, and O. D. McMasters, *J.Appl.Phys.* **50**, 1674, (1979).

57. P. Westwood, J. S. Abell, and K. C. Pitman, Phase relationships in the Tb-Dy-Fe ternary system, *J.Appl.Phys.* **67**, 4998, (1990).

58. G. F. Clark, B. K. Tanner, and H. T. Savage, *Phil.Mag.* **B46**, 331, (1982).

59. D. G. Lord, H. T. Savage, and R. G. Rosemeier, X-ray topography observation of a DyTbFe crystal, *J. Magn. Magn. Mat.* **29**, 684, (1982).

60. D. Y. Parpia, B. K. Tanner, and D. G. Lord, Direct observation of ferromagnetic domains, *Nature* **303**, 684, (1983).

61. A. L. Janio, A. Branwood, R. Dudley, and A. R. Piercy, *J.Phys.D.* **20**, 24, (1987).

62. D. G. Lord, V. Elliott, A. E. Clark, H. T. Savage, J. P. Teter, and O. D. McMasters, Optical observation of closure domains in TERFENOL-D single crystals, *IEEE Trans. Magn.* **24**, 1716, (1988).

63. J. P. Teter, K. Mahoney, M. Al-Jiboory, D. G. Lord, and O. D. McMasters, Domain observations and magnetostriction in TbDyFe twinned single crystals, *J. Appl. Phys.* **69**, 5768 (1991).

64. A. P. Holden, D. G. Lord, and P. J. Grundy, Surface deformations and domains in TERFENOL-D crystals by scanning probe microscopy, *J. Appl. Phys.* **79**, 6070, (1996).

65. P. Grutter, H. J. Mamin, and D. Rugar, *in* "Magnetic Force Microscopy (MFM) in Scanning Tunnelling Microscopy", Vol. II (H. -J. Guntherodt and R. Wiessendanger, (Eds). Springer-Verlag, Berlin, 1992.

66. A. Hubert, W. Rave, and S. Tomlinson, Imaging magnetic charges with magnetic force microscopy, *Physica Stat. Sol.*

67. D. G. Lord, A. P. Holden, and P. J. Grundy, Magnetic force microscopy of Terfenol-D fracture surfaces, *J. Appl. Phys.* **81**, 6200, (1996).

68. O. D. McMasters, Method of forming magnetostrictive rods from rare earth-iron alloys, U.S. Patent No. 4, 609, 402 (issued September 2, 1986).

69. E. D. Gibson, J. D. Verhoeven, F. A. Schmidt, and O. D. McMasters, Continuous method for manufacturing grain-oriented magnetostrictive bodies, U.S. Patent No. 4,770,704 (issued July 18, 1988).

70. J. D. Verhoeven, E. D. Gibson, O. D. McMasters, and J. E. Ostenson, Directional solidification and heat treatment of TERFENOL-D magnestostrictive materials, *Met. Trans. A* **21A**, 2249–2255, (1990).

71. J. D. Verhoeven, O. D. McMasters, and E. D. Gibson, Thermal treatment for increasing magnetostrictive response of rare earth-iron alloy rods, U.S. Patent No. 4,849,034 (issued July 18, 1989).

72. A. M. H. Hwang, P. Westwood, Y. J. Bi, and J. S. Abell, "Effect of Heat Treatment on the Magnestostriction and Microstructure of TERFENOL-D."

73. L. Sandlund, and T. Cedell, TERFENOL-D powder composite with high frequency performance, *in* "Third International Workshop on Transducers for Sonics and Ultrasonics," May 6–8, 1992. Orlando, FL. Hosted by the Naval Research Laboratory. conference proceedings, 1992 (ISBN 0-87762-993-5, 113–118).

74. L. Sandlund, M. Fahlander, T. Cedell, A. E. Clark, J. B. Restorff, M. Wun-Fogle, Magnetostriction, elastic moduli and coupling factors of composite TERFENOL-D, *J. Appl. Phys.* **75** (10), 15. May 1994.

75. M. Malekzadeh, R. Milton, Sintered rare earth-iron laves phase magnetostrictive alloy product and preparation thereof, U.S. Patent No. 4,152,178.

76. D. C. Jiles, "Introduction to Magnetism and Magnetic Materials." Chapman & Hall, New York, 1991, [ISBN 0-412-38630-5].

77. B. D. Cullity, "Introduction to Magnetic Materials." Addison–Wesley, Reading, MA, 1972. [ISBN 0-201-01218-9].

78. J. P. Teter, A. E. Clark, and O. D. McMasters, Anisotropic magnetostriction in $Tb_{0.27}Dy_{0.73}Fe_{1.95}$, *J. Appl. Phys.* **61**, 1987.

79. F. E. Baker, S. H. Brown, J. R. Brauer, and T. R. Gerhardt, Comparison of magnetic fields computed by finite element and classical series methods, *Int. J. Numerical Methods Eng.* **19**, 271–280, (1983).

80. D. T. Petterson, J. D. Verhoeven, O. D. McMasters, and W. A. Spitzig, Strength of TERFENOL-D, *J. Appl. Phys.* **65**(9) (1989).

## CHAPTER 2

# Modeling of Giant Magnetostrictive Materials

Göran Engdahl
*Royal Institute of Technology*
*Stockholm, Sweden*

## LINEAR MODELS

Since the discovery of magnetostriction in ferromagnetic metals such as Fe and Ni and various alloys, models have been used to describe this phenomenon in a macroscopic way. The first models were linear (1) due to the fact that ordinary magnetostriction at moderate exitations shows reasonable linearity and that the linear mathematical treatment leads to effective algorithms. Another assumption in these models is that the material shows no hysteresis effects, i.e., the material shows reversible properties.

### Definitions of Stress and Strain

In order to facilitate the understanding of the magnetoelastic processes in magnetostrictive materials and their device applications the definition of stress and strain usually adopted in literature (2) is first reviewed. A volume element of a magnetostrictive material is shown in Fig. 2.1. It is assumed to be subjected to stress. Two kinds of forces can act on the element. First, there are body forces which act throughout the body with a magnitude proportional to the volume of the element. Such a body force can, for example, be due to gravity or electromagnetic interaction. Second, there are forces acting on the surfaces of the element due to interaction with surrounding material. In this case, the forces per unit area defines the stress of the element. The stresses shown in Fig. 2.1 can be described as a tensor $[T_{ij}]$:

ISBN: 0-12-238640-X/$30.00.

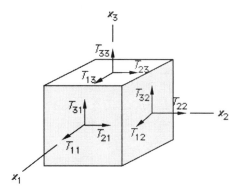

**FIGURE 2.1.**  The forces on the faces of a unit cube in a stressed body.

$$\left[T_{ij}\right] = \begin{bmatrix} T_{11} & T_{12} & T_{13} \\ T_{21} & T_{22} & T_{23} \\ T_{31} & T_{32} & T_{33} \end{bmatrix} \tag{2.1}$$

Normally, the body forces can be assumed to be small compared to the forces due to externally applied mechanical forces. Nye (2) shows that the stress tensor must be symmetric in the absence of body forces, i.e., $T_{ij} = T_{ji}$. This even holds for an inhomogeneously stressed body. Hence, there are six independent elements in the stress tensor. It is therefore convenient to express the stress in the form of a six-component vector. The relation between the indices in the tensor and vector notations is as follows:

| Tensor notation | 11 | 22 | 33 | 23,32 | 31,13 | 12,21 |
|---|---|---|---|---|---|---|
| Vector notation | 1 | 2 | 3 | 4 | 5 | 6 |

The strain definition can be illustrated in two dimensions (Fig. 2.2). From the deformation of the body in Fig. 2.2, the following quantities are defined:

$$e_{11} = \frac{\partial u_1}{\partial x_1}; e_{12} = \frac{\partial u_1}{\partial x_2}; e_{21} = \frac{\partial u_2}{\partial x_1}; e_{22} = \frac{\partial u_2}{\partial x_2}; \tag{2.2}$$

Writing the same quantities in three dimensions by using the Einstein's summation convention gives

$$e_{ij} = \frac{\partial u_i}{\partial x_j} (i, j = 1, 2, 3) \tag{2.3}$$

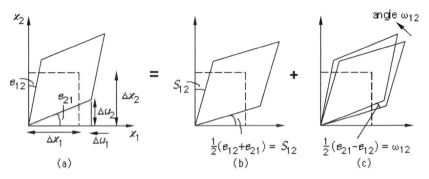

**FIGURE 2.2.** Undeformed (dashed) and deformed (solid) body. The general deformation shown in (a) can be represented by a strain (b) plus a rotation (c).

This convention will be used throughout this chapter.

The tensor $[e_{ij}]$ is not a satisfactory measure of the strain of a deformed body because a pure rotation of the body without deformation gives $e_{ij} \neq 0$ for $i \neq j$. Therefore, $[e_{ij}]$ is decomposed into a pure strain and a pure rotation, as shown in Figs. 2.2b and 2.2c. The symmetrical part of $[e_{ij}]$ is defined as the strain tensor $[S_{ij}]$:

$$[S_{ij}] = \begin{bmatrix} S_{11} & S_{12} & S_{13} \\ S_{21} & S_{22} & S_{23} \\ S_{31} & S_{32} & S_{33} \end{bmatrix} = \begin{bmatrix} e_{11} & \frac{1}{2}(e_{12}+e_{21}) & \frac{1}{2}(e_{13}+e_{31}) \\ \frac{1}{2}(e_{12}+e_{21}) & e_{22} & \frac{1}{2}(e_{23}+e_{32}) \\ \frac{1}{2}(e_{13}+e_{31}) & \frac{1}{2}(e_{23}+e_{32}) & e_{33} \end{bmatrix} \quad (2.4)$$

Observe that the off-diagonal elements of $[S_{ij}]$, which are the shear strains, only correspond to one-half of the angle change between the deformed and the undeformed body in Fig. 2.2a. The antisymmetric part of $[e_{ij}]$ gives the rotation tensor $[\omega_{ij}]$:

$$\begin{bmatrix} 0 & \omega_{12} & \omega_{13} \\ \omega_{21} & 0 & \omega_{23} \\ \omega_{31} & \omega_{32} & 0 \end{bmatrix} = \begin{bmatrix} 0 & -\frac{1}{2}(e_{21}-e_{12}) & \frac{1}{2}(e_{13}-e_{31}) \\ \frac{1}{2}(e_{12}-e_{21}) & 0 & -\frac{1}{2}(e_{32}-e_{23}) \\ -\frac{1}{2}(e_{13}-e_{31}) & \frac{1}{2}(e_{32}-e_{23}) & 0 \end{bmatrix} \quad (2.5)$$

so

$$[e_{ij}] = [S_{ij}] + [\omega_{ij}]$$

The strain tensor $[S_{ij}]$ is symmetric and therefore has six independent elements. Like the stress, the strain can be expressed as a six-component vector using the same relations between the indices in the matrix and vector notations. In the vector notation the shear strains correspond to the whole change in the angle between the deformed and the undeformed body. This gives $(S_1, S_2, S_3, S_4, S_5, S_6) = (S_{11}, S_{22}, S_{33}, 2S_{23}, 2S_{31}, 2S_{12})$, where $S_{23} = S_{32}$, $S_{13} = S_{31}$, and $S_{12} = S_{21}$. $S_i$, $i = 1, \cdots 6$, are sometimes called engineering strains.

## Energy and Work of Magnetostrictive Materials

When a unit volume of a magnetostrictive material is magnetized by the magnetic flux density $d\mathbf{B}$, a magnetic work $dW$,

$$dW = H_m dB_m \quad (m = 1, 2, 3; \text{refers to the } x_1, x_2, \text{and } x_3 \text{ directions}) \quad (2.6)$$

will be added to it.

If at the same time this volume is subjected to a mechanical strain $dS_i$ and this deformation process is isothermal and reversible, the mechanical work delivered to the same volume is

$$T_i dS_i \quad (i = 1, 2, \cdots 6; \text{refers to the components of engineering strains}) \tag{2.7}$$

In this reversible system the change of the internal energy then is $dU$:

$$dU = T_i dS_i + H_m dB_m \tag{2.8}$$

The Gibbs energy for adiabatic conditions can be written as

$$G = U - T_i S_i - H_m B_m \tag{2.9}$$

By differentiation, one gets

$$dG = dU - dT_i S_i - dH_m B_m - T_i dS_i - H_m dB_m \tag{2.10}$$

and for reversible processes

$$dG = -S_i dT_i - B_m dH_m \tag{2.11}$$

which gives

$$S_i = \left. \frac{-\partial G}{\partial T_i} \right|_H ; \quad B_m = \left. \frac{-\partial G}{\partial H_m} \right|_T \tag{2.12}$$

Taking the partial derivative of $S_i$ with respect to $H_m$ and the partial derivative of $B_m$ with respect to $T_i$ gives

$$\left. \frac{\partial S_i}{\partial H_m} \right|_T = \left. \frac{\partial B_m}{\partial T_i} \right|_H = d_{mi} \tag{2.13}$$

In the most common mode of operation of magnetostrictive

materials, the longitudinal mode, the applied magnetic field and stress have components only in the $x_3$ direction. In this case,

$$\left.\frac{\partial S_3}{\partial H_3}\right|_T = \left.\frac{\partial B_3}{\partial T_3}\right|_H = d_{33} \qquad (2.14)$$

where $d_{33}$ or $d$ is known as the magnetostrictive constant.

## Magnetomechanical Coupling

A characteristic property of magnetostrictive materials is that a mechanical strain will occur if they are subjected to a magnetic field in addition to strains originated from pure applied stresses. Also, their magnetization changes due to changes in applied mechanical stresses in addition to the changes caused by changes of the applied magnetic field.

These dependencies can be described by mathematical functions if one assumes that the material shows reversible properties

$$\mathbf{S} = \mathbf{S}(\mathbf{T}, \mathbf{H}) \qquad (2.15a)$$

$$\mathbf{B} = \mathbf{B}(\mathbf{T}, \mathbf{H}) \qquad (2.15b)$$

Differentiation of the previous equations and using the previous defined notations give

$$dS_i = \left.\frac{\partial S_i}{\partial T_j}\right|_H dT_j + \left.\frac{\partial S_i}{\partial H_k}\right|_T dH_k, i = 1, \cdots 6 \qquad (2.16a)$$

$$dB_i = \left.\frac{\partial B_m}{\partial T_j}\right|_H dT_j + \left.\frac{\partial B_m}{\partial H_k}\right|_T dH_k, m = 1, 2, 3 \qquad (2.16b)$$

In this equation system the magnetostrictive constant $d_{mi}$, sometimes called the piezomagnetic constant, can be identified ($d_{mi} = d_{im}$ by symmetry).

Moreover, it is common to denote constants of compliances and permeabilities according to

$$\left.\frac{\partial S_i}{\partial T_j}\right|_H = s_{ij}{}^H, \text{elastic compliances at const.} H \tag{2.17a}$$

$$\left.\frac{\partial B_m}{\partial H_k}\right|_T = \mu_{mk}{}^T, \text{magnetic permeabilities at const.} T \tag{2.17b}$$

For small variations in $d\mathbf{T}$ and $d\mathbf{H}$ the linearized constitutive equations can be formulated as

$$S_i = s_{ij}{}^H T_j + d_{ki} H_k, i = 1, \cdots 6 \tag{2.18a}$$

$$B_m = d_{mj} T_j + \mu_{mk}{}^T H_k, m = 1, 2, 3 \tag{2.18b}$$

Several coefficients in Eq. (2.18) may be zero and some may have the same value due to symmetry. For a polycrystalline ferromagnetic material with $x_3$ chosen as the direction of the magnetic polarization (magnetic bias) and prestress (mechanical bias), Eq. (2.18) turns out to be (3, 4)

$$\begin{bmatrix} S_1 \\ S_2 \\ S_3 \\ S_4 \\ S_5 \\ S_6 \end{bmatrix} = \begin{bmatrix} s_{11}^H & s_{12}^H & s_{13}^H & 0 & 0 & 0 \\ s_{12}^H & s_{11}^H & s_{13}^H & 0 & 0 & 0 \\ s_{13}^H & s_{13}^H & s_{33}^H & 0 & 0 & 0 \\ 0 & 0 & 0 & s_{44}^H & 0 & 0 \\ 0 & 0 & 0 & 0 & s_{44}^H & 0 \\ 0 & 0 & 0 & 0 & 0 & s_{66}^H \end{bmatrix} \cdot \begin{bmatrix} T_1 \\ T_2 \\ T_3 \\ T_4 \\ T_5 \\ T_6 \end{bmatrix} + \begin{bmatrix} 0 & 0 & d_{31} \\ 0 & 0 & d_{31} \\ 0 & 0 & d_{33} \\ 0 & d_{15} & 0 \\ d_{15} & 0 & 0 \\ 0 & 0 & 0 \end{bmatrix} \cdot \begin{bmatrix} H_1 \\ H_2 \\ H_3 \end{bmatrix} \tag{2.19}$$

$$\begin{bmatrix} B_1 \\ B_2 \\ B_3 \end{bmatrix} = \begin{bmatrix} 0 & 0 & 0 & 0 & d_{15} & 0 \\ 0 & 0 & 0 & d_{15} & 0 & 0 \\ d_{31} & d_{31} & d_{33} & 0 & 0 & 0 \end{bmatrix} \cdot \begin{bmatrix} T_1 \\ T_2 \\ T_3 \\ T_4 \\ T_5 \\ T_6 \end{bmatrix} + \begin{bmatrix} \mu_{11}^T & 0 & 0 \\ 0 & \mu_{11}^T & 0 \\ 0 & 0 & \mu_{33}^T \end{bmatrix} \cdot \begin{bmatrix} H_1 \\ H_2 \\ H_3 \end{bmatrix} \tag{2.20}$$

Assuming linear relationship between $\mathbf{B}$ and $\mathbf{H}$ and $\mathbf{S}$ and $\mathbf{T}$ gives the internal energy as (5)

$$U = \frac{1}{2} S_i T_i + \frac{1}{2} H_m B_m \tag{2.21}$$

Substitution of $S_i$ and $B_m$ in Eq. (2.21) by Eqs. (2.19) and (2.20) gives

$$U = \frac{1}{2} T_i s_{ij} T_j + \frac{1}{2} T_i d_{mi} H_m + \frac{1}{2} H_m d_{mi} T_i + \frac{1}{2} H_m \mu_{mk} H_k \tag{2.22}$$

$$= U_e + U_{em} + U_{me} + U_m = U_e + 2U_{me} + U_m$$

where $U_e$ is the pure elastic energy, $U_{me} = U_{em}$ is the mutual magnetoelastc energy, and $U_m$ is the pure magnetic energy of the system. One important single figure of merit for magnetostrictive materials is the magnetomechanical coupling coefficient $k$, which is defined as the ratio of the magnetoelastic energy to the geometric mean of the elastic and magnetic energy (5),

$$k = \frac{U_{me}}{\sqrt{U_e U_m}} \tag{2.23}$$

In cases in which the applied magnetic field and the stress have components only in the direction of $x_3$. $H_1 = H_2 = 0$, $H_3 \neq 0$ and $T_1 = T_2 = 0$, $T_3 \neq 0$. In eqs. (2.19) and (2.20) one then gets $B_1 = B_2 = 0$, $B_3 \neq 0$, and $S_4 = S_5 = S_6 = 0$, $S_1 = S_2 \neq 0$, $S_3 \neq 0$. $S_1$ and $S_2$, however, do not contribute to the energy of the system since $T_1 = T_2 = 0$, so the square of the longitudinal coupling coefficient becomes

$$k_{33}{}^2 = \frac{d_{33}{}^2}{\mu_{33}{}^T s_{33}{}^H} \tag{2.24}$$

## Longitudinal Coupling

The most important mode of operating magnetostrictive materials is the longitudinal. Let us assume this direction along the $x_3$ axis. Therefore, stresses, strains, and magnetic field quantities are directed parallel to this axis. It is then convenient to omit all subscripts of 3.

In this case the constitutive linearized equations that describe the magnetostriction or piezomagnetism appear as follows:

$$S = s^H T + dH \tag{2.25a}$$

$$B = dT + \mu^T H \tag{2.25b}$$

An interesting alternative form can be obtained by eliminating $H$ from the first equation and then eliminating $T$ from the second equation:

$$S = s^B T + \frac{d}{\mu^T} B \tag{2.26a}$$

$$B = \frac{d}{s^H} S + \mu^S H \tag{2.26b}$$

where

$$s^B = \frac{dS}{dT}\Big|_B = s^H \left(1 - \frac{d^2}{s^H \mu^T}\right) = s^H (1 - k^2) \tag{2.27a}$$

$$\mu^S = \frac{dB}{dH}\Big|_S = \mu^T \left(1 - \frac{d^2}{s^H \mu^T}\right) = \mu^T (1 - k^2) \tag{2.27b}$$

The factor $k$ can easily be identified as the coupling factor in the longitudinal case, where all indices of 3 are dropped.

Note that for constant magnetic induction, i.e., under short-circuit conditions $(B = 0)$, it is evident that the elastic modulus is reduced through the coupling factor. Similarly, under blocked or clamped conditions $(S = 0)$ the magnetic permeability is also reduced through the coupling factor.

Division of the previous equations gives

$$\mu^S s^H = \mu^T s^B \qquad (2.28)$$

The constitutive magnetostriction equations can, after elimination of $S$ in Eq. (2.27), be written as

$$S = s^B T + gB \qquad (2.29a)$$

$$H = -gT + \nu^T B \qquad (2.29b)$$

where $g = d/\mu^T$ and $\nu^T = 1/\mu^T$ is designated the reluctivity at constant stress. In this form, one can observe the antireciprocal nature of the description since the coefficient $g$ has a different sign in the constitutive equations.

The ratio of the cross-coefficient products yields a different measure of the coupling $\kappa$:

$$\kappa = -\frac{g^2}{s^B \nu^T} = -\frac{k^2}{(1 - k^2)} \qquad (2.30)$$

which evidently is imaginary and therefore seldom used.

By making an additional elimination of the stress in Eq. (2.29b) in the constitutive magnetostriction equations, one finally gets

$$T = \frac{1}{s^B} S - \lambda B \qquad (2.31a)$$

$$H = -\lambda S + \nu^S B \qquad (2.31b)$$

where the "classical" magnetostrictive constant $\lambda$ is given by

$$\lambda = \frac{d}{\mu^T s^B} = \frac{d}{\mu^S s^H} \quad \text{and} \quad \nu^S = \frac{1}{\mu^S} \qquad (2.32)$$

The ratio of the cross-product of the coefficients now gives $\lambda^2 \mu^S s^B$. With $\lambda^2 = d^2/\mu^T \mu^S s^B s^H$ one can evaluate $\lambda^2 \mu^S s^B$:

$$\lambda^2 \mu^S s^B = \frac{d^2}{\mu^T \mu^S s^B s^H} \mu^S s^B = \frac{d^2}{\mu^T s^H} = k^2 \qquad (2.33)$$

With additional algebraic manipulation, the relation between the classical magnetostrictive constant $\lambda$ and the magnetostrictive constant $d$ is

$$\lambda d = \frac{k^2}{(1 - k^2)} \quad \text{or} \quad k^2 = \frac{\lambda d}{(1 + \lambda d)} \tag{2.34}$$

where the product $\lambda d$ is dimensionless.

## Equivalent Circuits

### Lumped Parameters

A rod of magnetostrictive material can be modeled in its longitudinal direction as a network of discrete components, i.e., springs and dampers with lumped masses $m$ and spring and damping constants $k_i$ and $r_i$, respectively (Fig. 2.3). For a system that is moderately and harmonically excited with a frequency such that no elastic wave propagation occurs inside the rod, the linearized model with lumped parameters can be used to achieve a good understanding of the transduction process.

At higher frequencies, where elastic wave propagation occurs inside the rod due to its distributed mass, the lumped parameters must be replaced by distributed parameters. If we also in this case can assume linear conditions, the distributed parameters can be calculated analytically from the wave equation. Both these models are easy to use and need only a few input parameters.

As previously noted, the constitutive equations of piezomagnetism can have many different forms depending on which two quantities of the set of magnetizing field $H$, magnetic induction $B$, stress $T$, and strain $S$ are

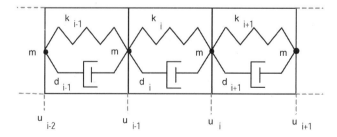

**FIGURE 2.3.** Mechanical model in the longitudinal direction of magnetostrictive rod.

chosen as independent (3). The last presented form, where $T = T(S, B)$ and $H = H(S, B)$, will be used to achieve a model for quasi-static harmonical operation of piezomagnetic materials in which the electrical voltage $U$ and current $I$ are the "in quantities" and the mechanical force $F$ and velocities $v_1$ and $v_2$ are the "out quantities" (see Fig. 2.4 for definitions) (6):

$$T = \frac{1}{s^B} S - \lambda B \tag{2.35a}$$

$$H = -\lambda S + \frac{1}{\mu^S} B \tag{2.35b}$$

The strain $S$ is constant along the rod under quasi-static conditions and is related to the phase velocities by

$$S = \frac{v_2 - v_1}{j\omega l_r} \tag{2.36}$$

Observe that sinusodial variation of all variables with an angular frequency $\omega$ is assumed, which results in a multiplication by $1/j\omega$ instead of a integration with respect to time and a multiplication by $j\omega$ instead of a derivation with respect to time. The force thus can be obtained by combining Eqs. (2.35a) and (2.36):

$$F = -TA_r = -\frac{A_r}{s^B} \frac{v_2 - v_1}{j\omega l_r} + \lambda A_r B \tag{2.37}$$

Note the minus sign before $T$, which indicates that $F$ is defined as positive for compressive stresses.

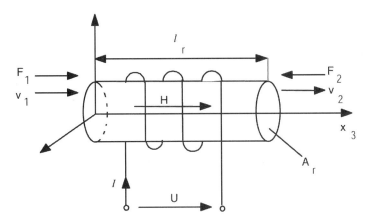

**FIGURE 2.4.** TERFENOL rod of radius $r_r$ with magnetizing coil carrying the current $I$.

It is convenient to define compliances referring to the whole rod at constant induction and constant magnetizing fields, respectively, as

$$C^B = s^B \frac{l_r}{A_r} \tag{2.38a}$$

$$C^H = s^H \frac{l_r}{A_r} \tag{2.38b}$$

The magnetizing field from a long and thin solenoid with $N$ number of turns and length $l_{coil}$ is

$$H = \frac{NI}{l_r} \tag{2.39}$$

It is assumed that $l_r >> r_r$ and $l_{coil} \approx l_r$.

By substituting $B$ in Eq. (2.37) by use of Eq. (2.35b) and using Eqs. (2.27), (2.32), (2.38), and (2.39), the force can be expressed as

$$F = -\frac{v_2 - v_1}{j\omega C^H} + \frac{dA_r N}{s^H l_r} I \tag{2.40}$$

The induction through the solenoid with a cross section $A_c (= A_r)$ is related to the voltage across the terminals by

$$B = \frac{U}{j\omega N A_r} \tag{2.41}$$

which is obtained by Faraday's law. Inserting this into Eq. (2.35b) gives

$$\frac{NI}{l_r} = -\lambda \frac{v_2 - v_1}{j\omega l_r} + \frac{U}{j\omega N A_r \mu^S} \tag{2.42}$$

The clamped and free inductance, respectively, of the magnetizing coil are defined as

$$L^S = \frac{\mu^S N^2 A_r}{l_r} \tag{2.43a}$$

$$L^T = \frac{\mu^T N^2 A_r}{l_r} \tag{2.43b}$$

Rewriting Eq. (2.42) and using Eq. (2.32) gives

$$U = j\omega L^S I + \frac{dA_r N}{s^H l_r} (v_2 - v_1) \tag{2.44}$$

Equations (2.40) and (2.44) constitute the coupling between the mechanical and electrical quantities by means of the factor

$$\theta = \frac{dA_{\mathrm{r}}N}{s^{H}l_{\mathrm{r}}} \tag{2.45}$$

Under clamped conditions, when $v_2 - v_1 \to 0$, one can check whether the voltage over the terminals will be defined by the clamped inductance $L^S$:

$$v_2 - v_1 \to 0 \Rightarrow U = j\omega L^S I \tag{2.46}$$

For no mechanical load the voltage should be defined by the free inductance $L^T$. If Eq. (2.40) with $F = 0$ is inserted into Eq. (2.44),

$$U = j\omega L^S I + j\omega C^H \left(\frac{dA_{\mathrm{r}}N}{s^{H}l_{\mathrm{r}}}\right)^2 I = j\omega L^S (1 - \frac{k^2}{1 - k^2})I = j\omega L^T I \tag{2.47}$$

Figure 2.5 shows a representation of Eqs. (2.40) and (2.44) as an equivalent electrical circuit, where $\theta{:}1$ acts as the turns ratio of an electromechanical transformer which relates the force $F$ to the current $I$ and the velocity $v$ to the voltage $U$. The circuit is usually referred to as the mobility (or admittance) representation of magnetostriction.

An alternative representation of piezomagnetism, which relates the force $F$ to the voltage $U$ and the velocity $v$ to the current $I$, referred to as the impedance representation of magnetostriction (6), can be derived by starting from Eqs. (2.37) and (2.42). By substituting the expression of $B$ in Eq. (2.41), one gets

$$F = -\frac{v_2 - v_1}{j\omega C^B} + \frac{d}{j\omega N\mu^T s^B} \cdot U \tag{2.48}$$

$$I = -\frac{d}{j\omega N\mu^T s^B}(v_2 - v_1) + \frac{U}{j\omega L^S} \tag{2.49}$$

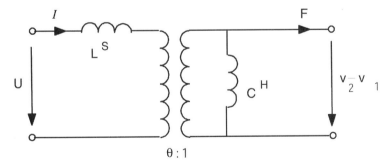

FIGURE 2.5.   Equivalent electrical circuit in mobility representation with lumped parameters.

In this case the electromechanical coupling is carried out by the factor

$$|\Phi| = \frac{d}{j\omega N \mu^T s^B} \qquad (2.50)$$

Here, $v_2 - v_1 \to 0$ gives $I = U/j\omega L^S$ for clamped conditions. $F = 0$ in Eq. (2.48) inserted in Eq. (2.49) gives

$$I = -j\omega C^B \left(\frac{d}{j\omega N \mu^T s^B}\right)^2 U + \frac{U}{j\omega L^S} = \frac{1 - k^2}{j\omega L^S} U = \frac{U}{j\omega L^T} \qquad (2.51)$$

i.e., no mechanical load conditions.

Equations (2.48) and (2.49) cannot exactly be represented by an equivalent circuit comprising passive elements and an ideal transformer due to the antireciprocal nature of this description of magnetomechanical transducers (5, 7, 8), which is reflected by the different signs in front of $d/j\omega N \mu^T s^B$ in Eqs. (2.48) and (2.49). In the equivalent circuit in Fig. 2.6, the mechanical side corresponds to Eq. (2.48).

By multiplying the first term in Eq. (2.49) by the complex operator $-j$ (i.e. subtracting $90°$ from the phase of the velocity $v_2 - v_1$), the antireciprocal feature will virtually disappear. The impedance representation with an ideal transformer gives reasonable results where one does not consider the phase shift between electrical and mechanical quantities.

When the rod is connected to a mechanical load comprising a mass $M$ and a mechanical resistance $R$, the equivalent circuits in the impedance and mobility representations will look like those in Fig. 2.7. Observe that the circuit in Fig. (2.7b) comprises the previous inconsistency due to the antireciprocal nature of the impedance description comprising a transformer.

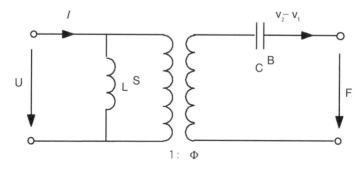

FIGURE 2.6.   Equivalent circuit in the impedance representation.

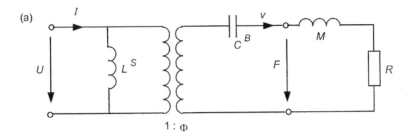

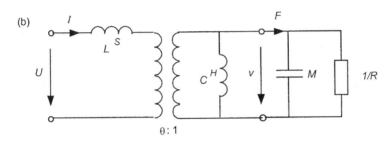

**FIGURE 2.7.** Equivalent circuits of the mechanical system in Fig. 2.12 in (a) impedance and (b) mobility representation.

Alternate mobility and impedance descriptions can be obtained by using the previous stated identities:

$$s^B = s^H(1 - k^2) \tag{2.52a}$$

$$\mu^S = \mu^T(1 - k^2) \tag{2.52b}$$

in Eqs. (2.40), (2.44), (2.48), (2.49).
     We then get

$$\frac{F}{1 - k^2} = -\frac{v_2 - v_1}{j\omega C^B} + \frac{dA_r N}{s^B l_r} I \tag{2.53}$$

$$\frac{U}{1 - k^2} = j\omega L^T I + \frac{dA_r N}{s^B l_r}(v_2 - v_1) \tag{2.54}$$

and

$$F(1-k^2) = -\frac{v_2 - v_1}{j\omega C^H} + \frac{d}{j\omega N\mu^T s^H} U \tag{2.55}$$

$$I(1-k^2) = -\frac{d}{j\omega N\mu^T s^H}(v_2 - v_1) + \frac{U}{j\omega L^T} \tag{2.56}$$

With $\phi' = (dA_r N)/(s^B l_r)$ and $\theta' = d/j\omega N\mu^T s^H$, the corresponding equivalent circuits will look as shown in Fig. 2.8.

Observe that current sources are cut off and voltage sources are short-circuited in impedance estimations of branches in the electrical impedance description, and that force sources are cut off and motion sources are short-circuited in impedance estimations of branches in the mechanical mobility description. Also observe that force sources are short-circuited and motion sources are cut off in impedance estimations of branches in the mechanical impedance description.

To obtain a correct impedance representation, a so-called gyrator [8] must be used for the coupling between electrical and mechanical branches. The difference between an ideal transformer and an ideal gyrator is given below:

| Ideal transformer | Ideal gyrator | |
|---|---|---|
| $\dfrac{U_1}{N_1} = \dfrac{U_2}{N_2}$ | $\dfrac{U_1}{G_1} = \dfrac{I_2}{G_2}$ | (2.57) |
| $I_1 N_1 = I_2 N_2$ | $I_1 G_1 = U_2 G_2$ | |

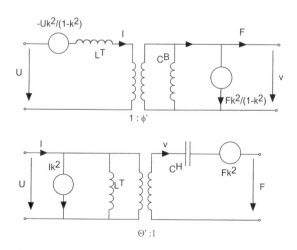

FIGURE 2.8.   Alternate mobility and impedance equivalent circuit descriptions.

$N_1$ and $N_2$ are the turns of the primary and secondary side of the transformer, respectively, and $G_1$ and $G_2$ are the correspondings turns of the the gyrator.

The gyrator concept can directly be applied to Eqs. (2.40) and (2.44), leading to

$$F = -\frac{v_2 - v_1}{j\omega C^H} + GI \tag{2.58}$$

$$U = j\omega L^S I + G(v_2 - v_1) \tag{2.59}$$

Comparison with the gyrator defintion gives

$$G_1 = G = \frac{dA_r N}{s^H l_r} \tag{2.60a}$$

$$G_2 = 1 \tag{2.60b}$$

An equivalent circuit of magnetostriction with the mechanical side in impedance representation and the mechanical coupling performed by a gyrator is shown in Fig. 2.9.

There is no obvious choice for which representation to use. This modified impedance representation perhaps seems better since the velocity is the flowing quantity and the force is the potential quantity in the mechanical branch which intuitively seems natural. Another advantage might be the similarity to equivalent circuits of electrical transformers and rotating machines. For a more thorough investigation of equivalent circuits of piezomagnetic systems, see Berlincourt *et al.* (5) and Katz (7).

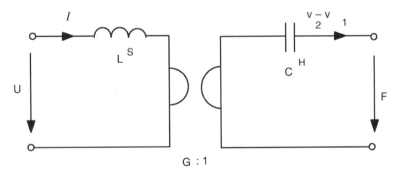

**FIGURE 2.9.** Equivalent circuit in impedance representation with the coupling performed by a gyrator.

## *Distributed Parameters*

The lumped parameter approach is valid for "low frequencies" when no wave propagation occurs inside a magnetostrictive rod due to its distributed mass and elasticity. In this case, it is sufficient to model the rod only with its compliances $C^B$ or $C^H$. For "higher frequencies" its mass must also be considered.

Longitidinal wave propagation in a bar of piezoelectric material when the applied field is parallel has been studied in literature. For an oscillating magnetostrictive rod magnetized along its symmetry axis $x_3$ we use a similar treatment.

Refer to at Fig. 2.4. First, we use Newton's second law and denote the $x_3$ by $x$.

$$\frac{\partial T}{\partial x} = \rho \frac{\partial^2 u}{\partial t^2} \tag{2.61}$$

where $\rho$ is the mass density and $u$ is the particle displacement. Substituting the stress with Eq. (2.35a), and using the definition of strain (see Eq. 2.2),

$$S = \frac{\partial u}{\partial x} \tag{2.62}$$

leads to

$$\frac{1}{s^B} \frac{\partial^2 u}{\partial x^2} - \lambda \frac{\partial B}{\partial x} = \rho \frac{\partial^2 u}{\partial t^2} \tag{2.63}$$

Since we assumed only a longitudinal component of $B$, and div $B = 0$ according to Maxwell's equations, we have $\partial B/\partial x = 0$. Equation (2.63) is further simplified for harmonic oscillation, i.e., $u = u_x(x) \, e^{j\omega t}$, giving

$$\frac{\partial^2 u_x}{\partial x^2} = K^2 u_x \tag{2.64}$$

where the wave number $K$ is defined as

$$K = \omega \sqrt{\rho s^B} = \frac{\omega}{c^B} \tag{2.65}$$

$c^B$ is the sound velocity obtained at constant induction. The general solution to Eq. (2.64) is

$$u_x(x) = A \cos Kx + B \sin Kx \tag{2.66}$$

The boundary conditions can be given in terms of the face velocities and are as follows:

$$v_1 = j\omega u_x(0) \tag{2.67a}$$

$$v_2 = j\omega u_x(l_r) \tag{2.67b}$$

which gives the displacement

$$u(x) = \frac{v_1}{j\omega}\cos Kx + \frac{v_2 - v_1 \cos Kl_r}{j\omega \sin Kl_r}\sin Kx \tag{2.68}$$

The strains at the faces are given by applying Eq. (2.62) to Eq. (2.68):

$$S(0) = \frac{v_2 - v_1 \cos Kl_r}{jc^B \sin Kl_r} \tag{2.69a}$$

$$S(l_r) = \frac{v_2 \cos Kl_r - v_1}{jc^B \sin Kl_r} \tag{2.69b}$$

The stresses at the faces are obtained by inserting Eq. (2.69) into Eq. (2.35a) and the forces at the faces are as previous given by $F = -AT$. Equations (2.32), (2.45), and (2.65) have also been used to arrive at the following:

$$F_1 = \frac{j\rho c^B A_r}{\sin Kl_r}v_2 - jv_1\rho c^B A_r \cot Kl_r + \frac{d}{\mu^T s^B}\frac{1}{j\omega N}U \tag{2.70a}$$

$$F_2 = -\frac{j\rho c^B A_r}{\sin Kl_r}v_1 + jv_2\rho c^B A_r \cot Kl_r + \frac{d}{\mu^T s^B}\frac{1}{j\omega N}U \tag{2.70b}$$

The system in Fig. 2.4 can then be modeled by a circuit with two mechanical ports and one electrical port by means of the previous equation system. Kirchow's second law for the two mechanical branches has also been used. By use of the identity $\tan(\alpha/2) = (1 - \cos\alpha)/\sin\alpha$ equation (2.70) can be written as

$$F_1 = (v_1 - v_2)Z_2 + v_1 Z_1 + \Phi U \tag{2.71a}$$

$$F_2 = (v_1 - v_2)Z_2 - v_2 Z_1 + \Phi U \tag{2.71b}$$

$$Z_1 = j\rho c^B A_r \tan\frac{Kl_r}{2} \tag{2.72a}$$

$$Z_2 = -\frac{j\rho c^B A_r}{\sin Kl_r} \tag{2.72b}$$

where $Z_1$ and $Z_2$ can be identified in Eq. (2.70).

In Fig. 2.10 the circuit is shown in an impedance representation comprising the mechanical impedances $Z_1$ and $Z_2$. One can also see that $\Phi$ has the same expression as for the lumped parameter model (Eq. 2.50). Figure 2.11 shows the input electrical impedance of the system in Fig. 2.12

when modeled in the impedance representation with lumped and distributed parameters, respectively, when wave propagation occurs in the rod. The difference in the two impedance curves indicates the need for distributed parameter modeling.

For wavelengths much larger than the rod length (i.e.; when $1/K >> l_r$), the distributed parameter model degenerates to the lumped parameter model, as expected. Taylor expansion of the impedances gives

$$Z_1 \approx j\rho c^B A_r \frac{Kl_r}{2} = \frac{j\omega M_r}{2} \tag{2.73a}$$

$$Z_2 \approx -j\rho c^B A_r \left( \frac{1}{Kl_r} + \frac{Kl_r}{6} \right) = \frac{1}{j\omega C^B} - \frac{j\omega M_r}{6} \tag{2.73b}$$

If the rod mass $M_r$ is small we return to the circuit shown in Fig. 2.7a. Observe that the previous impedance representation shows the same antireciprocal features as these in Fig. 2.6. To remove this inconsistency it is possible to use a gyrator in a similar way as for the lumped parameter description.

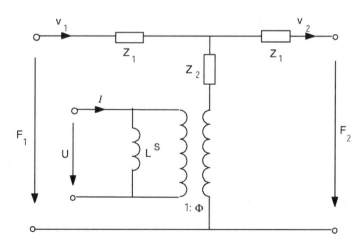

**FIGURE 2.10.** Equivalent circuit in impedance representation of magnetostrictive rod with distributed parameters.

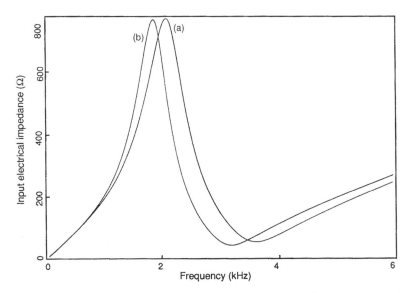

**FIGURE 2.11.** Input electrical impedance of the mechanical system in Fig. 2.12, with $M = 0.5\,\text{kg}$, $R = 2000\,\text{Ns/m}$, and $R_{\text{coil}} = 1\,\Omega$ for a rod with $l_r = 100\,\text{mm}$ and $A_r = 300\,\text{mm}^2$, modeled (a) with lumped and distributed parameters and (b) distributed parameters, respectively (24).

## System Interaction

So far the magnetomechanical transduction has been considered only from the view of the magnetostrictive material. The magnetizing coil and the magnetic circuit have been assumed to function ideally. Also, the mechanical transmission to the external environment has been assumed to be ideal.

In real life this situation never exists. The magnetostrictive rod needs to have a physical coil with a certain resistance and the magnetic circuit will show flux leakage and a certain amount of reluctance. The rod end has to be connected to a mechanical transmission link with some compliance that transfers the mechanical power to the mechanical load.

All these things will affect the magnetomechanical performance. Therefore, the described parameters of a physical actuator comprising magnetostrictive materials will be affected. A very useful concept to describe this influence of the surrounding systems is the effective coupling factor.

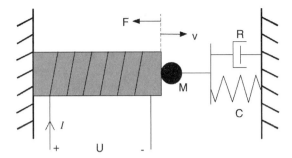

FIGURE 2.12.  TERFENOL rod with magnetizing coil carrying current $I$. One end of the rod is attached to a mass $M$, a mechanical resistance $R$, while the other end $i$ is fixed ($v_1 = 0, v_2 = v$).

The square of the coupling factor was previous derived to be

$$k^2 = \frac{d^2}{\mu^T s^H} \tag{2.74}$$

in the case of longitudinal coupling along the $x_3$ axis, where all indices 3 are dropped.

The square of the coupling coefficient is, as will be shown, also defined as the fraction of maximum stored magnetic energy in the material that can be transformed into elastic energy or alternatively the fraction of maximum elastic energy that can be transformed into magnetic energy.

We have for the magnetic $\rightarrow$ elastic coupling:

$$\frac{E_{el}}{E_{mag,max}} = \frac{E_{mag,max} - E_{mag,min}}{E_{mag,max}} = \frac{\frac{1}{2}\mu^T H^2 - \frac{1}{2}\mu^S H^2}{\frac{1}{2}\mu^T H^2} = \frac{\mu^T - \mu^S}{\mu^T} = \frac{d^2}{\mu^T s^H} = k^2 \tag{2.75}$$

and for the elastic $\rightarrow$ magnetic coupling:

$$\frac{E_{mag}}{E_{el,max}} = \frac{E_{el,max} - E_{el,min}}{E_{el,max}} = \frac{\frac{1}{2}\frac{S^2}{s^B} - \frac{1}{2}\frac{S^2}{s^H}}{\frac{1}{2}\frac{S^2}{s^B}} = \frac{s^H - s^B}{s^H} = \frac{d^2}{\mu^T s^H} = k^2 \tag{2.76}$$

It was previous shown that the internal energy $U$ of a magnetostrictive material can always be divided according to

$$U = U_e + U_{em} + U_{me} + U_m \tag{2.77}$$

The material is assumed to be linear, i.e., the coupling does not depend on the excitation level. Therefore, one can find numbers $k_1$ and $k_2$ such that the following hold:

$$k^2 = \frac{k_1 E_{el}}{k_1 E_{mag,max}} = \frac{U_{me}}{U_m} \tag{2.78}$$

$$k^2 = \frac{k_2 E_{mag}}{k_2 E_{el,max}} = \frac{U_{em}}{U_e} \tag{2.79}$$

Multiplication of the previous equations gives

$$k^4 = \frac{U_{me} U_{em}}{U_m U_e} \Rightarrow k^2 = \frac{U_{me}}{\sqrt{U_m U_e}} \tag{2.80}$$

where $U_{em} = U_{me}$. This alternative definition will be used to investigate system interaction on the magnetic/electrical and the mechanical side of an magnetostrictive actuator.

In a real magnetostrictive actuator there is a magnetic circuit comprising a driving coil that imposes the magnetizing field on the magnetostrictive material. One can describe this magnetizing equipment by means of reluctances or inductances (Fig. 2.13).

An effective coupling factor $k_{eff}$ that takes care of the influence of the magnetizing equipment can now be written as

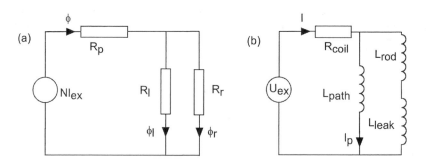

FIGURE 2.13.   The magnetic circuit of a magnetostrictive actuator in electrical and magnetic representations.

$$k_{\text{eff}}{}^2 = \frac{E_{\text{mag,total,max}} - E_{\text{mag,total,min}}}{E_{\text{mag,total,max}}}$$

$$= \frac{(E_{\text{mag,mat,max}} + E_{\text{mag,leak}} + E_{\text{mag,path}}) - (E_{\text{mag,mat,min}} + E_{\text{mag,leak}} + E_{\text{mag,path}})}{E_{\text{mag,mat,max}} + E_{\text{mag,leak}} + E_{\text{mag,path}}}$$

$$= \frac{(\frac{1}{2}\mu^T H^2 l_r A_r - \frac{1}{2}\mu^S H^2 l_r A_r)}{\frac{1}{2}\mu^T H^2 l_r A_r + \frac{1}{2}\mu_l H^2 l_l A_l + \frac{1}{2}\mu_p H_p{}^2 l_p A_p}$$

$$= \frac{\frac{1}{2}\mu^T H^2 l_r A_r}{\frac{1}{2}\mu^T H^2 l_r A_r + \frac{1}{2}\mu_l H^2 l_l A_l + \frac{1}{2}\mu_p H_p{}^2 l_p A_p} \frac{(\frac{1}{2}\mu^T H^2 l_r A_r - \frac{1}{2}\mu^S H^2 l_r A_r)}{\frac{1}{2}\mu^T H^2 l_r A_r} \quad (2.81)$$

$$= \frac{\frac{1}{2}\mu^T H^2 l_r A_r}{\frac{1}{2}\mu^T H^2 l_r A_r + \frac{1}{2}\mu_l H^2 l_l A_l + \frac{1}{2}\mu_p H_p{}^2 l_p A_p} \frac{\mu^T - \mu^S}{\mu^T}$$

$$= \frac{L^T}{L^T + L_{\text{leak}} + L_{\text{path}}\left(\frac{l_p}{l-l_p}\right)^2} k^2 \quad \text{or}$$

$$= \frac{1}{1 + \frac{\mu_l A_l}{\mu^T A_r} + \frac{\mu_p H_p{}^2 l_p A_p}{\mu^T H^2 l_r A_r} k^2}$$

It has here been assumed that the equivalent leakage length $l_l$ is equal to the rod length. The relative magnetic permeability of the leakage flux path with cross-section $A_l$ normally is $\mu_0$. A more detailed analysis of the coil leakage will be given later. It is also assumed that there is an equivalent magnetic permeability $\mu_p$ in the flux feedback path with an equivalent length $l_p$ and cross-section $A_p$. $H_p$ is the equivalent magnetizing field across the flux feedback path.

From the previous equations it is evident that the coil flux leakage and reluctance in the flux feedback path reduce the effective coupling factor of an magnetostrictive actuator. It is common to assign this reduction a magnetic coupling coefficient $k_M$ such that

$$k_{\text{eff}}{}^2 = k_M{}^2 k^2 \quad (2.82)$$

It is also possible to define an effective coupling factor $k_{\text{eff}}$ that handles the influence of a mechanical transmission device with compliance $C^T$ and prestress device with compliance $C^P$ in a magnetostrictive actuator.

In a similar way this effective coupling factor is written as

$$
\begin{aligned}
k_{\text{eff}}{}^2 &= \frac{E_{\text{el,total,max}} - E_{\text{el,total,min}}}{E_{\text{el,total,max}}} \\[2mm]
&= \frac{(E_{\text{el,mat,max}} + E_{\text{el,trans}} + E_{\text{el,prestress}}) - (E_{\text{el,mat,min}} + E_{\text{el,trans}} + E_{\text{el,prestress}})}{E_{\text{el,mat,max}} + E_{\text{el,trans}} + E_{\text{el,prestress}}} \\[2mm]
&= \frac{\frac{1}{2}\frac{S^2}{C^B}}{\frac{1}{2}\frac{S^2}{C^B} + \frac{1}{2}\frac{S_T{}^2}{C^T} + \frac{1}{2}\frac{S^2}{C^P}} \quad \frac{\left(\frac{1}{2}\frac{S^2}{C^B} - \frac{1}{2}\frac{S^2}{C^H}\right)}{\frac{1}{2}\frac{S^2}{C^B}} \\[2mm]
&= \frac{\frac{1}{2}\frac{S^2}{C^B}}{\frac{1}{2}\frac{S^2}{C^B} + \frac{1}{2}\frac{S_T{}^2}{C^T} + \frac{1}{2}\frac{S^2}{C^P}} \quad \frac{\frac{1}{C^B} - \frac{1}{C^H}}{\frac{1}{C^B}} \\[2mm]
&= \frac{\frac{1}{C^B}}{\frac{1}{C^B} + \frac{C^T}{(C^B)^2} + \frac{1}{C^P}} \quad \frac{\frac{1}{C^B} - \frac{1}{C^H}}{\frac{1}{C^B}} \\[2mm]
&= \frac{\frac{1}{C^B}}{\frac{1}{C^B} + \frac{C^T}{(C^B)^2} + \frac{1}{C^P}} k^2 = k_{\text{E}}{}^2 k^2
\end{aligned}
\tag{2.83}
$$

Here, it has been assumed that the force in the mechanical transmission $S_T/C^T$ is equal to the force $S/C^B$ in the magnetostrictive rod, where $S_T$ is the strain of the mechanical transmission device.

For a magnetostrictive actuator the total effective coupling factor can now be written as

$$
k_{\text{eff}}{}^2 = k_{\text{M}}{}^2 k_{\text{E}}{}^2 k^2
\tag{2.84}
$$

where $k_M$ and $k_E$ are defined according to the above equations.

It is evident from the expression of the effective coupling factor that the influence of an additional mechanical load comprising springs can be handled by using effective values of the resulting compliances $C_{\text{eff}}{}^S$ and $C_{\text{eff}}{}^P$ in series of or parallel with the mechanical outlet of the actuator:

$$
\frac{1}{C_{\text{eff}}{}^S} = \frac{1}{C^P} + \frac{1}{C_{\text{mech load in parallel}}}
\tag{2.85a}
$$

$$
C_{\text{eff}}{}^P = C^T + C_{\text{mech load in series}}
\tag{2.85b}
$$

## Resonance Phenomena

In a magnetostrictive rod the internal energy is mechanical, magnetic, and magnetomechanical. In a time-varying situation the fractions of these kind of energies vary. Then, it is inevitable that oscillation or resonance phenomena will occur.

This can easily be studied by using equivalent circuits. If we connect the mechanical branch as in Fig. 2.7a with a mechanical load comprising a mass $M$ and a viscous damper $R$, the electrical input admittance $Y_e$ of the circuit is

$$Y_e = \frac{1}{j\omega L^S} + \frac{|\Phi|^2}{R + j\omega M + 1/(j\omega C^B)} = \frac{1}{j\omega L^S} + \frac{j\omega C^B |\Phi|^2}{1 - \omega^2 M C^B + j\omega R C^B} \quad (2.86)$$

With the square of the transformer ratio in Fig 2.7a, $|\Phi|^2 = d^2/(\omega N\mu^T s^B)^2 = k^2/(\omega^2 L^S C^B)$, one derives, after some rearrangements,

$$Y_e = \frac{1-k^2}{L^S} \frac{1 - \omega^2 M C^H + j\omega R C^H}{j\omega(1 - \omega^2 M C^B + j\omega R C^B)} \quad (2.87)$$

where $C^B = C^H(1 - k^2)$ has been used. At resonance the input admittance $Y_e$ is purely resistive (i.e, it is a real number). For $\omega \neq 0$, the ratio between real and imaginary components in the numerator and the dominator then is equal, which gives

$$\omega^4 + \omega^2\left(-\frac{1}{MC^H} - \frac{1}{MC^B} + \frac{R^2}{M^2}\right) + \frac{1}{M^2 C^H C^B} = 0 \quad (2.88)$$

This equation has the solutions

$$\omega^2 = \frac{1}{2}\left(\frac{1}{MC^H} + \frac{1}{MC^B} - \frac{R^2}{M^2}\right) \pm \sqrt{\frac{1}{4}\left(\frac{1}{MC^H} + \frac{1}{MC^B} - \frac{R^2}{M^2}\right)^2 - \frac{1}{M^2 C^H C^B}} \quad (2.89)$$

When series resonance with $R = 0$ occurs in the mechanical branch in Fig 2.7a, this corresponds to minimum input electrical impedance occurring at

$$\omega_a^2 = \frac{1}{2}\left(\frac{1}{MC^H} + \frac{1}{MC^B}\right) - \frac{1}{2}\left(\frac{1}{MC^H} - \frac{1}{MC^B}\right) \Rightarrow \omega_a = \sqrt{\frac{1}{MC^B}} \quad (2.90)$$

When parallel resonance with $R = 0$ occurs in the mechanical branch in Fig 2.7b, this corresponds to maximum input electrical impedance occurring at

$$\omega_r^2 = \frac{1}{2}\left(\frac{1}{MC^H} + \frac{1}{MC^B}\right) + \frac{1}{2}\left(\frac{1}{MC^H} - \frac{1}{MC^B}\right) \Rightarrow \omega_r = \sqrt{\frac{1}{MC^H}} \quad (2.91)$$

The resonance frequencies $\omega_r$ and $\omega_a$ can be interpreted as follows. At the frequency $\omega_r$ the maximum impedance occurs with $Y_e \rightarrow 0$ for $R = 0$. This this corresponds to the situation in which the current is constant or when the actuator is fed by a constant current source. This frequency is called the electrical resonance frequency or the electrical open circuit resonance because this resonance will occur if the electrical feeding is done by a current source. When $R = 0$ and no losses are present, this frequency is denoted

$$f_{r,i}^E = \frac{1}{2\pi}\sqrt{\frac{1}{MC^H}} \tag{2.92}$$

At the frequency $\omega_a$ the minimum impedance occurs with $Y_e \to \infty$ for $R = 0$. This corresponds to the situation in which the induction or voltage is constant or when the actuator is fed by a constant voltage source. This frequency is called the electrical antiresonance frequency or the electrical short-circuit resonance because this resonance will occur if the electrical feeding is done by a voltage source. When $R = 0$ and no losses are present, this frequency is denoted

$$f_{a,i}^E = \frac{1}{2\pi}\sqrt{\frac{1}{MC^B}} \tag{2.93}$$

The index i indicates that we are considering an ideal circuit with no losses and that the energy is stored only in the magnetostrictive rod. The index E stands for electrical.

If the expressions for the two resonance frequencies are divided, one gets

$$\frac{f_{r,i}^E}{f_{a,i}^E} = \sqrt{\frac{C^B}{C^H}} \tag{2.94}$$

It also hold that $C^B = C^H(1 - k^2)$. Combining the frequency ratio with this expression gives

$$k = \sqrt{1 - \left(\frac{f_{r,i}^E}{f_{a,i}^E}\right)^2} \tag{2.95}$$

As shown, it is possible to determine the coupling coefficient by investigation of the electrical resonance and antiresonances of a magnetostrictive actuator.

Figure 2.14 shows both situations with regard to the mechanical side in the equivalent circuits.

For a mechanical prestress device with compliance $C^P$ and a mechanical transmission device with compliance $C^T$, the circuits at the mechanical branches will appear as shown in Fig. 2.15. This indicates that an ideal prestress mechanism has an infinite compliance and an ideal transmission device zero compliance.

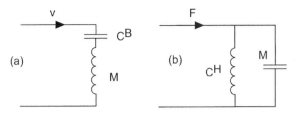

FIGURE 2.14. The impedances in the mechanical branch under (a) series and (b) parallel resonance.

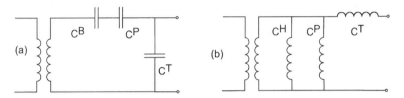

FIGURE 2.15. Equivalent circuit representations of a prestress device and a mechanical transmission in the (a) impedance and (b) mobility mechanical description.

In a similar way, we can examine the mechanical input admittance in Fig 2.7b when the electrical branch comprises an electrical resistance $R_e$ and capacitance $C$ in series:

$$Y_m = \frac{1}{j\omega C^H} + \frac{\Theta^2}{R_e + j\omega L^S + 1/(j\omega C)}; \quad \Theta^2 = \left(\frac{dA_r N}{sl_r}\right)^2 = \frac{k^2 L^T}{C^H} \quad (2.96)$$

gives

$$Y_m = \frac{1}{j\omega C^H}\left(1 - \frac{k^2 L^T \omega C}{1 - \omega^2 L^S C + j\omega R_e C}\right) \quad (2.97)$$

and

$$Y_m = \frac{1}{j\omega C^H}\left(\frac{1 - \omega^2 C L^T + j\omega R_e C}{1 - \omega^2 C L^S + j\omega R_e C}\right) \quad (2.98)$$

At resonance, $Y_m$ must be real number. For $\omega \neq 0$ the ratio between the real and imaginary components in the numerator dominator is equal, which gives

$$\frac{\omega R_e C}{1 - \omega^2 L^S C} = \frac{1 - \omega^2 C L^T}{-\omega R_e C} \tag{2.99}$$

and

$$\omega^4 + \omega^2 \left( -\frac{1}{L^S C} - \frac{1}{L^T C} + \frac{R_e^2}{L^S L^T} \right) + \frac{1}{L^S L^T C^2} = 0 \tag{2.100}$$

For $R_e = 0$,

$$\omega^2 = \frac{1}{2}\left(\frac{1}{L^S C} + \frac{1}{L^T C}\right) \pm \frac{1}{2}\left(\frac{1}{L^S C} - \frac{1}{L^T C}\right) \tag{2.101}$$

When series resonance occurs in the electrical branch in Fig 2.7b, this corresponds to a minimum input mechanical impedance occurring at

$$\left(\omega_a^M\right)^2 = \frac{1}{2}\left(\frac{1}{L^S C} + \frac{1}{L^T C}\right) + \frac{1}{2}\left(\frac{1}{L^S C} - \frac{1}{L^T C}\right) = \frac{1}{L^S C} \Rightarrow \omega_a^M = \sqrt{\frac{1}{L^S C}} \tag{2.102}$$

When parallel resonance occurs in the electrical branch, this corresponds to maximum input mechanical impedance occurring at

$$\left(\omega_r^M\right)^2 = \frac{1}{2}\left(\frac{1}{L^S C} + \frac{1}{L^T C}\right) - \frac{1}{2}\left(\frac{1}{L^S C} - \frac{1}{L^T C}\right) = \frac{1}{L^T C} \Rightarrow \omega_r^M = \sqrt{\frac{1}{L^T C}} \tag{2.103}$$

If the expressions for the two resonance frequencies are divided, one gets

$$\frac{\omega_{r,i}^M}{\omega_{a,i}^M} = \sqrt{\frac{L^S}{L^T}} \tag{2.104}$$

It also hold that $L^S = L^T(1 - k^2)$ because $\mu^S = \mu^T(1 - k^2)$. Combining the previous frequency ratio with this expression gives

$$k = \sqrt{1 - \left(\frac{f_{r,i}^M}{f_{a,i}^M}\right)^2} \tag{2.105}$$

The index M stands for mechanical.

As shown, it is also possible to determine the coupling coefficient by investigation of the mechanical resonance and antiresonances of a magnetostrictive actuator. From Eq. (2.98) it is also obvious that $\omega_{r,i}^M$ implies maximum mechanical impedance and $\omega_{r,a}^M$ implies minimal mechanical impedance in the mechanical mobility representation, where force is analogous to electrical current and velocity to electrical voltage.

Maximum mechanical impedance in the mobility description is then analogous to the situation in which the force is held constant or when the actuator is driven by a constant force source. Minimum mechanical impedance in the mobility description is then analogous to the situation in which the velocity is held constant or when the actuator is driven by a constant velocity source. Figure 2.16 shows the situations at the electrical side in the equivalent circuits. For a nonideal magnetic circuit the electrical branches will appear as shown in Fig. 2.17.

To summarize, the electrical and mechanical resonance and antiresonance frequencies can be described in terms of series and parallel resonances in the mechanical and electrical branches, respectively in an equivalent circuit diagram.

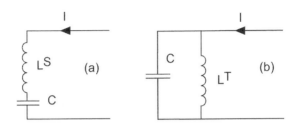

**FIGURE 2.16.** The impedances in the electrical branch under (a) series and (b) parallel resonance.

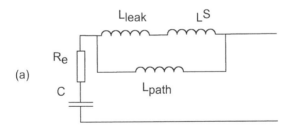

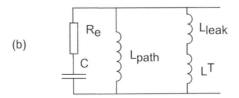

**FIGURE 2.17.** The impedances in the electrical branch under (a) series and (b) parallel resonance for a nonideal magnetic circuit.

It is illustrative to note that the magnetizing equipment and a mechanical prestress and transmission device can be described as discrete energy-storing components such as inductors and springs. From the equivalent circuits one can also see how imperfections in the magnetizing device and how the prestress and transmission device affect the coupling factor.

## Influence of Wave Propagation

The equation of motion for a magnetostrictive rod was previous found to be (Eq. 2.66)

$$u_x(x) = A\cos Kx + B\sin Kx \tag{2.106}$$

We assume that the rod is magnetized with an ideal magnetic circuit and that *it is* fixed at $x = 0$ and free at $x = l_r$. These boundary conditions give $A = 0$ and the stress at the free end equal to zero. Equation (2.25a) can then be written

$$S\big|_{x=l_r} = dH \Rightarrow \frac{\partial u_x}{\partial x}\Big|_{x=l_r} = dH \Rightarrow BK\cos Kl_r = dH \tag{2.107}$$

which leads to

$$B = \frac{dH}{K\cos Kl_r} \tag{2.108}$$

and

$$u_x = \frac{dH}{K}\frac{\sin Kx}{\cos Kl_r} \tag{2.109}$$

Equation (2.25a) yields $H = \frac{S - s^H T}{d}$, and together with Eq.(2.25b) we obtain

$$H = \frac{B\,s^H - dS}{\mu^T s^H - d^2} \tag{2.110}$$

where $B$ is independent of $x$ because there is no flux leakage. We then have

$$B = \mu^T H(l_r) \tag{2.111}$$

$H(x)$ will be a function of just $S(x)$:

$$H(x) = \frac{\mu^T s^H H(l_r) - dS(x)}{\mu^T s^H - d^2} \tag{2.112}$$

where $S(x)$ can be calculated from Eq. (2.109):

$$S(x) = \frac{\partial u_x}{\partial x} = \frac{dH(l_r) \cos Kx}{\cos Kl_r} \qquad (2.113)$$

Introducing the average magnetizing field $\overline{H}$,

$$\overline{H} = \int_0^{l_r} \frac{H(x)}{l_r} dx \qquad (2.114)$$

gives, after evaluating the integral,

$$\overline{H} = \frac{1}{1-k^2} H(l_r) \left(1 - \frac{k^2}{Kl_r} \tan(Kl_r)\right) \qquad (2.115)$$

As noted previously, the ideal magnetic circuit gives

$$I = \frac{\overline{H}l_r}{N} \quad \text{and} \quad U = Nj\omega BA_r \qquad (2.116)$$

The input electrical impedance $Z$ is then given by

$$Z = \frac{U}{I} = \frac{N^2 j\omega BA_r}{\overline{H}l_r} \qquad (2.117)$$

As in the lumped parameter description, we can now study under which conditions $Z \to 0$ and $Z \to \infty$.

When feeding with a current source under resonant condition we have an open-circuit resonance with $Z \to \infty$ (resonance). Therefore, when $\overline{H} \to 0$, Eq. (2.115) implies that

$$k^2 = \frac{\frac{\omega_r l_r}{c^B}}{\tan\left(\frac{\omega_r l_r}{c^B}\right)} \qquad (2.118)$$

When feeding with a voltage source under resonant condition, we have a short-circuit resonance with $Z \to 0$ (antiresonance). Therefore

$$\overline{H} \to \infty \Rightarrow \frac{\omega_a}{c^B} l_r = \frac{\pi}{2} n, n = 1, 3, 5, \cdots \qquad (2.119)$$

For $n = 1$, one can recognize the first resonance. We get

$$\frac{2\pi f_a l_r}{c^B} = \frac{\pi}{2} \Rightarrow l_r = \frac{\lambda}{4}$$

where

$$f_a \lambda = c^B \qquad (2.120)$$

and $f_a$ is identified as the quarter wavelength resonance frequency when feeding from a voltage source. Equation (2.119) for $n = 1$, together with Eq. (2.118), gives

$$k^2 = \frac{\pi f_r}{2 f_a} \frac{1}{\tan\left(\frac{\pi}{2}\frac{f_r}{f_a}\right)} \tag{2.121}$$

Figure 2.18 shows the ratio between the square of the coupling coefficient determined with and without wave propagation. Observe that a resonant system often is a combination of the two cases. Equation (2.121) can be approximated with different expansions with respect to $f_r/f_a$. The most common approximation is

$$k^2 = \frac{\pi^2}{8}\left(1 - \left(\frac{f_r}{f_a}\right)^2\right) \tag{2.122}$$

## Eddy Current Effects

The effects of eddy currents will increase by increasing frequency (17). The induced eddy currents give rise to ohmic losses and consequently decrease the performance of the magnetostrictive material and the actuator for a specific externally imposed magnetic field.

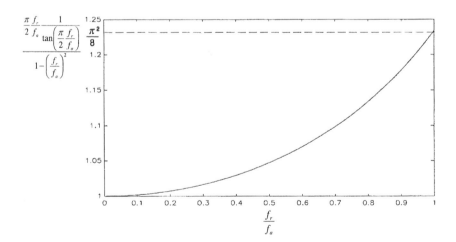

FIGURE 2.18.  The ratio between the square of the coupling coefficient determined with and without wave propagation (18).

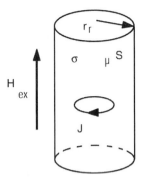

**FIGURE 2.19.** Magnetostrictive rod of radius $r_r$, permeability $\mu^S$, and conductivity $\sigma$ magnetized by the external magnetic field $H_{ex}$.

Consider a rod of a magnetic material with radius $r_r$ and electrical conductivity $\sigma$ shown in Fig. 2.19, magnetized by a longitudinal external field $H_{ex} = H_0 e^{j\omega t}$. Appendix A shows that the expression for the internal magnetizing field as a function of the radial coordinate inside the rod, assuming small signals, is given by

$$H_{\text{inside}} = H_0 \frac{\text{ber}(\gamma' r) + j\text{bei}(\gamma' r)}{\text{ber}(\gamma' r_r) + j\text{bei}(\gamma' r_r)} e^{j\omega t} \tag{2.123}$$

where $\text{ber}(\gamma' r)$ and $\text{bei}(\gamma' r)$ are the real and imaginary parts of the Bessel function $J_0(\sqrt{-1}\gamma' r)$. $\gamma'$ is the inverse of the penetration depth,

$$\gamma' = \sqrt{\omega \mu^S \sigma} \tag{2.124}$$

The reason why $\mu^S$ is used here instead of $\mu^T$ is that the effects of eddy currents are most pronounced at high frequencies, at which the rod can be considered clamped rather than free. In reality, a value between $\mu^S$ and $\mu^T$ shall be used. The flux penetration can be regarded as complete when $\gamma' r_r \leq 1$. The cut-off frequency $f_c$ is defined as $\gamma' r_r = 1$ and can be calculated from Eq. (2.124).

$$f_c = \frac{1}{2\pi\mu\sigma r_r^2} \tag{2.125}$$

If one integrates the expression for $H_{\text{inside}}$ over the rod cross section it is possible to estimate an effective internal magnetic field $H_{\text{eff}}$ as

$$H_{\text{eff}}(\gamma' r) \cdot e^{j\omega t} = \left( \frac{1}{\pi r_r^2} \int \int H_{\text{inside}}(\gamma' r) r \, dr \, d\phi \right) e^{j\omega t} \tag{2.126}$$

Azimutal symmetry is assumed, which implies that the integration with respect to $\phi$ is equivalent to a multiplication by $2\pi$.

Integration then yields

$$H_{\text{eff}}(\gamma'r) = \chi H_0 = (\chi_r - j\chi_i)H_0 \tag{2.127}$$

where

$$\chi_r = \frac{2}{\gamma'r_r} \frac{\text{ber}(\gamma'r_r)\text{bei}'(\gamma'r_r) - \text{bei}(\gamma'r_r)\text{ber}'(\gamma'r_r)}{\text{ber}^2(\gamma'r_r) + \text{bei}^2(\gamma'r_r)} \tag{2.128a}$$

$$\chi_i = \frac{2}{\gamma'r_r} \frac{\text{ber}(\gamma'r_r)\text{ber}'(\gamma'r_r) + \text{bei}(\gamma'r_r)\text{bei}'(\gamma'r_r)}{\text{ber}^2(\gamma'r_r) + \text{bei}^2(\gamma'r_r)} \tag{2.128b}$$

where $\chi_r$ and $\chi_i$ are the real and imaginary parts of the eddy current loss factor, respectively, and ber'( ) denotes the derivative of ber with respect to the argument $\gamma'r$. The variation of $\chi$ with frequency, normalized to the eddy current cut-off frequency, is shown in Fig. 2.20.

It is convenient to define a complex permeability $\mu_c$ by multiplication of the static permeability with the eddy current factor. This complex

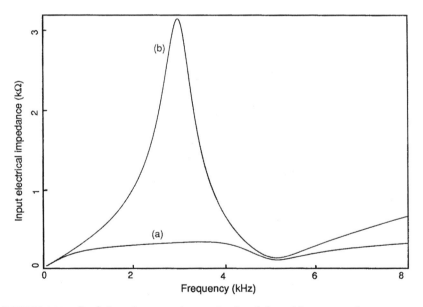

**FIGURE 2.20.**   Real, imaginary, and magnitude of the eddy current factor $\chi$ as a function of frequency normalized by $f_c$ (24).

permeability relates the effective magnetic flux density $B_{\text{eff}}$ to the external magnetizing field $H_0$:

$$B_{\text{eff}} = \mu_c H_0 = \chi \mu H_0 = (\chi_r - j\chi_i)\mu H_0 \qquad (2.129)$$

Eddy currents can now be taken into account for in the linearized models of magnetostriction by this eddy current factor $\chi$. In Eq. (2.35) the eddy current factor affects only the clamped permeability, and hence the term $1/\chi\mu^S$ rather than $1/\mu^S$ should be entered into Eq. (2.35b). It can easily be shown that the consequence of this action changes Eqs. (2.44) and (2.49) in the following way:

$$U = j\omega\chi L^S I + \chi \frac{dA_r N}{s^H l_r} (v_2 - v_1) \qquad (2.130)$$

$$I = -\frac{d}{j\omega N\mu^T s^B} (v_2 - v_1) + \frac{U}{j\omega\chi L^S} \qquad (2.131)$$

Hence, the series impedance in the mobility representation (Fig. 2.5) and the shunt impedance in the impedance representation (Fig. 2.6) will be $j\omega\chi L^S$, which can be seen as an inductance in series with a resistance when eddy currents are present. Moreover, the transformer ratio in the mobility representation will be $\chi\Theta : 1$ when eddy currents are present.

For small losses one can perform the approximation

$$\mu_c = (\chi_r - j\chi_i)\mu \approx \chi_r\mu \qquad (2.132)$$

An effective coupling factor

$$k_{\text{eff}} = \sqrt{\chi_r}k_{\text{eff}}(0) \qquad (2.133)$$

can then be derived on the basis of the electrical and mechanical resonance and antiresonance frequencies as discussed under Resonance Phenomena, where $k_{\text{eff}}(0)$ is the coupling factor when no eddy currents are present.

It also follows that $k_{\text{eff}}$ can be obtained experimentally from the electrical or mechanical resonance frequencies for a magnetostrictive material comprising eddy currents according to

$$k_{\text{eff}} = \sqrt{1 - \left(\frac{f_r^M}{f_a^M}\right)^2} \qquad (2.134a)$$

$$k_{\text{eff}} = \sqrt{1 - \left(\frac{f_r^E}{f_a^E}\right)^2} \qquad (2.134b)$$

When account is taken for the magnetizing circuit and the mechanical prestress and transmission device, $k_{eff}$ is given by

$$k_{eff} = \sqrt{\chi_r} k_e k_m k \qquad (2.135)$$

where $k_e$ and $k_m$ are the previously defined elastic and magnetic coupling coefficients. The DC coupling factor $k(0)$ for the magnetostrictive rod depends on the degree of wave propagation within it (see previous sections).

The influence of eddy currents on the input electrical impedance is shown in Fig. 2.21. Note how the resonance peak vanishes when large eddy currents are present.

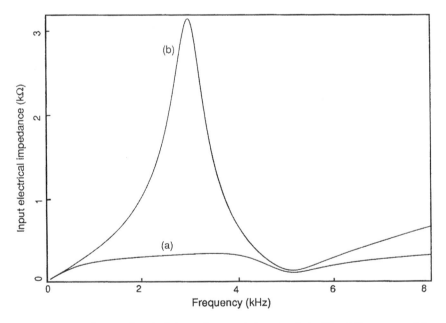

FIGURE 2.21.   Input electrical impedance of the circuit in Fig. 2.7 connected in series on the electrical branch with an equivalent coil resistance $R_{coil} = 1\Omega$, and $M = 0.5$ kg and $R = 2000$ Ns/m for a rod with $l_r = 50$ mm and $A_r = 300$mm$^2$, with (a) and without (b) taking eddy currents into account. No flux leakage and transmission compliance is taken into account (24).

## Hysteresis Effects

In linear descriptions of magnetostrictive materials, magnetic hysteresis can be described in terms of a complex permeability, in which the imaginary part represents the dissipation of hysteresis losses. Therefore, hysteresis can be taken into account by adding a term corresponding to the dissipated hysteresis loss in the expression for the magnetic permeabilty:

$$\mu = \mu_{\text{r}} - j\mu_{\text{H}} \tag{2.136}$$

Previously, it was stated that eddy currents influence the real part $\mu_r$, which has to be replaced by $(\chi_{\text{r}} - j\chi_{\text{i}})\mu_{\text{r}}$. This gives

$$\mu = \chi_{\text{r}}\mu_{\text{r}} - j(\chi_{\text{i}}\mu_{\text{r}} + \mu_{\text{H}}) \tag{2.137}$$

From the previous expression, it is evident that hysteresis losses result in an additional $\mu_{\text{H}}/\mu_r$ to $\chi_i$, giving

$$\chi_{\text{i,tot}} = \chi_{\text{i}} + \frac{\mu_{\text{H}}}{\mu_{\text{r}}} \tag{2.138}$$

In equivalent circuit diagrams hysteresis losses are manifested by an additional resistive component (Fig. 2.22) that should be inserted wherever the impedance $j\omega L^S$ is present in the electrical impedance representation.

## Mechanical Losses

Mechanical material losses inside a magnetostrictive material can be regarded as mechanical hysteresis losses, which are only affected by the purely elastic strain. Therefore $s^B$ has to be replaced with

$$s_{\text{c}}^B = s_{\text{r}}^B - js_{\text{i}}^B \tag{2.139}$$

No account for eddy currents
and hysteresis

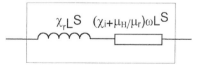

Account for eddy currents
and hysteresis

**FIGURE 2.22.** Representation of eddy currents and hysteresis in equivalent circuit descriptions of magnetostrictive actuators.

in analogy with the treatment of the magnetic permeability for the magnetic hysteresis. In electric equivalent circuits, a dissipative component $C_i^B \omega$ has to be added to $C^B$ in the mechanical mobility representation wherever it appears (Fig. 2.23).

## Skin Effect in the Driving Coil

If a low-frequency current passes through the coil wire, the current density is equally distributed over the cross section of the wire. At higher frequencies the current density will be reduced at the center of the wire. This implies a smaller effective cross section of the wire and therefore a higher coil wire resistance.

If we assume that there is a negligible flux leakage in the magnetic circuit, it is possible to separately estimate the impedance of the magnetizing coil related to its conducting wire.

A frequency-dependent resistor and a frequency-dependent inductor in series can then replace the equivalent coil resistance (Fig. 2.24). Appendix B shows that the lower relation is valid for a straight wire of

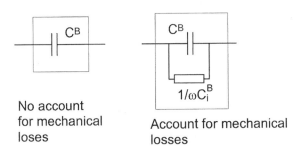

No account
for mechanical
loses

Account for mechanical
losses

FIGURE 2.23.   Representation of mechanical losses in equivalent circuit descriptions of magnetostrictive actuators.

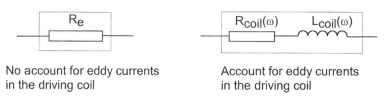

No account for eddy currents
in the driving coil

Account for eddy currents
in the driving coil

FIGURE 2.24.   Representation of eddy currents in the drive coil winding in equivalent circuit descriptions of magnetostrictive actuators.

length $l_w$ and radius $r_w$. Such an assumption is valid if the wire radius is small compared to the radius of the coil, which is usually the case. Therefore,

$$R_{\text{coil}} + j\omega L_{\text{coil}} = \frac{l_w \rho}{\pi r_w^2} \frac{\gamma r_w}{2} \frac{I_0(\gamma r_w)}{I_1(\gamma r_w)} \qquad (2.140)$$

where

$$\gamma = \sqrt{\frac{j\omega \mu_{\text{air}}}{\rho}} = \sqrt{\frac{1}{2} \frac{\omega \mu_{\text{air}}}{\rho}}(1+j) \qquad (2.141)$$

and $I_0$ and $I_1$ are modified Bessel functions of the first kind and order 0 and 1, respectively.

If $\gamma r_w$ is small, series expansion of $I_0$ and $I_1$ gives

$$R_{\text{coil}} + j\omega L_{\text{coil}} = \frac{l_w \rho}{\pi r_w^2}\left(1 + \frac{1}{8}\gamma^2 r_w^2\right) = R_{\text{coil}}(0) + j\omega \frac{\mu_{\text{air}}}{8\pi} l_w \qquad (2.142a)$$

$$\Rightarrow R_{\text{coil}}(\omega) = R_{\text{coil,DC}} \quad \text{and} \quad L_{\text{coil}}(\omega) = \frac{\mu_{\text{air}} l_w}{8\pi} \qquad (2.142b)$$

For large $\gamma r_w$, it is possible to make use of asymptotic expressions for the Bessel functions, which give

$$\frac{R_{\text{coil}}(\omega) + j\omega L_{\text{coil}}(\omega)}{R_{\text{coil,DC}}} = \frac{1}{2}\gamma r_w + \frac{1}{4} = \frac{1}{2}\kappa r_w + \frac{1}{4} + j\frac{1}{2}\kappa r_w \qquad (2.143)$$

where

$$\kappa = \sqrt{\frac{1}{2}\frac{\omega \mu_{\text{air}}}{\rho}} \qquad (2.144)$$

or

$$R_{\text{coil}}(\omega) = \left(\frac{1}{2}\kappa r_w + \frac{1}{4}\right) \cdot R_{\text{coil,DC}} \qquad (2.145)$$

and

$$L_{\text{coil}}(\omega) = \sqrt{\frac{\mu_{\text{air}}}{8\omega\rho}} r_w R_{\text{coil,DC}} \qquad (2.146)$$

In Fig. 2.25, the normalized impedance is visualized as a function of $\kappa r_w$.

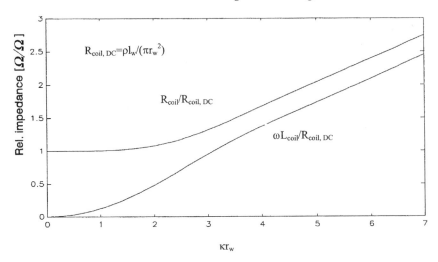

**FIGURE 2.25.**   The real and imaginary part of the inner impedance per unit length of a long, straight wire normalized with respect to the static wire resistance.

## The Quality Factor or the $Q$ Value

The quality factor or the system magnification is related to loss dissipation processes in the magnetoelastic transduction. The following is one definition of the $Q$ value:

$$Q = \frac{W_e}{W_l} \tag{2.147}$$

where $W_e$ is the maximum stored elastic energy and $W_l$ is the total energy loss during a periodic cycle $T$ of operation.

### The Q Value of a Mechanical Free Rod

For a magnetostrictive rod in electrical resonance the stored magnetic energy compared to the kinetic energy is small, i.e., the actual magnetoelastic coupling is weak. Therefore, the quality factor determined by electrical resonance measurements is influenced mostly by pure elastic losses.

It can be proved that under resonant conditions for a free magnetostrictive rod, the maximum value of the potential energy is equal to the maximum value of the kinetic energy for moderate elastic losses (18). The kinetic energy $dW_k$ in section $dx$ can then be expressed as

$$dW_k = d\left(\frac{mv^2}{2}\right) = \left(\frac{\partial u}{\partial t}\right)^2 \frac{\rho A_r}{2} dx \tag{2.148}$$

where $u(x)$ is the velocity of the rod along its length $1_r$. The rod cross section is $A_r$. Integration gives

$$W_k = \frac{\rho A_r}{2} \int_0^{l_r} \left(\frac{\partial u}{\partial t}\right)^2 dx \tag{2.149}$$

The total power loss in section $dx$ can be written

$$\partial P_l = \partial^2 W_l = dF_l \cdot dl = \underbrace{A_r \nu \frac{\partial u}{\partial t} dx}_{dF_l} \cdot \underbrace{\frac{\partial u}{\partial t} dt}_{dl} \tag{2.150}$$

where $\nu$ is the viscous attenuation coefficient of the material. Integration gives

$$W_l = A_r \nu \int_0^T \int_0^{l_r} \left(\frac{\partial u}{\partial t}\right)^2 dx dt \tag{2.151}$$

The rod velocity $u(x,t)$ can be separated according to

$$u(x,t) = u_x(x) \cdot u_t(t) \tag{2.152}$$

which gives

$$W_l = A_r \nu \int_0^{l_r} u_x^2 dx \int_0^T \left(\frac{\partial u_t}{\partial t}\right)^2 dt \tag{2.153}$$

$$W_k = \frac{\rho A_r}{2} \left(\frac{\partial u_t}{\partial t}\right)^2 \int_0^{l_r} u_x^2 dx \tag{2.154}$$

Now we can write the ratio:

$$Q = \frac{W_e}{W_l} = \frac{W_k}{W_l} = \frac{\frac{\rho A_r}{2} \left(\frac{\partial u_t}{\partial t}\right)^2 \Big|_{max} \int_0^{l_r} u_x^2 dx}{A_r \nu \int_0^{l_r} u_x^2 dx \int_0^T \left(\frac{\partial u_t}{\partial t}\right)^2 dt} = \frac{\frac{\rho A_r}{2} \left(\frac{\partial u_t}{\partial t}\right)^2 \Big|_{max}}{A_r \nu T \left(\frac{\partial u_t}{\partial t}\right)^2} \tag{2.155}$$

In resonance, $u_t$ is sinusoidal, which gives

$$\frac{\left(\frac{\partial u_t}{\partial t}\right)^2\Big|_{\max}}{\left(\frac{\partial u_t}{\partial t}\right)^2} = \frac{\left(\left(\frac{\partial u_t}{\partial t}\right)\Big|_{\max}\right)^2}{\sqrt{\left(\frac{\partial u_t}{\partial t}\right)^2}\sqrt{\left(\frac{\partial u_t}{\partial t}\right)^2}} = \frac{\left(\left(\frac{\partial u_t}{\partial t}\right)\Big|_{\max}\right)\cdot\left(\left(\frac{\partial u_t}{\partial t}\right)\Big|_{\max}\right)}{\sqrt{\left(\frac{\partial u_t}{\partial t}\right)^2}\cdot\sqrt{\left(\frac{\partial u_t}{\partial t}\right)^2}} = \sqrt{2}\sqrt{2} = 2$$

$$(2.156)$$

and

$$Q = \frac{\rho}{\nu T} = \frac{\rho \omega_{\text{res}}}{\nu 2\pi} \qquad (2.157)$$

which relates the $Q$ value to material constants as density and the viscous attenuation constant for a mechanical free magnetostrictive rod in electrical resonance.

### The Q Value for a Rod with a Mechanical Load

If the wave propagation in the rod is neglected and the mechanical load comprises a mass $M$ and a viscous damper $R$, then Eq. (2.25a) results in

$$S = s^H T + d\,H = s^H\left(-\frac{F}{A_r}\right) + d\,H = s^H \cdot (-1) \cdot \overbrace{\frac{(j\omega)^2 u_x \cdot M + (j\omega)u_x \cdot R}{A_r}}^{F} + d\,H$$

$$\frac{u_x}{l_r} = -s^H\frac{(j\omega)^2 u_x \cdot M + (j\omega)u_x \cdot R}{A_r} + d\,H$$

$$d\,H\frac{A_r}{Ms^H} = u_x\left(\frac{A_r}{s^H M l_r} + \frac{R}{M}j\omega - \omega^2\right)$$

$$u_x = \frac{d\,H\frac{A_r}{Ms^H}}{\left(-\omega^2 + \frac{R}{M}j\omega + \frac{A_r}{s^H M l_r}\right)}$$

$$(2.158)$$

The absolute value of the normalized frequency response $G$, will then be

$$|G(j\omega)| = \left|\frac{u_x(\omega)}{dHl_r}\right| = \frac{1}{\sqrt{\left(1 - \left(\frac{\omega}{\omega_0}\right)^2\right)^2 + \left(\frac{2\zeta\omega}{\omega_0}\right)^2}} \qquad (2.159)$$

where

$$\zeta = \frac{R}{2M}\sqrt{\frac{Ml_r s^H}{A_r}} \tag{2.160}$$

$$\omega_0{}^2 = \frac{A_r}{s^H Ml_r} \tag{2.161}$$

The maximal response corresponds to the stationary, normalized amplitude of forced vibrations in resonance. This maximum of the response at resonance can easily be calculated by investigating where the derivative of $|G|$ with respect to $\omega$ is zero, i.e.,

$$\frac{d}{dw}(|G(j\omega)|) = 0 \Rightarrow \omega_{res} = \omega_0\sqrt{1 - 2\zeta^2} \tag{2.162}$$

By inserting this resonance frequency in the expression for $|G|$, one can obtain the system magnification at resonance or the quality factor at resonance:

$$|G(j\omega)|_{max} = Q = \frac{1}{2\zeta\sqrt{1 - \zeta^2}} \tag{2.163}$$

If the damping is small, the quality factor can be approximated:

$$Q \approx \frac{1}{2\zeta} = \frac{M\omega_0}{R} \approx \frac{M\omega_{res}}{R} \tag{2.164}$$

This result is analogous to the $Q$ value of an inductor at any frequency with self-inductance $L$ and electric resistance $R_e$, i.e., $Q_{inductor} = \omega L / R_e$.

The quality factor or $Q$ value can be deduced from the resonance curve since the width of the resonance peak is governed by the decrease in magnification. For this purpose, the frequencies can be determined where the amplitudes of displacements correspond to half of the maximum absorbed power, i.e., when

$$\frac{|G(j\omega)|_{max}}{\sqrt{2}} \approx \frac{1}{2\zeta\sqrt{2}} = \frac{1}{\left(1 - \left(\frac{\omega}{\omega_0}\right)^2\right)^2 + \left(\frac{2\zeta\omega}{\omega_0}\right)^2} \tag{2.165}$$

This fourth-order equation can be solved with respect to $\omega$:

$$\omega(\zeta)_{1,2} = \pm\omega_0\sqrt{1 - 2\zeta^2 \pm 2\zeta\sqrt{\zeta^2 + 1}} \tag{2.166}$$

Neglecting the negative frequencies, a series expansion around $\zeta = 0$ gives

$$\omega(\zeta)_{1,2} = \omega_0 \left( 1 + \frac{d\omega(0)_{1,2}}{d\zeta} \zeta + R(\zeta)\zeta^2 \right) \tag{2.167}$$

The residual, $R(\zeta)$, is limited and small if $\zeta << 1$, $\Rightarrow$ $\begin{cases} \omega_1 \approx \omega_0(1+\zeta) \\ \omega_2 \approx \omega_0(1-\zeta). \end{cases}$
This leads to

$$\omega_1 - \omega_2 = \omega_0 2\zeta \Rightarrow \frac{\omega_0}{\omega_1 - \omega_2} \approx \frac{\omega_{res}}{\omega_1 - \omega_2} = \frac{1}{2\zeta} \approx Q \tag{2.168}$$

The difference between $\omega_1$ and $\omega_2$ is called the bandwidth of the system. The points at which the amplitude of $|G|$ is reduced to $Q/\sqrt{2}$ are called half power points due to the fact that the dissipated power is proportional to the square of the amplitude of displacement.

The previous equation indicates the possibility to estimate $Q$ or $\zeta$ by determination of the half power points and the resonance frequency. Another way to estimate the $Q$ value is to determine the ratio between the amplitudes at resonance and at quasi-static excitation when feeding from a constant current source (see Eq. 2.163).

In a magnetostrictive actuator $Q$ values can be associated to all involved components. $Q$ values for eddy currents and hystereses, for example, can be defined in analogy with $Q$ values for inductive components as follows:

$$Q_{eddy\ currents} = \frac{\chi_r \omega L^s}{\chi_i \omega L^s} = \frac{\chi_r}{\chi_i} \tag{2.169}$$

$$Q_{hysteresis} = \frac{\omega L^s}{\frac{\mu_H}{\mu_r}\omega L^s} = \frac{\mu_r}{\mu_H} \tag{2.170}$$

or when both eddy currents and hysteresis are present

$$Q_{eddy\ currents+hysteresis} = \frac{\chi_r}{\chi_i + \frac{\mu_H}{\mu_r}} = \frac{\mu_r \chi_r}{\chi_i \mu_r + \mu_H} \tag{2.171}$$

The mechanical and magnetic and other $Q$ values contribute to the overall $Q$ value of the system in a complex way. The $Q$ value is therefore normally evaluated experimentally.

## Efficiency

A more straightforward quality concept related to magnetostrictive materials and magnetostrictive actuators is their power efficiency. In linear modeling, however, it is difficult to estimate this quantity because the occurring losses merely are described in terms of complex permeabilities or compliances. This description is in most cases insufficient because the elastic, magnetic, and magnetoelastic losses are caused by hysteresis phenomena.

Nevertheless, the efficiency is easy to define according to

$$\eta_{\text{mec}} = \frac{P_{\text{mec}}}{\sum P_i} \tag{2.172}$$

where $\sum P_i$ stands for the sum of all dissipated power including the delivered mechanical power $P_{\text{mec}}$.

A similar electric efficiency,

$$\eta_{\text{electric}} = \frac{P_{\text{electric}}}{\sum P_i} \tag{2.173}$$

can also be defined in the case in which the magnetostrictive material is used as an electric generator.

Mechanical and electrical efficiencies, are of the previously mentioned reason, much easier to measure experimentally but can also be estimated by nonlinear modeling (see Nonlinear Modeling).

## Conversion Ratio and Conversion Efficiency

One way to describe the efficiency of the energy transformation in a magnetostrictive material is to examine the incremental mechanical energy (18),

$$\Delta W_{\text{e}}(t) = \int_{s(t-\Delta t)}^{s(t)} T dS \approx T(S(t) - S(t - \Delta t)) \tag{2.174}$$

and the magnetic energy

$$\Delta W_{\text{m}}(t) = \int_{B(t-\Delta t)}^{B(t)} H dB \approx \frac{((B(t) - B(t - \Delta t)) \cdot ((H(t) + H(t - \Delta t))}{2} \tag{2.175}$$

at every time interval.

Figure 2.26 shows the ratio $\Delta W_{\text{e}}/\Delta W_{\text{m}}$ during a half cycle at quasistatic operation. The bottom line shows the energy ratio for the expanding phase, and the top line shows the corresponding ratio for the contraction phase. The dashed line shows the magnetostriction.

For $H = 10\,\text{kA/m}$ in the expansion phase in Fig. 2.26, the mechanical energy output is 1.7 times higher than the supplied magnetic energy. At

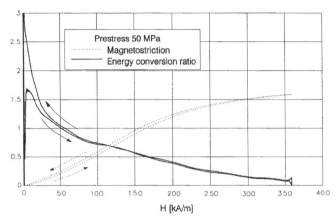

**FIGURE 2.26.** Energy conversion ratio and magnetostriction as a function of magnetizing field at prestress of $50\,\mathrm{N/mm^2}$ (*18*).

$H > 50\,\mathrm{kA/m}$, input and output energies are equal. Therefore, the conversion efficiency then is

$$\frac{\left.\dfrac{W_e}{W_m}\right|_{\text{expansion}}}{\left.\dfrac{W_e}{W_m}\right|_{\text{contraction}}} \approx 100\% \qquad (2.176)$$

It is also obvious that the conversion efficiency is considerably lower below $30\,\mathrm{kA/m}$. The correspondning losses are due to nonreversible losses which arise from domain wall movements i.e. from magnetic hysteresis. A conclusion is that low magnetizing levels ($< 30\,\mathrm{kA/m}$) give high conversion ratios at the expense of low conversion efficiency, i.e. high magnetic hysteresis losses.

## FINITE ELEMENT MODELING

A variational formulation using the Hamilton principle for the equations associated to electromechanically coupled structures has been developed. Appendix C provides a detailed description of this algorithm.

The method can describe the three-dimensional dynamic behavior of structures including magnetostrictive parts and compute its overall relevant characteristics (4). The method is based on an original variational formulation of electromechanical coupling equations and is developed within the framework of the finite element (FE) method.

The model obtained allows the dynamic behavior of complex-geometry and heterogeneous three-dimensional structures to be described. Due to the consideration of current sources using a reduced scalar potential formulation of the magnetic field, strong algorithmic analogies have been used with respect to the method implemented within the ATILA[1] computation code for piezoelectric transducers. The ATILA code is the computation code of DCN (French navy) for piezoelectric transducers. This commercially available program has been efficiently extended to the computation of magnetostrictive structures.

The essential feature of the ATILA code structure is identical to that of the FE method applied to the computation of piezoelectric transducers. Consequently, rather than developing a specific code representing many engineering years, DCN deemed that it was more efficient to implement the necessary computation modules so that the ATILA code could be used for processing both piezoelectric and magnetostrictive transducers. After a successful implementation of the method (4, 9), this code was first made available to design complex magnetostrictive sonar transducers such as flextensional structures (10). An agreement between DCN and Institut Supérieur d'Electronic du Nord (ISEN) has allowed the distribution of the ATILA as a CAD software package without any restriction for civilian application fields, such as actuators, motors, industrial transducers, and sensors (11).

The FE model is based on the following assumptions:

1. The time variations of all quantities are harmonic (i.e., sinusoidal).

2. The magnetic phenomena are described in a quasistatic approximation (i.e., no rate-dependent magnetization and magnetostriction are assumed).

3. The sources are directly expressed by a finite number of excitation currents.

4. No magnetic part of infinite permeability is included in the analyzed domains and therefore does not interleave with any current loop.

---

[1]ATILA (Analyse des Transducteurs par Integration de l'équation de Laplace) is the finite element code originally described by J. N. Decarpigny (see Decarpigny *et al.*, *J. Acoust. Soc. Am.* 78, 1499–1507, 1985), now commercially available worldwide through Cedrat and Magsoft.

5. The material behavior is linear and nonhysteretic.

6. The system energy is conservative.

All these assumptions coincide with the assumptions made regarding the previously described nonlinear models. Therefore, the requirements regarding material data, excitation sources, and mechanical load essentially are the same in a magnetoelastic FE analysis by ATILA.

There are, however, some important differences:

- In a cylindrical 2D FE model, the model requires numerical values of $s_{11}^H$, $s_{13}^H$, $d_{13}$, $\mu_{11}^T$, and $d_{15}$ in addition to the required $s_{33}^H$, $d_{33}$, and $\mu_{33}^T$ in the longitudinal model.

- In a full 3D FE model, the model requires numerical values of all nonzero elements in the material matrixes in Eqs. (2.19) and (2.20).

Experimental determination of the magnetoelastic shear compliance, $s_{13}^H$ and shear magnetostriction constant, $d_{15}$, is not straightforward. Cedrat Recherch, however, has worked out a method based on a specially designed actuator (12) which is able to determine these parameters, including the transverse permeability, $\mu_{11}$, and $s_{11}^H$.

The remaining required material constants for a full 3D analysis could be estimated on the basis of the 2D constants. In the ATILA program all the required material data for many commercial qualities of magnetostrictive material are available. Table 2.1 provides examples of numerical values of some of these constants. One must be aware, however, that such values differ considerably between different material qualities.

## NONLINEAR MODELS

One characteristic of linear models is that they are only valid for small signal excitations, in which bias magnetization level and prestress are accounted for by adjusting the magnetostrictive linear parameters as $d$, $s^H$, $s^B$, $\mu^T$, and $\mu^S$ in an appropriate way. Also, linear models cannot give an appropriate description of hysteresis effects that can be of significant importance even at low excitations.

To overcome these obvious limitations connected with linear models, efforts have been made since the mid-1980s to develop nonlinear models (19). These are inherently formulated in the time domain and provide the possibility to analyze devices and apparatuses based on

**Table 2.1.** Numerical Values of Magnetostrictive Constants

| Constant | Value |
|---|---|
| $s_{11}{}^{H}$ | $12.5 \times 10^{-11a}$ |
| $s_{12}{}^{H}$ | $-1.8 \times 10^{-12a}$ |
| $s_{13}{}^{H}$ | $-1.7 \times 10^{-11a}$ |
| $s_{33}{}^{H}$ | $4.0 \times 10^{-11a}$ |
| $s_{44}{}^{H}$ | $1.8 \times 10^{-10a}$ |
| $s_{66}{}^{H}$ | $5.4 \times 10^{-11a}$ |
| $s_{11}{}^{B}$ | $3.5 \times 10^{-11}$ |
| $s_{12}{}^{B}$ | $-1.3 \times 10^{-11}$ |
| $s_{13}{}^{B}$ | $-8.4 \times 10^{-12}$ |
| $s_{33}{}^{B}$ | $2.1 \times 10^{-11}$ |
| $s_{44}{}^{B}$ | $1.2 \times 10^{-10}$ |
| $s_{66}{}^{B}$ | — |
| $d_{31}$ | $-5.3 \times 10^{-9a}$ |
| $d_{15}$ | $2.8 \times 10^{-8a}$ |
| $d_{33}$ | $1.1 \times 10^{-8a}$ |
| $\mu_{11}^{T}$ | $1.3 \times 10^{-5a}$ |
| $\mu_{33}^{T}$ | $5.3 \times 10^{-6a}$ |
| $\mu_{11}^{B}$ | $5.7 \times 10^{-6a}$ |
| $\mu_{33}^{B}$ | $2.7 \times 10^{-6a}$ |

[a]Values are taken from the ATILA database.

magnetostrictive materials, which interact with surrounding systems in specific applications.

## The Lumped Element Approach

The nonlinear models are based on the lumped element approach, which means that parts of the geometry that have similar potential, current, mass, magnetic flux, mechanical stress, strain, or other relevant properties are lumped together and represented by one discrete component (20). This approach is often used when modeling electric systems, but it can also be used for various physical systems, including hydraulic, mechanical, thermal, and magnetic. Examples of components or components include resistors, capacitors, inductors, reluctances, masses, springs, and viscous dampers.

The elements in a system can be considered to have through and across variables. The through variables must sum to zero at connection points (nodes) between components. In electrical networks the current is the through variable, in mechanical systems in the mobility representation the through variable is the force, and in thermal systems the heat flux is the through variable. The across variable is generally the driving force, i.e., the potential difference in electrical systems and temperature difference in thermal systems.

A system described by the lumped element approach comprises many components that are linked in a network, in which the relations between through and across variables for each component are defined by constitutive equations. In Fig. 2.27, this is illustrated for a resistor and an inductor.

In principle, any phenomenon, material, or process that can be described by algebraic and/or differential equations can be described by this approach, which results in a system of algebraic and/or differential equations with respect to time. Making an analysis is similar to obtaining a solution to the resulting equation system. This can be done in several ways, but one common equation solver is based on the Newton–Raphsson method.

There are many software packages, including PSPICE, Saber, Simplorer, Dymola, and Sandys, that comprise such an equation solver. The mentioned packages for dynamic simulation are principally similar

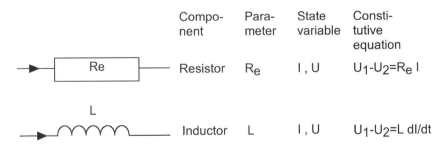

| Component | Parameter | State variable | Constitutive equation |
|---|---|---|---|
| Resistor | $R_e$ | $I$, $U$ | $U_1 - U_2 = R_e I$ |
| Inductor | $L$ | $I$, $U$ | $U_1 - U_2 = L \, dI/dt$ |

**FIGURE 2.27.**   Lumped circuit components with constitutive equations.

except for different syntaxes and user interfaces. One important difference is their different ability to handle changes of state in the simulated system. For example, in electric power engineering, different equations are valid during conduction and blocking of a semiconductor. Also, in hydraulics, the valid equations change completely depending on the state of the valves in a system.

It is also important that the program that is used for the lumped element approach can manage different physical quantities such as electrical, magnetic, mechanical, and hydraulic in a convenient way. This implies that PSPICE only can be used in cases in which all quantities can be translated into electrical terms, which is exactly what occurs when translating the physical situation into equivalent circuits. It is therefore preferable to use PSPICE in linear modeling of magnetostrictive materials by use of equivalent circuits. The lumped element approach can be regarded as an extension or generalization of the equivalent circuit approach.

Table 2.2 summarizes the features of potential packages suitable for the lumped element approach regarding nonlinear modeling of systems comprising magnetostrictive materials.

## The Axial Model of Magnetostrictive Rods

The governing idea in the axial model is to delimit the state variables to stress and strain in a finite number of longitudinal sections (19) (Fig. 2.28). The state variables $T_i$ and $u_i$ are assumed to be valid in the whole volume of each segment. Let the rod in Fig 2.28 have section length $l_s$, cross-sectional area $A_r$, and mass density $\rho$. Discretization of Eq. (2.36) gives the strain in section $i$ as

$$S_i = \frac{u_i - u_{i-1}}{l_s} \tag{2.177}$$

Discretization of Newton's second law (Eq. 2.61) gives

$$-\frac{T_{i+1} - T_i}{l_s} = \rho \frac{d^2 u_i}{dt^2}, \quad i = 1, \ldots, n-1 \tag{2.178}$$

where it is assumed that the mass is concentrated to the section boundary. The additional minus sign on the left-hand side in Eq. (2.178) is due to the fact that the compressive stress is regarded as positive.

The strain and the induction are nonlinear functions of the magnetizing field and the stress. They are taken from static magnetostriction and

**Table 2.2.**   Features of Programs for Use in Dynamic Simulation of Magnetostrictive Systems

| Program | Platform[a] | System input | State handling | Component library | Full physics capability |
|---|---|---|---|---|---|
| Pspice | Sun- and, HP-UNIX, PC | Circuit schematics, text | No | Yes (a big one) | No |
| Saber | Sun- and, HP-UNIX, Solaris, PC | Circuit schematics, text | Yes, but not easy to use | Yes (a big one) | Yes |
| Sandys | IBM- and, HP- UNIX | Text | Yes | A small one | Yes |
| Dymola | HP-UNIX, PC | Circuit schematics, text | No | Yes | Yes |
| Simplorer | HP-UNIX, PC-Dos, PC | Circuit schematics, text | No | Yes | Yes |

[a] The codes in general also support other platforms. A minimum capability is presented.

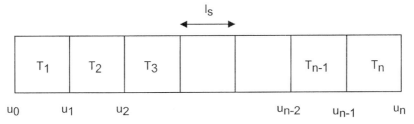

FIGURE 2.28. Magnetostrictive rod cut in a finite number of sections, each with the state variables $u_i$ and $T_i$.

magnetization curves obtained from experimental measurements. These curves comprise hysteresis. Therefore, the magnetostriction and magnetization curves have to be "dehysterized". In this context, dehysterization is determined by calculating average strain and induction values from the "up" and "down" curves corresponding to the same magnetizing field. Figure 2.29 shows examples of dehysterized magnetostriction and magnetization curves.

The obtained dehysterized curves represent the constitutive relations $S = S_{\exp}(H, T)$ and $B = B_{\exp}(H, T)$ between strain, stress, induction, and magnetizing field. For each section of the rod,

$$S_i = S_{\exp}(H_i, T_i) \qquad i = 1, \dots, n \tag{2.179a}$$
$$B_i = B_{\exp}(H_i, T_i) \tag{2.179b}$$

For reversible processes the condition

$$\left.\frac{\partial B}{\partial T}\right|_H = \left.\frac{\partial S}{\partial H}\right|_T \tag{2.180}$$

must be fulfilled [compare with Eq. (2.14) for the linearized model]. However, it is not obvious that the dehysterization method used here always ensures that this condition is fulfilled. Moreover, since the slope of the dehysterized material characteristics is approximately the same as the differential slope of the major loops, the model is restricted to high excitation levels.

So far, the rod has been modeled only by means of springs and masses, without any intrinsic material damping. To model the rod with intrinsic material damping, one can consider each section of the rod divided into a nondamped rod connected with a viscous damper. This situation is the same as that described previously for the $Q$ value of a

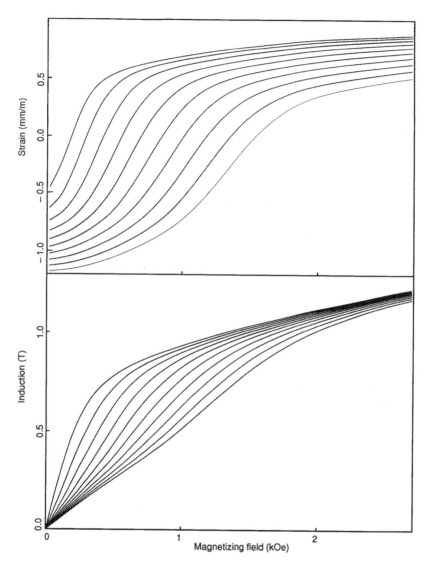

**FIGURE 2.29.** Examples of dehysterized magnetostriction and magnetization curves (24).

magnetostrictive rod, when one assumes that the velocity $u_x$ is the velocity difference between the two rod ends.

Control theory states that critical damping occurs if $\zeta = 1$. This fact makes it possible to determine the value of the equivalent viscous damper in a rod section which implies critical damping. Therefore,

$$\zeta = 1 = \frac{R_c}{2M}\sqrt{\frac{Ml_c s^H}{A_r}} = \frac{R_c}{2\sqrt{\rho l_c A_r}}\sqrt{\frac{l_c s^H}{A_r}} \quad \Rightarrow \quad R_c = 2A_r\sqrt{\frac{\rho}{s^H}} \quad (2.181)$$

The viscous counteracting force/area in section $i$ corresponding to the fraction $\zeta$ of the critical damping is

$$2\zeta\sqrt{\frac{\rho}{s_i^H}}\left(\frac{du_i}{dt} - \frac{du_{i+1}}{dt}\right) \quad (2.182)$$

The material compliance is dependent on the magnetization and stress and is defined by

$$s_i^H = \frac{S_{0i} - S_i}{T_i} \quad (2.183)$$

where $S_{0i} = S_{\exp}(H_i, 0)$ is the magnetostrictive strain that would be obtained for zero stress. The difference, $S_{0i} - S_i$, gives the pure elastic compression of the rod caused by the compressive stress $T_i$ for the constant magnetic field $H_i$.

The dynamic electromechanical behavior can then be described by means of the following equation system:

$$S_i = \frac{u_i - u_{i-1}}{l_s} \quad (2.184a)$$

$$T_i - T_{i+1} + 2\zeta\sqrt{\frac{\rho}{s_i^H}}\left(\frac{du_{i-1}}{dt} - \frac{du_i}{dt}\right) - 2\zeta\sqrt{\frac{\rho}{s_{i+1}^H}}\left(\frac{du_i}{dt} - \frac{du_{i+1}}{dt}\right) = \rho l_s \frac{d^2 u_i}{dt^2} \quad (2.184b)$$

$$s_i^H = \frac{S_{0i} - S_i}{T_i} \quad (2.184c)$$

$$S_i = S_{\exp}(H_i, T_i) \quad (2.184d)$$

$$B_i = B_{\exp}(H_i, T_i) \quad (2.184e)$$

This can be solved by one of the previously mentioned programs for simulation of systems described by differential and algebraic equations. For each time step in the simulation, actual values of $S_i$ and $B_i$ are accessed from a database comprising material data obtained from experimental material characterization.

*Initial Conditions*

For initial conditions it is often assumed that the rod is in rest and has a well-defined length corresponding to the bias magnetization $H_{bias}$ and pre-stress $T_p$. Therefore,

$$\frac{du_i}{dt} = 0 \tag{2.185a}$$

$$\frac{u_i - u_{i-1}}{l_s} = S_{exp}(H_{bias}, T_p) \tag{2.185b}$$

*Boundary Conditions*

The boundary conditions define the mechanical and magnetic load/excitation of the rod. In the case of magnetic excitation and an ideal magnetic circuit, one can use

$$H_i = H_{ex}$$

where $H_{ex}$ is a prescribed imposed field along the rod sections, or

$$B_i = B_{ex}$$

where $B_{ex}$ is a prescribed imposed magnetic flux density through the rod sections.

A mechanical passive load at one rod end is defined by

$$T_n A_r = T_{n+1} A_l = M_1 \frac{d^2 u_n}{dt^2} + R_1 \frac{du_n}{dt} + \frac{u_n}{C_1} \tag{2.186}$$

where $M_1, R_1$, and $C_1$ are the mass, viscous damping, and compliance of the load. In this 1-D description the area of the transmission can be considered to be the same as the rod cross-sectional area, i.e., $A_r = A_l$, which gives $T_n = T_{n+1}$. If the other rod end is fixed, then $u_0 = 0$, but it can in principle be attached to another mechanical passive load ($M_0$, $R_0$, and $C_0$), which results in the boundary condition

$$T_0 A_l = T_1 A_r = M_0 \frac{d^2 u_0}{dt^2} + R_0 \frac{du_0}{dt} + \frac{u_0}{C_0} \tag{2.187}$$

In general, the mechanical load does not need to be passive. In fact, it can be any function of the position and/or its derivatives

$$T_0 A_r = f(t, u_0, \dot{u}_0, \ddot{u}_0, \dddot{u}_0, \ldots) \tag{2.188a}$$

$$T_n A_r = g(t, u_n, \dot{u}_n, \ddot{u}_n, \dddot{u}_n, \ldots) \tag{2.188b}$$

In the case of mechanical excitation, $f$ and $g$ can also be arbitrary functions of time. Therefore, the model can also be used in studies of the transduction from mechanical to electrical energies.

## Electrical Feeding

The two boundary conditions when an $H$ or a $B$ field excites the rod assume the same imposed field in all consecutive sections. In the case in which $H$ is the driving quantity that corresponds to a current source, $B_i$ will differ between the sections if wave propagation effects occur. The equation that represents the boundary condition, if there is no flux leakage out of the rod, is

$$H_{ex} = \frac{NI_{ex}}{l_{coil}} = \frac{1}{n}\sum_{j=1}^{n} H_i \tag{2.189}$$

if the reluctance of the return flux path is negligible.

Similarly, in the case in which $B$ is the driving quantity which corresponds to a voltage source, the equation that represents the boundary condition is

$$B_i = B_{ex} \text{ for } i = 1 \cdots n \tag{2.190}$$

In this case, the material database gives the actual magnetizing field:

$$H_i = H_{exp}(B_i, T_i) \tag{2.191}$$

and corresponding $H_{ex}$ from Eq. (2.189).

## The Magnetic Circuit

As long as one considers only $H_{ex}$ and/or $B_{ex}$ as the driving quantity, one does not need to specify the magnetic circuit because if $H_{ex}$ is specified it is always possible to obtain a corresponding $B_{ex}$ from the material database and vice versa by using

$$B_{ex} = B_{exp}(H_{ex}, T_i) \tag{2.192a}$$

or

$$H_{ex} = H_{exp}(B_{ex}, T_i) \tag{2.192b}$$

In real applications, however, one is interested in driving quantities as imposed current $I_{ex}$ and/or imposed voltage $U_{ex}$. To make it possible to excite the rod with input currents and/or voltages in the model, it is necessary to specify the magnetizing system or the magnetic circuit.

In principle, such a magnetizing system involves a reluctance and a coil flux leakage (Fig. 2.30). In this 1-D model it is sufficient to estimate $R_p$ and $R_l$ in order to take the magnetic circuit into account. Assuming equivalent cross-sectional areas and lengths of the flux return and leakage paths can yield an estimate of $R_p$ and $R_l$ according to

$$R_p = \frac{l_p}{\mu_p A_p} \tag{2.193a}$$

$$R_l = \frac{l_l}{\mu_l A_l} \tag{2.193b}$$

The relation between imposed current $I_{ex}$ and imposed magnetic field $H_{ex}$ can then be described by the following equations:

$$NI_{ex} = R_p \phi + \frac{R_l R_r}{R_l + R_r} \phi \tag{2.194a}$$

$$\phi = \phi_l + \phi_r \tag{2.194b}$$

$$H_{ex} = \frac{\phi_r R_r}{l_r} = \frac{\phi_l R_l}{l_r} \tag{2.194c}$$

The rod reluctance $R_r$ can be estimated as $R_r = l_r/\mu_r A_r$, where $\mu_r$ is the effective magnetic permeability of the magnetostrictive rod. In real operation, $\mu^S \leq \mu_r \leq \mu^T$, where $\mu^S$ and $\mu^T$ represent the extreme conditions clamped and free rod, respectively.

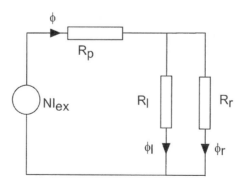

FIGURE 2.30.   Reluctance description of the magnetic circuit of a magnetostrictive actuator.

Similarly, the relation between imposed current $U_{ex}$ and imposed magnetic field $B_{ex}$ can be described by the following equations:

$$U_{ex} = R_{coil}I + L_{leak}\frac{d}{dt}(I - I_p) + L_{rod}\frac{d}{dt}(I - I_p) \qquad (2.195a)$$

$$U_{ex} = R_{coil}I + L_{path}\frac{d}{dt}I_p \qquad (2.195b)$$

$$A_rB_{ex} = L_{rod}(I - I_p) \qquad (2.195c)$$

The inductances are

$$L_{leak} = \frac{\mu_1 N^2 A_1}{l_1} \qquad (2.196a)$$

$$L_{path} = \frac{\mu_p N^2 A_p}{l_p} \qquad (2.196b)$$

$$L_{rod} = \frac{\mu_r N^2 A_r}{l_r} \qquad (2.196c)$$

where $\mu_1$ and $\mu_p$ are the effective magnetic permeabilities of the flux leakage and flux return path, respectively.

In the model, appropriate equations should be added to the equation system if one intends to account for the magnetic circuit.

## Bias Magnetization

Magnetostriction is a phenomenon that mathematically can be characterized by an even function of the magnetizing field. Therefore, the strain of a magnetostrictive rod is independent of the sign of the applied longitudinal magnetizing field. Thus, the frequency of the strain will be doubled when the rod is magnetized sinusoidally, as illustrated in Fig. 2.31. To obtain "linear" strain, an additional bias magnetization with a magnitude that is at least equal to the amplitude of the dynamic field must be imposed on the rod (Fig. 2.31b).

The bias magnetization can be generated by a DC current in the magnetizing coil or by permanent magnets in the magnetic circuit. In both cases it is possible to estimate an equivalent current $I_{bias}$ corresponding to an applied bias field $H_{bias}$ or $B_{bias} = B_{exp}(H_{bias}, T_p)$ by solving the following equation system:

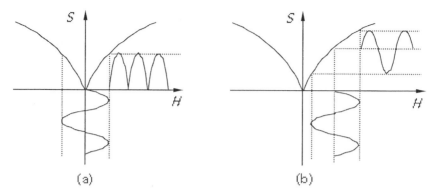

(a)                                    (b)

**FIGURE 2.31.**   Illustration of the magnetostrictive strain without (a) and with (b) a bias magnetic field.

$$NI_{bias} = R_p \phi + \frac{R_l R_r}{R_l + R_r} \phi \qquad\qquad (2.197a)$$

$$\phi = \phi_l + \phi_r \qquad\qquad (2.197b)$$

$$H_{bias} = \frac{\phi_r R_r}{l_r} = \frac{\phi_l R_l}{l_r} \qquad\qquad (2.197c)$$

In the case of permanent magnets, $I_{bias}$ should not be added to $I$ in terms comprising estimations of coil losses or resistive voltage drops over the coil resistance, which refers to real conductive currents. If, as occurs in many cases, the permanent magnets are located in the flux return path, their reluctances must be added to $R_p$. The reluctance of each permanent magnet can be estimated by

$$R_{magnet} = \frac{l_m}{\mu_m A_m} \qquad\qquad (2.198)$$

where $l_m$, $A_m$, and $\mu_m$ are the magnet thickness, cross-sectional area, and magnetic permeability during operation. $\mu_m$ is near $\mu_0$ for NdFeB magnets.

## Eddy Currents

Previously, the complex eddy current factor was derived for harmonic excitation under the assumption that the permeability is constant within the whole rod. By incorporating this factor into each section in the axial model for harmonic magnetization, the influence of eddy currents can be taken into account.

This can be done by multiplying the external magnetization field generated by the coil with the eddy current factor. $H_{ex}$ in Eq. (2.194c) should then be substituted by $\chi H_{ex}$ and $B_{ex}$ in Eq. (2.195c) by $\chi B_{ex}$. Notice that the eddy current factor is complex. Therefore, $H_{ex}$ and $B_{ex}$ will be altered with respect to amplitude and phase.

The amplitude alteration is $|\chi|$, and the phase alteration is $\arctan(\chi_i/\chi_r)$. Also observe that $\chi$ is a function of frequency, whereas the model operates in the time domain. Therefore, another interpretation is that a sinusoidal externally applied field quantity will be split into a sinus and a cosinus function with aplitudes $\chi_r$ and $\chi_i$, respectively.

At low frequencies, $\chi_r \approx 1$ and $\chi_i \approx 0$, which implies that the internal field is nearly equal to the external field. The eddy current factor approach to manage the influence of eddy currents can be regarded as a hybrid between small signal linear and nonlinear modeling.

When using the eddy current factor approach, it is important to notice that significant errors occur as soon as the mechanical stresses vary significantly in the radial direction. The different mechanical stresses influence the magnetic permeability of the material and therefore affect the field penetration in the rod (13).

## Modeling of Laminated Magnetostrictive Rods

If the actuator rods are laminated in such way that the field inside each laminate can be regarded as constant, it is possible to define an effective magnetizing field $H_{eff}$ as follows (14):

$$H_{eff} = f(\bar{B}) + k\frac{\partial \bar{B}}{\partial t} \tag{2.199}$$

where $\bar{B}$ is interpreted as a geometrical average value of $B$ across the whole rod cross-sectional area.

If one assumes that $f(\bar{B}) = (1/\mu)\bar{B} = \nu \cdot \bar{B}$, then in the frequency plane one derives

$$\tilde{H}_{eff} = \nu\tilde{\bar{B}} + j\omega k\tilde{\bar{B}} \tag{2.200}$$

The parameter $k$ can then be estimated by means of a transversal magnetic FE analysis by applying a harmonic $H$ field on the circular periphery of the laminated rod cross section and calculating the average $B$ field through the different laminates of the rod.

If a time average of $\nu$ is assumed or estimated from experimental data, it is easy to estimate $k$ on the basis of the calculated average $B$ field from the FE analysis. The $k$ value is easily calculated from the phase shift between the H- and B-fields because Eq. (2.200) gives

$$k = \frac{\nu_r \tan(\varphi)}{\omega} \tag{2.201}$$

where $\varphi$ is the phase shift of the applied H-field relative to the resulting B-field.

With estimated values of $k$ and $\nu$ it is possible to use the expressions in Eqs. (2.199) and (2.200) for the H-field in equivalent circuit analysis, lumped element analysis, and FE analysis. The relation between $\bar{B}$ and $\bar{H}$, however, is still given by Eq. (2.184e) or a hysteresis model.

## Hysteresis

The influence of hysteresis can, in the axial model, be treated by means of the eddy current factor. The algorithm for inclusion of hysteresis used for the previously described linear model should be used here. When applying the eddy current factor to describe hysteresis, one must be aware that the errors in the simulated performance can be significant, especially when the domain wall motion dominates.

## The Radial–Axial Model of Magnetostrictive Rods

The presented axial model is not capable of resolving radial variations of, e.g., mechanical stress and field quantities. Therefore, at frequencies over the previously defined cut-off frequency $f_c$, it is necessary to also make a radial division of the sections/lumped elements. Such a division is shown in Fig. 2.32.

The constitutive equation for the field distribution inside the rod when it is exposed to a longitudinal external field is derived in Appendix A and is

$$\frac{\partial^2 H}{\partial r^2} + \frac{1}{r}\frac{\partial H}{\partial r} = \sigma \frac{\partial B}{\partial t} \tag{2.202}$$

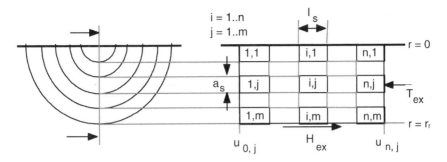

**FIGURE 2.32.** A magnetostrictive rod with radius $r_r$ sectionized in axial and radial directions.

This equation is known as the heat equation. In each element in Fig. 2.32 it is now possible to define state variables $u_{i,j}$, $T_{i,j}$, $B_{i,j}$, $H_{i,j}$, and $S_{i,j}$, $u_{i,j}$ is the displacement of the section boundary, $T_{i,j}$ is the mechanical stress, and $H_{i,j}$ and $B_{i,j}$ are the magnetizing field and the flux density in the middle of section $(i,j)$. By combining the discretized Newton's law and the heat equation with magnetostriction and magnetization experimental data, we can set up the following differential algebraic equations for each section (21):

$$S_{i,j} = \frac{u_{i,j} - u_{i-1,j}}{l_s} \tag{2.203a}$$

$$T_{i,j} - T_{i+1,j} + 2\xi \sqrt{\frac{\rho}{s_{i,j}^H}} \left( \frac{du_{i-1,j}}{dt} - \frac{du_{i,j}}{dt} \right) - 2\xi \sqrt{\frac{\rho}{s_{i+1,j}^H}} \left( \frac{du_{i,j}}{dt} - \frac{du_{i+1,j}}{dt} \right)$$

$$= \rho l_s \frac{d^2 u_{i,j}}{dt^2} \tag{2.203b}$$

$$s_{i,j}^H = \frac{S_{0i,j} - S_{i,j}}{T_{i,j}} \tag{2.203c}$$

$$S_{i,j} = S_{\exp}(H_{i,j}, T_{i,j}) \tag{2.203d}$$

$$\frac{H_{i,j+1} - 2H_{i,j} + H_{i,j-1}}{a_s^2} + \frac{1}{a_s(j - 1/2)} \frac{H_{i,j+1} - H_{i,j-1}}{2a_s} = \sigma \frac{dB_{i,j}}{dt} \tag{2.203e}$$

$$B_{i,j} = B_{\exp}(H_{i,j}, T_{i,j}) \tag{2.203f}$$

## Initial Conditions

The initial conditions are in analogy with the axial model given by

$$\frac{du_{i,j}}{dt} = 0 \tag{2.204a}$$

$$\frac{u_{i,j} - u_{i-1,j}}{l_s} = S_{exp}(H_{bias}, T_p) \tag{2.204b}$$

which corresponds to the situation in which the rod is at rest and has a well-defined length for the bias external magnetizing field $H_{bias}$ and mechanical prestress $T_p$.

## Boundary Conditions

The mechanical boundary conditions defines the strain or stress at the rod ends, i.e., $x = 0$ and $x = l_r$.

$$\left.\begin{array}{lll} x = 0 & : & u_{0,j} = u_{ex,x=0} \\ x = l_r & : & u_{n,j} = u_{ex,x=l_r} \end{array}\right\} \text{ prescribed strain at the rod ends} \tag{2.205}$$

$$\left.\begin{array}{lll} x = 0 & : & \frac{1}{m}\sum_{j=1}^{m} T_{1,j} = T_{load,x=0} \\ \\ x = l_r & : & \frac{1}{m}\sum_{j=1}^{m} T_{n,j} = T_{load,x=l_r} \end{array}\right\} \text{ prescribed stress at the rod ends}$$

$$\tag{2.206}$$

In principle, the different rod ends can be subjected to either prescribed stress or strain independently. One special case is

$$\left.\begin{array}{lll} x = 0 & : & u_{0,j} = 0 \\ \\ x = l_r & : & \frac{1}{m}\sum_{j=1}^{m} T_{n,j} = T_{load,x=l_r} \end{array}\right\} \tag{2.207}$$

where one rod end is fixed and the other end is subjected to prescribed force $A_r T_{load,x=l_r}$. Another common situation occurs when the rod ends are loaded with passive mechanical loads represented by $(M_0, R_0, C_0)$ and $(M_1, R_1, C_1)$, giving

$$T_{\text{load},x=0} = M_0 \frac{d^2 u_0}{dt^2} + R_0 \frac{du_0}{dt} + \frac{u_0}{C_0} \qquad (2.208a)$$

$$T_{\text{load},x=l_r} = M_1 \frac{d^2 u_1}{dt^2} + R_1 \frac{du_1}{dt} + \frac{u_1}{C_1} \qquad (2.208b)$$

The magnetic boundary conditions define magnetic field quantities along the rod envelope surface. There are several ways to do this. The following are often used.

The prescribed external H-field condition:

$$\left. \begin{array}{l} H_{i,m+1} = H_{\text{ex}} \\ \frac{\partial H}{\partial r}\big|_{r=0} = 0 \end{array} \right\} \qquad (2.209)$$

The continuous flux condition:

$$\left. \begin{array}{l} \frac{1}{m}\sum_{j=1}^{m} B_{i,j} = B_{\text{ex}} \\ \frac{\partial B}{\partial r}\big|_{r=0} = 0 \end{array} \right\} \qquad (2.210)$$

The derivative of external field and flux condition:

$$\left. \begin{array}{l} \frac{\partial H}{\partial r}\big|_{r=r_1} = \frac{\sigma}{2\pi r_1}\frac{d\phi_r}{dt} \\ \frac{\partial H}{\partial r}\big|_{r=r_0} = 0 \end{array} \right\} \qquad (2.211)$$

The first two conditions represent cases in which external H-and B-fields excite the rod, respectively. The third condition defines the applied magnetizing coil voltage represented by $-N\frac{d\phi_r}{dt}$, where $\phi_r$ is the magnetic flux through the rod, $N$ is the number of coil turns, $\sigma$ is the electric conductivity of the coil wire material, and $r_1$ is the inside radius of the magnetizing coil.

## Electrical Feeding

A magnetostrictive actuator is normally excited by a voltage or current source. In a real situation there is also a magnetic circuit with path and leakage reluctances $R_p$ and $R_1$. Therefore, when feeding the actuator from a voltage source one should use the derivative of external field and flux condition, for which $B_{\text{ex}} = \phi_r/A_r$ and Eq. (2.195) should be used. When feeding the actuator from a current source, one must use the external field condition for which $H_{ex}$ and Eq. (2.194) must be used.

## The Magnetic Circuit and Bias Magnetization

There is no difference between the axial and radial–axial model with respect to accounting for the magnetic circuit and bias magnetization. Therefore, $R_p$, $R_l$ and $I_{bias}$, as for the axial model, are calculated by means of Eqs. (2.193a), (2.193b) and (2.197), respectively.

## Eddy Current Losses

In the radial–axial model the eddy current losses can easily be calculated. In the model it is assumed that the magnetic field quantities are directed along the rod axis. Therefore, the azimutal eddy current density $j_\phi$ is given by

$$\left( \underbrace{\frac{\partial H_r}{\partial z}}_{=0} - \frac{\partial H_z}{\partial r} \right)_\phi = -\frac{\partial H_z}{\partial r} = j_\phi \tag{2.212}$$

Thus, the total dissipated eddy current losses can be obtained by the expression

$$P_{\text{eddy}} = \rho \sum_{j=1}^{m} \left( \sum_{i=1}^{n} j_{\phi,i,j}^2 V_{i,j} \right) \tag{2.213}$$

where $\rho$ is the electrical resistivity of the rod, $V_{i,j}$ is the volume of segment $(i,j)$, and $j_{\phi,i,j}$ is the corresponding current density in that segment.

In the case of laminated rods, the dissipated eddy current losses are given by

$$P_{\text{eddy}} = k \sum_{i=1}^{n} \left( \frac{\partial B_i}{\partial t} \right)^2 V_i \tag{2.214}$$

where $k$ is given by Eq. (2.201) and $V_i$ and $B_i$ are the volume and flux density of segment $i$ in the axial model, respectively.

## Hysteresis Effects

One feasible way to account for magnetoelastic hysteresis effects is to use a hysteresis model. Such a model should have the following properties:

- Agree well with experiments
- Can be applied to arbitrary processes (not limited to harmonic or periodic processes)
- Admit simultaneous mechanical and magnetic inputs/outputs
- Computationally simple and efficient
- Reasonable easy to determine material parameters

A model that meets the previous demands is based on thermo-dynamics (23). A brief description of this model is given as follows: A basic assumption in this model approach is that magnetic/magnetostrictive hysteresis is analogous to dry mechanical friction (so-called Coulomb friction). An object on a curved surface, which is subjected to gravity, normal force, and a controllable force $F$, can illustrate this phenomenon. The potential energy $W$ of this object is proportional to its height above the ground and is indirectly a function of the horizontal coordinate $x$. The physical principle of virtual work states that the sum of the gravitational and normal force is $-\partial W/\partial x$. Without friction there would be a single-valued relation between applied force $F$ and position $x$ given by the condition that the total force $F - \partial W/\partial x$ is zero. However, if there is Coulomb friction, the object will not move unless $|F - \partial W/\partial x|$ reaches some threshold value $F_F$. In this case, the $F - x$ relation can be decribed by a hysteresis curve with a width $2F_F$. Figure 2.33 shows a typical example.

We now adopt the Coulomb friction concept for the description of magnetomechanical hysteresis. Suppose that the magnetostrictive material we study can be divided into a finite number of fractions, and each of these shows the same magnetomechanical hysteresis. We call these fractions pseudoparticles. Suppose we have one pseudoparticle with magnetization $m$ and strain $s$. There is a clear analogy between such a pseuodoparticle and a mechanical system with an object on a curved surface.

For the mechanical system it holds that $\mathbf{x}$ strives toward a position in which

$W(\mathbf{x}) - \mathbf{F} \cdot \mathbf{x}$ is minimized. $\mathbf{F}$ and $\mathbf{x}$ are assumed to be vectors in space.

For a magnetostrictive pseudoparticle its Gibbs potential energy

$g = f(m,s) - Hm - Ts$ strives toward a minimum due to the principle

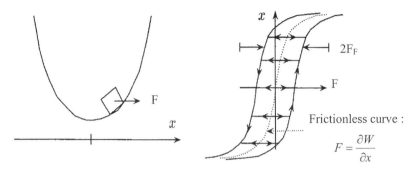

**FIGURE 2.33.** An object on a surface subjected to an applied force $F$ and Coulomb friction. (Right) The resulting relation between $F$ and $x$.

of entropy maximation. The quantity $f$ is the free energy at a certain temperature of the pseudoparticle that depends only on $m$ and $s$, which are its corresponding magnetization and strain.

Now we assume that the relation between $(H,T)$ and $(m,s)$ is analogous to the relation between $\mathbf{F}$ and $\mathbf{x}$ in the presence of Coulomb friction. In other words, for a given $(m,s)$, there exists a region such that as long as $(H,T)$ remain within the region there is no change in $(m,s)$.

Such behavior can be reproduced by stating that $f$ is the sum of a convex function $f_{an}(m, s)$ corresponding to the anhysteretic behavior and a small ripple $f_F(m, s)$ and then applying the principle of minimization of $g$. The ripple introduces many densely distributed local minima of $g$. These local minima can be found by derivation of $g$ and setting this derivative equal to zero. This gives

$$\frac{\partial g}{\partial m} = \frac{\partial f}{\partial m} - H = 0 \tag{2.215a}$$

$$\frac{\partial g}{\partial s} = \frac{\partial f}{\partial s} - T = 0 \tag{2.215b}$$

which will be valid for many metastable states lying near each other. The transition conditions between these states will be exactly analogous to those that are valid for the change of states in the described Coulomb dry friction case. The situation is shown in Fig. 2.34 for a purely magnetic case, with $f_F = -(H_F/N) \cos(Nm)$, where $N$ is large.

Now we introduce the auxiliary quantities

$$\tilde{H} = \frac{\partial f_{an}}{\partial m} \tag{2.216a}$$

and

$$\tilde{T} = \frac{\partial f_{\mathrm{an}}}{\partial s} \tag{2.216b}$$

Since without friction we would have

$$H = \frac{\partial f_{\mathrm{an}}}{\partial m} \tag{2.217a}$$

$$T = \frac{\partial f_{\mathrm{an}}}{\partial s} \tag{2.217b}$$

it is natural to write the anhysteric curves as

$$m = M_{\mathrm{an}}(\tilde{H}, \tilde{T}) \tag{2.218a}$$

$$s = S_{\mathrm{an}}(\tilde{H}, \tilde{T}) \tag{2.218b}$$

where $M_{\mathrm{an}}$ and $S_{\mathrm{an}}$ are the inverse of $\partial f_{\mathrm{an}}/\partial m$ and $\partial f_{\mathrm{an}}/\partial s$, respectively.

The shapes of the metastable regions in the $(H,T)$ space can in principle be estimated from experiments. For simplicity, here they are assumed to be elliptical and can be expressed by the inequality

$$H_{\mathrm{F}}^{-2}(H - \tilde{H})^2 + T_{\mathrm{F}}^{-2}(T - \tilde{T})^2 \leq 1 \tag{2.219}$$

$H_{\mathrm{F}}$ and $T_{\mathrm{F}}$ are adjustable parameters corresponding to the half-width of purely magnetic and purely mechanical hysteresis loops, respectively.

Concerning the direction in which changes occur, for simplicity assume that a resulting change in $(\tilde{H}, \tilde{T})$ is parallel to $(H - \tilde{H}, T - \tilde{T})$ or, mathematically expressed,

$$(\Delta \tilde{H}, \Delta \tilde{T}) = \Delta \xi (H - \tilde{H}, T - \tilde{T}) \tag{2.220}$$

where $\Delta \xi$ is adjusted so that the left-hand side of Eq. (2.219) is exactly 1. It is straightforward to realize that this happens when

$$\Delta \xi = 1 - \left( H_{\mathrm{F}}^{-2}(H - \tilde{H})^2 + T_{\mathrm{F}}^{-2}(T - \tilde{T})^2 \right)^{-\frac{1}{2}} \tag{2.221}$$

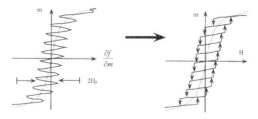

FIGURE 2.34.   Relation between $H$ and $m$ when the free energy $f(m)$ contains a small ripple. Dotted curves show behavior without ripple.

An incremental algorithm for determining $(m,s)$ for any variation of $(H,T)$ can now be formulated as follows: Suppose that at some moment we have known values $(H,T)$ and $(\tilde{H},\tilde{T})$ and that $(H,T)$ are changed by a small amount. With the new $(H,T)$ values and the old $(\tilde{H},\tilde{T})$ values, we first check to determine if Eq. (2.219) is fulfilled. If it is, there will be no change in $(\tilde{H},\tilde{T})$; otherwise, $(\tilde{H},\tilde{T})$ are updated by means of Eqs. (2.220) and (2.221). Finally, $(m,s)$ are found by use of Eq (2.218).

It is now possible to determine the energy loss $q$ associated with the hysteresis phenomenon for an arbitrary process. The change in energy loss is the difference between changes in total magnetic and mechanical work and stored energy, $\Delta q = H\Delta m + T\Delta s - \Delta f$. As mentioned previously, the energy ripple $f_F$ is small. A Taylor expansion of $f_{an}$ then gives $\Delta f \approx \Delta f_{an} \approx \tilde{H}\Delta m + \tilde{T}\Delta s$. Dividing by $\Delta t$ and letting $\Delta t \to 0$ yields

$$\dot{q} = (H - \tilde{H})\dot{m} + (T - \tilde{T})\dot{s} \qquad (2.222)$$

Therefore, for any cyclic process, the net power loss is the sum of the areas of the closed magnetic and mechanical hysteresis loops:

$$\oint (Hdm + Tds) \qquad (2.223)$$

It is also straightforward that $\dot{q} \geq 0$ at all times. This can be derived from the fact that $f_{an}$ is convex and that $(\tilde{H},\tilde{T}) \propto (H - \tilde{H}, T - \tilde{T})$.

A single pseudoparticle proportional exhibits some basic qualitative properties of magmetomechanical hysteresis. However, low-level excitations result in so-called minor loops, which consist of horizontal lines connecting major loop branches. This is not satisfactory. To achieve a better agreement, with real behavior, we consider an assembly of different pseudoparticles that differ only by the magnitude of the friction. Thus, they have the same nonhysteretic properties, the same ratio between magnetic and mechanical friction, etc. To express this, we associate each particle with a dimensionless positive number $\alpha$. For a given particle, $H_F$ and $T_F$ are multiplied by $\alpha$ in Eqs. (2.219) and (2.221). General expressions for the net magnetization and strain can then be written as

$$M = \int_0^\infty M_{an}(\tilde{H}(\alpha), \tilde{T}(\alpha))\varsigma(\alpha)d\alpha \qquad (2.224a)$$

$$S = \int_0^\infty S_{an}(\tilde{H}(\alpha), \tilde{T}(\alpha))\varsigma(\alpha)d\alpha \qquad (2.224b)$$

Here, $\tilde{H}(\alpha)$ and $\tilde{T}(\alpha)$ are the values of $\tilde{H}$ and $\tilde{T}$ for a particle with frictional multiplier $\alpha$, and $\varsigma(\alpha)$ is a weight distribution function giving the volume fraction for the corresponding particle. It satisfies the normalization

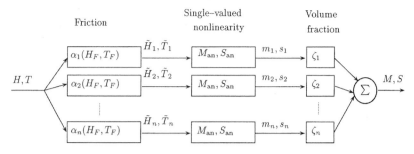

**FIGURE 2.35.** Block diagram showing the calculation procedure of the hyteresis model.

condition $\int_0^\infty \varsigma(\alpha)d\alpha = 1$. In a numerical implementation we must use a finite number of particles and approximate Eqs. (2.224a) and (2.224b) by sums. In the case of $n$ pseudoparticles, we then have

$$M = \sum_{i=1}^{n} m_i \varsigma_i = \sum_{i=1}^{n} M_{\mathrm{an}}(\tilde{H}_i, \tilde{T}_i)\varsigma_i \tag{2.225a}$$

$$S = \sum_{i=1}^{n} s_i \varsigma_i = \sum_{i=1}^{n} S_{\mathrm{an}}(\tilde{H}_i, \tilde{T}_i)\varsigma_i \tag{2.225b}$$

where $\sum_{i=1}^{n} \varsigma_i = 1$ and $\tilde{H}_i$ and $\tilde{T}_i$ correspond to the frictions $\alpha_i H_F$ and $\alpha_i T_F$, respectively. The procedure is illustrated in Fig. 2.35.

To use the model we need to determine the anhysteretic functions $M_{\mathrm{an}}(H, T)$ and $S_{\mathrm{an}}(H, T)$, the distribution $\varsigma(\alpha)$, and the friction constants.

First, $M_{\mathrm{an}}$ and $S_{\mathrm{an}}$ can be obtained experimentally. By using the fundamental relation between magnetic flux density $B$ and magnetization $B = \mu_0 H + M$, one can obtain $M^+(H, T)$ and $M^-(H, T)$ from the upper and lower branches of a saturation loop with respect to $H$ under constant $T$. An approximation that is quite accurate gives

$$M_{\mathrm{an}}(H, T) \approx \frac{1}{2}(M^+(H - H_\mathrm{c}, T) + (M^-(H + H_\mathrm{c}, T)) \tag{2.226a}$$

$$S_{\mathrm{an}}(H, T) \approx \frac{1}{2}(S^+(H - H_\mathrm{c}, T) + (S^-(H + H_\mathrm{c}, T)) \tag{2.226b}$$

where $H_\mathrm{c}$ is the coercive force of the corresponding major magnetization loop.

For convenience, $\varsigma(\alpha)$ is assumed to be a sum of a Gaussian and Dirac delta function:

$$\varsigma(\alpha) = A \exp\left(-(\alpha - 1)^2\right) + \lambda\delta(\alpha) \qquad (2.227)$$

where $A$ is given by normalization and $\lambda$ is an adjustable parameter. The second term in this expression gives a fully reversible contribution since $\alpha = 0$ corresponds to the frictionless case. In small minor loops, any changes of $M$ and $S$ occur without friction since particles with $\alpha > 0$ have minor loops that are only horizontal lines as seen in Fig. 2.33. Inserting Eq. (2.227) into Eq. (2.224) or Eq. (2.225) shows that $\lambda$ can be determined as the ratio between the slope of a minor loop for some arbitrary $(H,T)$ and the slope of the anhysteretic curve for the same $(H,T)$. Finally, $H_F$ and $T_F$ can be found from magnetic and mechanical losses. For a major $M,H$ loop under constant $T$, the loss contribution from a pseudoparticle is clearly $4\alpha H_F M_s$, where $M_s$ is the saturation magnetization. The total loss of a major loop is therefore

$$4H_F M_s \int_0^\infty \alpha\varsigma(\alpha)d\alpha \qquad (2.228)$$

which is proportional to $H_F$. $H_F$ can then be adjusted to give a measured magnetic loss. Similarly, $T_F$ can be estimated from a measured mechanical loss $\oint TdS$ under constant $H$.

In practice, approximately five pseudoparticles are sufficient for a satisfactory hystersis handling capability. Typical parameter values are $\lambda = 0.48, H_F = 2.56\text{kA/m}$, and $T_F = 1.6\text{MPa}$. Figure 2.36 shows some examples of output from the model together with experimental curves.

When using the magnetomechanical hysteresis model in simulations,

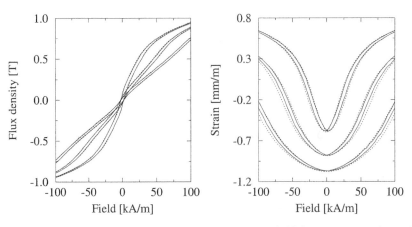

**FIGURE 2.36.**   Flux density and strain vs magnetic field for constant values of $T$. Solid lines, calculations; dotted lines, experiments.

strain $S$ and magnetic flux density $B$ values will be taken from the previous hysteresis model instead of from the other database comprising dehysterized data, i.e., the numerical values are given by

$$S_i = S_{\text{hyst model}}(H_i, T_i) \qquad\qquad (2.229)$$

$$B_i = \mu_0 H_i + M_{\text{hyst model}}(H_i, T_i)$$

The experimental data are instead used to estimate the anhysteretic curves and parameters required by the hysteresis model.

## System Integration

The presented nonlinear model approach has some important features regarding interaction with surrounding systems. There are no definite boundaries between the model of the active material, the magnetostrictive actuator, and the electric feed and mechanical load systems.

In fact, it is possible to model the electrical feeding and control systems and the mechanical process in which the actuator interacts in a totally integrated system description. In Chapter 3, this feature is treated more in detail. Conceptually, the nonlinear model can be described in terms of equivalent circuits (Fig. 2.37). The magnetostrictive actuator is characterized by its electrical across and through quantities actuator current $i_{\text{act}}$ and actuator voltage $u_{\text{act}}$ that are connected to a magnetostriction component via a coil resistance $R_{\text{coil}}$. The magnetostriction component comprises hysteresis behavior. The magnetic circuit that magnetizes the active material is represented by an internal flux loop short-circuited by a reluctance component equal to the magnetic circuit flux path reluctance $R_{\text{p}}$. The magnetic circuit reluctance $R_{\text{p}}$ in principle also shows hysteresis. In many cases, it is sufficient to regard $R_{\text{p}}$ as a pure reluctance. A relatively simple way to account for losses in the magnetic circuit is to regard $R_{\text{p}}$ as complex valued. Using this approach hysteresis and eddy current losses in the magnetic circuit can be handled. A thorough treatment, however, uses a FE method which can handle hysteresis. (25) (FE model of magnetic circuit).

The mechanical quantities $F$ and $v$ are connected via the internal mass $M_{\text{act}}$ and internal transmission components $1/k_{\text{trans}}$ to the magnetostriction component. The magnetostriction component has two electrical, two magnetic, and two mechanical pins. The magnetic bias field in the actuator is obtained by either permanent magnets modeled by a loss-free DC current $I_{\text{bias}}$ superimposed on $I_{\text{act}}$ or by a real DC current $I_{\text{bias}}$ with coil losses. The mechanical prestress $T_{bias}$ is obtained by applying a force $F_{bias}/A_r$, which can be achieved from a precompression of a spring in the mechanical load or internally by a precompressed spring connected between the actuator force outlets.

## A Tentative Extension to a Full 2D Nonlinear Model

The nonlinear models of magnetostrictive materials that have been presented are all restricted to handle only the longitudinal components of the magnetic and mechanical quantities. In many cases, this is sufficient since most applications are based on the longitudinal motion of the material at moderate frequencies. There are cases, however, in which a more general model is needed. In this section, a tentative extension of the radial-axial model to a full 2-D model is discussed in which the radial components of all quantities in the axisymmetric geometry are considered (24).

Consider the volume element of magnetostrictive material in Fig. 2.38. Here, $T_r$, $T_\phi$ and $T_z$ are the "ordinary" compressive stresses and $T_{zr}(= T_{rz})$ is the only shear stress that has to be considered in this axisymmetric

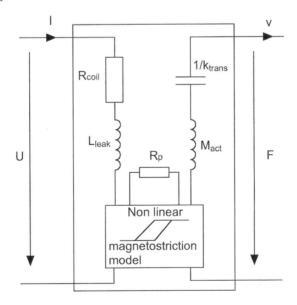

**FIGURE 2.37.**   Equivalent description of magnetostrictive actuator according to a nonlinear model.

problem. The mechanical equilibrium equations for the volume element are (15)

$$r - \text{component:} \quad \frac{\partial T_r}{\partial r} + \frac{\partial T_{zr}}{\partial r} + \frac{T_r - T_\phi}{r} = \rho \frac{\partial^2 u_r}{\partial t^2} \qquad (2.230a)$$

$$z - \text{component:} \quad \frac{\partial T_{zr}}{\partial r} + \frac{\partial T_z}{\partial z} + \frac{T_{zr}}{r} = \rho \frac{\partial^2 u_z}{\partial t^2} \qquad (2.230b)$$

where $u_z$ and $u_r$ are the axial and radial displacements, respectively. Note that no internal damping of the material will be considered here for simplicity. The strains corresponding to the stresses are as follows:

$$S_r = \frac{\partial u_r}{\partial r} \qquad (2.231a)$$

$$S_\phi = \frac{u_r}{r} \qquad (2.231b)$$

$$S_z = \frac{\partial u_z}{\partial z} \qquad (2.231c)$$

$$S_{zr} = \frac{1}{2} \left( \frac{\partial u_r}{\partial z} + \frac{\partial u_z}{\partial r} \right) \qquad (2.231d)$$

See the beginning of this chapter regarding the definition of $S_{zr}$.

The expressions for the magnetic field inside the volume element, when considering the effects of circulating eddy currents, can be derived following the procedure in Appendix A for the axisymmetric geometry:

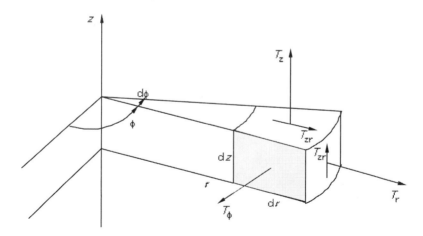

FIGURE 2.38.  Volume element of a magnetostrictive material in a cylindrical geometry.

$r$ − component:  $\qquad -\dfrac{\partial^2 H_r}{\partial z^2} + \dfrac{\partial^2 H_z}{\partial r \partial z} = -\sigma \dfrac{\partial B_r}{\partial t}$  (2.232a)

$z$ − component:  $\left(\dfrac{\partial^2}{\partial r \partial z} + \dfrac{1}{r}\dfrac{\partial}{\partial z}\right) H_r - \left(\dfrac{\partial^2}{\partial r^2} + \dfrac{1}{r}\dfrac{\partial}{\partial r}\right) H_z = -\sigma \dfrac{\partial B_z}{\partial t}$  (2.232b)

The magnetomechanical coupling is included in the form of the dependence of strain and induction on magnetizing field and stress, respectively. These dependencies can be achieved from static experimental characterization of the magnetostrictive material.

The problem becomes more difficult when increasing the geometrical dimension. Although only one component each of $S$, $T$, $B$, and $H$ was considered in the longitudinal case, four components of $S$ and $T$ and two components of $B$ and $H$ must be handled here. The room derivatives here will be expressed by finite differences. The compressive strains and stresses and the magnetic fields are defined in the middle of a section, whereas the shear strain and stress and the displacements are defined on the radial and axial boundaries of a section (compare with the sectionized rod in Fig. 2.32). The expressions obtained can easily be implemented in a program which handles ordinary differential equations (see Table 2.2):

$$a_s S_r^{i,j} = u_r^{i,j} - u_r^{i,j-1}$$

$$L_s S_z^{i,j} = u_z^{i,j} - u_z^{i,j-1}$$

$$r^j S_\phi^{i,j} = u_r^{i,j}$$

$$2l_s a_s S_{zr}^{i,j} = a_s\left(u_r^{i,j} - u_r^{i,j-1}\right) + l_s\left(u_z^{i,j} - u_z^{i,j-1}\right)$$

$$\rho l_s a_s r^j \frac{\partial^2 u_r^{i,j}}{\partial t^2} = r^j l_s\left(T_r^{i,j} - T_r^{i+1,j}\right) + r^j a_s\left(T_{zr}^{i,j} - T_{zr}^{i+1,j}\right) + a_s l_s\left(T_r^{i,j} - T_\phi^{i,j}\right)$$

$$\rho l_s a_s r^j \frac{\partial^2 u_z^{i,j}}{\partial t^2} = r^j l_s\left(T_{zr}^{i,j} - T_{zr}^{i+1,j}\right) + r^j a_s\left(T_z^{i,j} - T_z^{i+1,j}\right) + a_s l_s T_{zr}^{i,j}$$

$$-\frac{H_r^{i+1,j} - 2H_r^{i,j} + H_r^{i-1,j}}{l_s^2} + \frac{\left(H_z^{i+1,j+1} - H_z^{i+1,j-1}\right) - \left(H_z^{i-1,j+1} - H_z^{i-1,j-1}\right)}{4 l_s a_s} = -\sigma \frac{dB_r^{i,j}}{dt}$$

$$-\frac{H_z^{i,j+1} - 2H_z^{i,j} + H_z^{i,j-1}}{a_s^2} + \frac{\left(H_r^{i+1,j+1} - H_r^{i+1,j-1}\right) - \left(H_r^{i-1,j+1} - H_r^{i-1,j-1}\right)}{4 l_s a_s}$$

$$\frac{1}{r^j}\frac{H_r^{i+1,j} + H_r^{i-1,j}}{2 l_s} - \frac{1}{r^j}\frac{H_z^{i,j+1} + H_z^{i,j-1}}{2 a_s} = -\sigma \frac{dB_z^{i,j}}{dt}$$

(2.233)

In the previous equations, $r^j = a_s(j - 1/2)$.

In the most general case each component of the strains and the magnetizations depends on all the components of the applied magnetizing field and stress. As a first approximation one can assume the same dependencies as defined for a polycrystal rod with bias field and prestress applied in the $x3$ $(z)$ direction, given in Eqs.(2.19) and (2.20). From the definitions given at the beginning of this chapter, one can see that the shear components $(rz)$ correspond to the fifth component of the stress and strain vectors in Eqs. (2.19) and (2.20). Hence, the following constitutive material dependencies must be included in the model description for each section:

$$S_r^{i,j} = S_r\left(H_z^{i,j}, T_r^{i,j}, T_\phi^{i,j}, T_z^{i,j}\right)$$
$$S_\phi^{i,j} = S_\phi\left(H_z^{i,j}, T_r^{i,j}, T_\phi^{i,j}, T_z^{i,j}\right)$$
$$S_z^{i,j} = S_z\left(H_z^{i,j}, T_r^{i,j}, T_\phi^{i,j}, T_z^{i,j}\right)$$
$$S_{zr}^{i,j} = S_{zr}\left(H_r^{i,j}, T_{zr}^{i,j}\right)$$ 
$$B_r^{i,j} = B_r\left(H_r^{i,j}, T_{zr}^{i,j}\right)$$
$$B_z^{i,j} = B_z\left(H_z^{i,j}, T_r^{i,j}, T_\phi^{i,j}, T_z^{i,j}\right)$$

(2.234)

The greatest difficulty with this extended model is to perform the necessary experiments for characterization of the magnetostrictive material. Even with the simplified material behavior assumed previously, the characterization would be very difficult to perform. A suggested measurement setup has proved to give an estimation of $d_{15}$ and the elastic shear modulus at constant bias magnetization. Additional simplifications must be considered regarding the material dependencies to make it possible to make a numerical implementation of the tentative extension to a full 2-D nonlinear model.

## FE Modeling

Nonlinear FE analysis has been possible in ATILA for electrostrictive (piezoelectric) materials since 1996 and an extension to magnetostrictive materials is envisaged. Some attempts has already been made to treat the nonlinear properties of giant magnetostrictive materials in FE modeling by use of surface splines (16). With increasing speed of computing capabilities it is possible that nonlinear modeling in the future will be a useful design tool for applications based on magngetostrictive materials.

# REFERENCES

1. S. Butterworth and F. D. Smith, The equivalent circuit of the magnetostriction oscillator, *Phys. Soc. London Proc.* **43**, 165–185 (1931).

2. J. F. Nye, "Physical Properties of Crystals." Oxford University Press, London, 1957.

3. IEEE Standard on magnetostrictive materials: Piezomagnetic nomencature, IEEE Std 319-1990, February 1991.

4. F. Claeyssen, Design and construction of low-frequency sonar transducers based on rare earth-iron alloys. Doctoral thesis, INSA, Lyon, France, 1989. [translation from French].

5. D. A. Berlincourt, D. R. Curran, and H. Jaffe, Piezoelectric piezomagnetic materials and their function in transducers, *in* "Physical Acoustics, Principles and Methods," (W. P. Mason, Ed.), Vol. 1, Part A. Academic Press, New York, 1964.

6. J. L. Butler, "Highly Active Magnetostrictive Transducers." Image Acoustics, 1986.

7. H. W. Katz, "Solid State Magnetics and Dielectric Devices." Wiley, New York, 1959.

8. B. D. H. Tellegren, The gyrator, a new electric network component, *Philips Res. Rep.* **3**, 81–101 (1948).

9. F. Claeyssen, D. Boucher, K. Anifrani, R. Bossut, and J.N. Decarpigny, Analysis of magnetostrictive transducers by the ATILA finite element code, *J. Am. Soc. Am.* **85**, (pl. 1), 590 (1989).

10. B. Hamonic, 'Contribution à l'étude du rayonnement de transducteurs utilisant les vibrations de coque mince', thèse de doctorat, USTL, Lille, France, 1987.

11. Lhermet N., "ATILA V.5.1.1 User Manual—A 3D CAD Software for Piezoelectric & Magnetostrictive Structures," Isen, Lille, France, 1997. (Distributed by Cedrat, Meylan, France, and Magsoft, Troy, NY.)

12. P. Bouchilloux, N. Lhermet, and F. Claeyssen, Dynamic shear characterization in a magnetostrictive rare earth-alloy, *M.R.S. Symp. Proc.* **360**, 265–272 (1994).

13. L. Kvarnsjö and G. Engdahl, Examination of the interaction between eddy currents and magnetoelasticity in Terefenol-D, *J. Appl. Phys.* **69**, 5777–5779 (1991).

14. F. Stillesjö, G. Engdahl, and A. Bergqvist, A design technique for magnetostrictive actuators with laminated active material, paper presented at the 7th joint MMM-INTERMAG Conference, San Francicso, January 1998.

15. Formal sampling i Hallfasthetslävia, Department of Solid Mechanics, Royal Institute of Technology, 1990. (In Swedish).

16.  M. E. H. Benbouzid, G. Reyne, G. Meunier, L. Kvarnsjö, and G. Engdahl, Dynamic modeling of giant magnetostriction in Terfenol-D by the finite element method, paper presented at the Conference on Electromagnetic Field Computation, Grenoble, 1994.

17.  L. Kvarnsjö, G. Engdahl, Examination of eddy current influence on the behavior of a giant magnetostrictive functional unit, *J. of Appl. Phys.*, **67**, 5010-5012 (1990).

18.  T. Cedell, Magnetostrictive materials and selective applications, Doctoral Thesis, Lund University, Sweden 1996

19.  G. Engdahl, L. Svensson, Simulation of magnetostrictive performance of TERFENOL-D in mechanical devices, *J. Appl. Phys.* **63** (*8*), 15 April 1988

20.  P. Holmberg, A. Bergqvist, G Engdahl, Modelling a magnetomechanical drive by a coupled magnetic, electric and mechanical lumped circuit approach, *J. of Appl. Phys.*, **81**, (8), 1997

21.  G. Engdahl, L. Kvarnsjö, A time dependent radially resolved simulation model of giant magnetostrictive materials, in *Mechanical Modellings of New Electromagnetic Materials*, R.K.T. Hsieh Ed., Elsevier, p. 131-138 (1990).

22.  H. Tiberg, Design in Applied Magnetostriction, Licentiate Thesis, Royal Institute of Technology, Sweden, 1994

23.  A. Bergqvist, G. Engdahl, A model for magnetomechanical hysteresis and losses in magnetostrictive materials, *J. Appl. Phys.* 79 (8), (1996).

24.  L. Kvarnsjö, On Characterization, Modelling and Application of Highly Magnetostrictive Materials, Doctoral Thesis, Royal Institute of Technology, Sweden, 1993

25.  A. Bergqvist, H. Tiberg, G. Engdahl, Application of vector preisach model in a magnetic circuit , *J. of Appl. Phys.* **73**, 5839-5841 (1993).

# CHAPTER 3

# Magnetostrictive Design

Göran Engdahl
*Royal Institute of Technology*
*Stockholm, Sweden*

Charles B. Bright
*ETREMA Products, Inc.*
*Ames, Iowa*

## GENERAL

Magnetostrictive design and electrotechnical design are activities rather than disciplines of theoretical considerations. Instead one has to deal with a task that can be divided into many subtasks that in many cases have to be executed in an appropriate order.

In most design situations, several aspects have to be considered. These also have to be accounted for simultaneously. In electric power engineering, the following aspects are crucial:

- Magnetic
- Mechanical

- Electrical
- Thermal

A magnetostrictive actuator is an example in which

- Its magnetic circuit has to be designed to give an appropriate magnetization of the magnetostrictive material.
- It has to be properly designed with regard to rated voltage and current and with respect to the dielectric strength of comprising insulation materials.
- The stiffness of the force outlet, mechanical transmission, and actuator fixture and prestress system have to be designed to satisfy given demands.
- An appropriate cooling system has to match the overall heat generation and internal heat transfer during operation.

Most important, an appropriate compromise between the different aspects must be found in which the consequences of various designs are considered.

The starting point is in general a demand specification in terms of currents, voltages, strains, forces, power levels, frequency characteristics, response times, temperatures, etc. During the design phase all these data must be transformed to construction data such as principal design choices, material choices, length measures, geometric structures, and tolerances. This process is illustrated in Fig. 3.1.

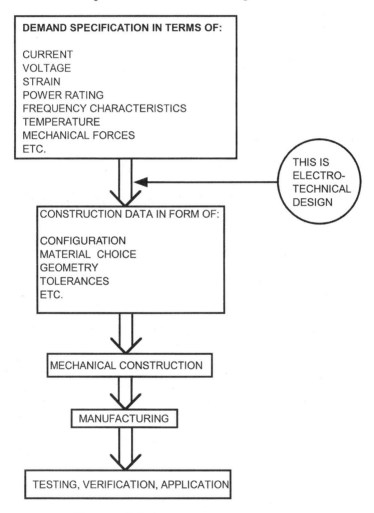

FIGURE 3.1.   The overall design process.

The conversion from performance demands to basic design data is done by means of

- Analytical methods

   Simplified calculations

- Finite element calculations

   Magnetic

   Mechanical

   Magnetomechanical

   Thermal

- Dynamic simulation

   Circuit simulation

   Full physics simulation

For a magnetostrictive actuator the activities can be structured according to Fig. 3.2.

## SOME RELEVANT PHYSICAL DATA OF TERFENOL-D

Figures 3.3 and 3.4 show room-temperature magnetostriction as a funcion of magnetic flux density and magnetizing field for different prestress levels of a representative single rod of $Tb_{0.3}Dy_{0.7}Fe_{1.92}$ stoichiometry TERFENOL-D produced by ETREMA Products, Inc. (Ames, Iowa). Figure 3.5 shows the corresponding magnetization. The effect of temperature on $Tb_{0.3}Dy_{0.7}Fe_{1.9}$ stoichiometry TERFENOL-D is shown in Fig. 3.6. Information not contained in these figures is summarized in Table 3.1, which is a composite from several published sources.

From Figs. 3.3–3.6, the following general observations can be noted:

- The magnetostriction $S$ of TERFENOL-D is a function of mechanical compressive prestress $T_p$, magnetic field intensity $H$, and temperature $t$. These fundamental relationships are illustrated in the

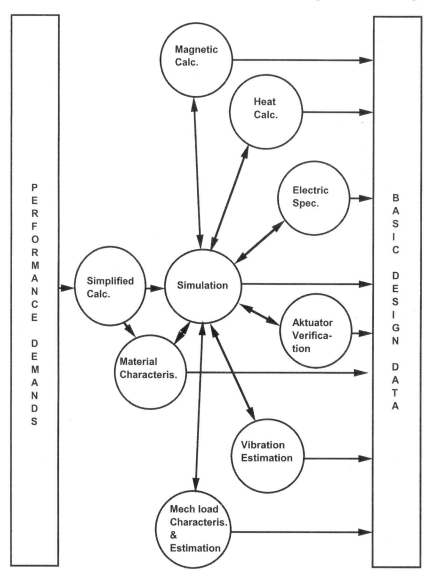

**FIGURE 3.2.**   The electrotechnical design activities regarding a magnetostrictive actuator.

following Figs. 3.3–3.6. At constant temperature, increasing $T$ decreases $\Delta S$ for constant $\Delta H$. At this point, it is convenient to introduce slope $d^T$, which is defined as $\Delta S/\Delta H$ at constant stress $T$.

**Table 3.1.**   TERFENOL-D Material Properties

| Property | Symbol | Unit | Value |
|---|---|---|---|
| Density | $\rho$ | $\mathrm{kg\,m^{-3}}$ | $9.21 - 9.25 \times 10^3$ |
| Young's modulus | | | |
|     at constant current | $Y^H$ | $\mathrm{N\,m^{-2}}$ | $1.8 - 5.50 \times 10^{10}$ |
|     at constant voltage | $Y^B$ | $\mathrm{N\,m^{-2}}$ | $5.0 - 9.0 \times 10^{10}$ |
| Sound speed | $c$ | $\mathrm{ms^{-1}}$ | 1395–2444 |
| Bulk modulus | | $\mathrm{N\,m^{-2}}$ | $9.00 \times 10^{10}$ |
| Tensile strength | | $\mathrm{N\,m^{-2}}$ | $2.8 - 4.0 \times 10^7$ |
| Compression strength | | $\mathrm{N\,m^{-2}}$ | $3.04 - 8.80 \times 10^8$ |
| Curie temperature | $T_c$ | °C | 357 |
| Coefficient of thermal expansion | CTE | $\mathrm{\mu mm^{-1}K^{-1}}$ | 11.0 @ 25°C |
| Specific heat | $c$ | $\mathrm{kJ\,kg^{-1}K^{-1}}$ | 0.33 |
| Thermal conductivity | $\lambda$ | $\mathrm{W/m^{-1}K^{-1}}$ | 13.5 @ 25°C |
| Electrical resistivity | $\rho_e$ | ohm m | $6.0 \times 10^{-7}$ |
| Energy density | | $\mathrm{kJ\,m^{-3}}$ | 4.9–25 |
| Relative magnetic permeability | $\mu_r/\mu_0$ | Dimensionless | 2–10 |
| Coupling factor | $k$ | Dimensionless | 0.7–0.8 |

- Each $H$–$S$ curve is "flatter" at the bottom ($d^T$ is low). Each curve also has a nearly linear section in which $d^T$ is greatest called the "burst zone," in which TERFENOL-D produces the most magnetostriction per unit magnetic field intensity increase. Each curve then flattens again due to saturation. TERFENOL-D does not exhibit polarity—$S$ increases with increasing $H$, regardless of magnetic field polarity.

- For any given stoichiometry, maximum magnetostriction occurs at its magnetization reorientation temperature, $t_{mr}$. Any temperature deviation from $t_{mr}$ decreases magnetostriction with the decrease being much steeper below $t_{mr}$ $Tb_{0.30}Dy_{0.70}Fe_{1.92}$ stoichiometry TERFENOL-D has a $t_{mr}$ near 0°C (1).

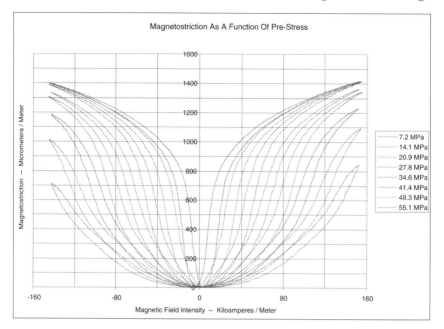

**FIGURE 3.3.**    Magnetostriction curves for various prestress levels as a function of an applied magnetic flux density of a representative single rod of $Tb_{0.3}Dy_{0.7}Fe_{1.92}$ stoichiometry TERFENOL-D.

## MAGNETIC AND MECHANICAL OPERATION RANGES

The first step in the design process is to make a rough estimation of the size of the required magnetostrictive material in relation to the demands regarding strain and forces of the actuator. In principle, the required length of the active material is proportional to the strain demand and the required cross-sectional area to the force demand.

Regarding the strain demand, a rule of thumb is that one should at least achieve 1 ms (peak to peak millistrain) effectively for a material with $\approx 2$ ms saturation strain. This figure is for quasi-static conditions at relatively low prestress, before dynamic conditions are considered. Regarding the force demand, the situation is complex because the prestress level, the magnetic excitation level, and the mechanical load situation all affect the resulting force. The effective force in terms of mechanical stress of the active material ranges, depending on various applications, from $\approx 5$ to $\approx 70$ MPa. Increasing stress bias allows larger excursions around the bias point without overcoming the stress bias.

It is a feature of TERFENOL-D that its strain is quadratic. That is, regardless of magnetic field polarity, it expands with increasing field. To

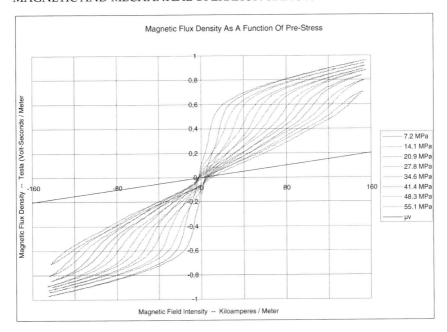

**FIGURE 3.4.** Magnetization curves for various prestress levels as a function of an applied magnetizing field of a representative single rod of $Tb_{0.3}Dy_{0.7}Fe_{1.92}$ stoichiometry TERFENOL-D.

maximize equal strain about a null point, TERFENOL-D is partially expanded with a bias magnetic field to the desired null. The alternating magnetic field superposed on the static bias field then enables positive and negative motion about the null. TERFENOL-D couples more energy per unit volume if prestress is increased, with a corresponding increase of the magnetic field bias. This is because larger stress excursions around null then can be accommodated without overcoming the stress bias.

The choice of effective length and cross-sectional area of the magnetostrictive material is for the previously discussed reason the same as that of the mechanical operation range of the actuator.

One way to illustrate the operation ranges of a magnetostrictive actuator is to visualize the regions in the magnetic field—strain *(H,S)* plane comprising magnetostrictive curves for different mechanical stress levels (Fig. 3.7). For a given material it is possible to specify operation areas in the *(H,S)* plane with different mechanical prestress levels, bias magnetization, and magnetic excitation levels.

Studies (2) addressed how to obtain a high total efficiency by choosing appropriate prestress and bias magnetization levels. The results reveal that there is a nearly linear relation between optimal prestress and

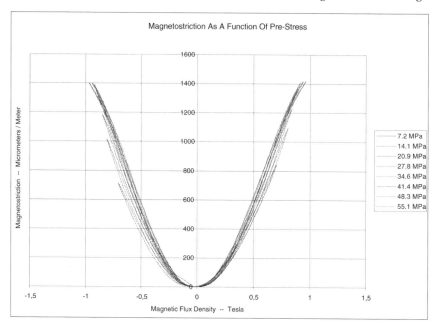

**FIGURE 3.5.** Normalized magnetostriction curves for various prestress levels as a function of an applied magnetizing field of a representative single rod of $Tb_{0.3}Dy_{0.7}Fe_{1.92}$ stoichiometry TERFENOL-D.

bias magnetization irrespective of the mechanical load situation. For a magnetostrictive material dedicated to hydroacoustic transducer applications, this linear relation is

$$T_p = 0.48H_{bias} + 1.0 \qquad (3.1)$$

where $T$ is expressed in MPa and $H$ in kOe. Figures 3.8 and 3.9 show examples of how the efficiency depends on the prestress and bias magnetization levels. One can see that a slightly higher efficiency is obtained at a lower mechanical prestress level. The efficiency decrease is nearly linear with respect to the magnetic bias and the mechanical prestress levels. The width of the efficiency peak is significantly broader at higher prestress levels. This is an advantage with regard to tolerances and temperature drift of the mechanical prestress system.

Another important element of a magnetostrictive actuator is its power transduction capability (i.e., its quantitative ability to deliver mechanical power from magnetic power). The product of the magnetic bias field and mechanical prestress level is then relevant because it is related to the

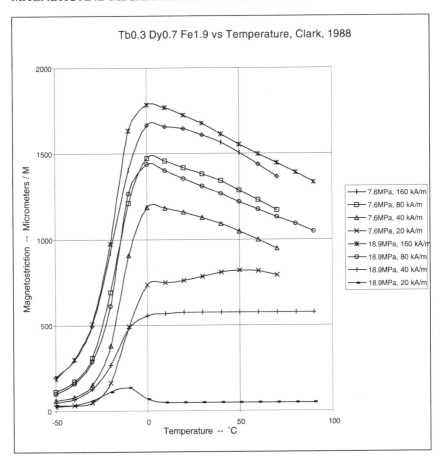

**FIGURE 3.6.** Temperature dependence of the saturation magnetostriction of a representative single rod of $Tb_{0.3}Dy_{0.7}Fe_{1.92}$ stoichiometry TERFENOL-D.

power conversion level $P$ of an actuator for a certain $B_{\text{peak to peak}}$ and $S_{\text{peak to peak}}$ at a given operation angular frequency according to

$$P \propto \omega l_r A_r \sqrt{H_{\text{bias}} T_p} \tag{3.2}$$

The output power of the actuator is

$$P \propto \omega l_r A_r \sqrt{\Delta S_{\text{rms}} \cdot \Delta B_{\text{rms}}} \sqrt{H_{\text{bias}} T_p} \eta_m \tag{3.3}$$

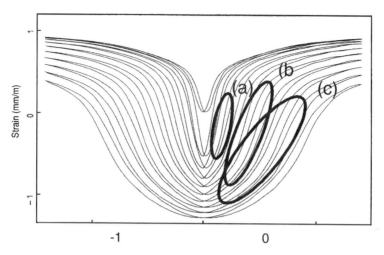

**FIGURE 3.7.**   Operation areas in the $H$–$S$ plane.

The efficiency $\eta_m$ relates to the performance of the active material and depends on $H_{bias}$ and $T_p$.

For nonlinear materials the material efficiency also depends on the magnetomechanical excitation levels (i.e., $\Delta S_{rms}$ and $\Delta B_{rms}$). For hysteretic material, the efficiency also depends on the phase angle between $\Delta S$ and $\Delta B$. The output power of a magnetostrictive actuator can be expressed as

$$P_{actuator} \propto \omega l_r A_r \sqrt{\Delta S_{rms} \cdot \Delta B_{rms}} \sqrt{H_{bias} T_p} \eta_{mag} \eta_{mech} \eta_m (H_{bias}, T_p, \Delta S_{rms}, \Delta B_{rms}, \arg(\Delta S, \Delta B), \omega)$$

$$(3.4)$$

where $\eta_{mag}$ and $\eta_{mech}$ are the efficiencies of the magnetization and mechanical systems of the actuator, respectively. The material efficiency is accounted for in the presented hysteresis model, which in the axial radial model also accounts for the presence of eddy currents.

In the previous equation, there are many significant design parameters. The influence of mechanical and magnetic bias levels, excitation levels, and active material volume is evident. It is instructive to see how efforts made in the design of the magnetization and mechanical systems influence the resulting output power of the actuator and also how an appropriate choice of the operation range (i.e., $H_{bias}$ and $T_p$) can improve the actuator performance. The expression also indicates the relevance of characterizing a magnetostrictive material in terms of its intrinsic material efficiency.

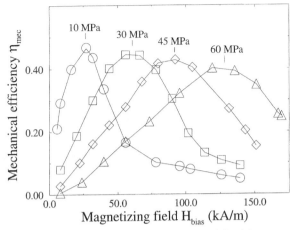

**FIGURE 3.8.** Mechanical efficiencies as a function of the bias magnetizing field for different mechanical prestress levels.

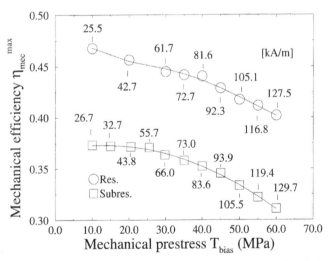

**FIGURE 3.9.** Mechanical efficiencies as a function of the mechanical prestress level for different bias magnetizing fields.

## MAGNETIC DESIGN

From the design equation (Eq. 3.4) the impact of the magnetic design is manifested by the magnetic efficiency $\eta_{mag}$, which expresses the fraction of the input electric power that is transferred to magnetic power. Therefore, if we had access to a coil wire material with superconducting properties, all input power would be transferred to magnetic power because there would be no resistive coil losses irrespective of the current level. In all practical cases one has to accept coil losses, e. g. copper or aluminum losses. To decrease the needed current level, one must design the magnetic circuit in such a way that the achieved magnetic power in the active material is maximized for a given coil loss level (7, 11). This can be formulated in terms of reluctances. The well-known relation for a coil with $N$ turns

$$NI = \phi R \tag{3.5}$$

states that, for a given current, magnetic flux increases with decreasing magnetic circuit reluctance. Another important feature of the magnetizing circuit is to direct as much of the obtained flux as possible through the active material. Normally, there is leakage flux $\phi_l$, which gives

$$\phi_{\text{total}} = \phi_r + \phi_l \tag{3.6}$$

The reluctances $R_p$ and $R_l$ corresponding to the magnetic flux path and the leakage, together with the reluctance of the magnetostrictive rod and the applied magnetomotive force, (MMF) determine the magnitudes of $\phi_r$ and $\phi_l$ (see Eq. 2.194) . In a good magnetic circuit, $R_p$ should be as low as possible and $R_l$ as high as possible.

For a magnetic circuit without permanent magnets, there is in principle no problem achieving control of $R_p$ and $R_l$ independently of each other. In many cases, it is desirable to incorporate permanent magnets to obtain a given bias magnetization level because there is then no need for a DC magnetization current. In such a magnetic circuit, however, there is an intrinsic conflict between low $R_p$ and high $R_l$ regarding the total AC and DC flux. Therefore, it is advisable to distinguish between permanent magnet circuits and nonpermanent magnet circuits.

## Nonpermanent Magnet Circuits

Nonpermanent magnet circuits are preferably used whenever the efficiency of the actuator has low priority, for instance, when it is used to produce single-force impulses to trigger a mechanical process or when there are few restrictions regarding accessible input power and/or temperature rise of the actuator.

With no permanent magnets involved it is possible achieve a low $R_p$

value because one can easily put a high-permeability material along the whole flux path between the rod ends with no interruption (Fig 3.10). For rods with low $l_r/r_r$ values it is in some cases enough to supply the rod with large pole pieces of soft iron. For rods with a high $l_r/r_r$ value the effective $R_l$ value will decrease relative to the effective reluctance of the magnetostrictive rod. For a homogeneous magnetizing of such rods it is therefore recommended to divide the rod and stack many separate rod elements with intermediate high-permeability material (see Fig 3.11, which shows the geometry of stacked and nonstacked permanent magnets). Figures 3.12 and 3.13 show magnetic finite element (FE) calculations that corresponds to Figs. 3.11a and 3.11b.

Figures 3.11–3.13 show how such a measure increases the effective leakage reluctance in relation to the rod stack reluctance and so reduces the flux leakage. There is evidently a trade-off between an effective magnetic circuit and its geometrical size. For long and thin actuator rods it is therefore advisable to stack many shorter rod elements which improve the homogeneity of the magnetic flux density along the rod elements.

TERFENOL-D can now be produced in other shapes besides rods (NKK, Japan), which makes it possible to arrange magnetic circuits with other topologies. One of these arrangements consists of inverting the geometry of the magnetic circuit of the actuator in such a way that the location of magnetostrictive and soft magnetic material are switched (Fig. 3.14). This can be done by use of hollow magnetostrictive rods, which will decrease the effective $l_r/r_r$ value and therefore the flux leakage and increase the homogeneity of the magnetic flux density in the magnetostrictive material. Such a configuration shows much promise because it can give quite compact actuator designs with the magnetizing device in the interior.

All the presented figures show actuators with magnetic flux return paths. Such an arrangement reduces flux leakage and the required drive current. This is especially valid for high $l_r/r_r$. For low $l_r/r_r$ (<4), it is in many cases sufficient to use only pole pieces of appropriate size.

## Permanent Magnet Circuits

When efficiency, limited power supply, temperature rise, continuous operation, etc. must be taken into consideration it is preferable to use permanent magnets in the magnetizing circuit. As mentioned previously, there is an intrinsic conflict between low flux path reluctance and a permanent magnet biasing flux. There are two principal approaches to achieve a bias magnetization—series and parallel magnetization.

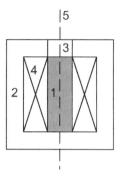

**FIGURE 3.10.** Magnetic circuit of a magnetostrictive actuator comprising a TERFENOL-D rod (1), a high-permeability flux path (2), a high-permeability and mechanically stiff force outlet (3), and a driving coil (4) with no permanent magnets and symmetry axis (5).

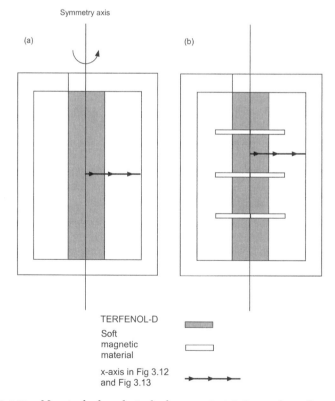

**FIGURE 3.11.** Nonstacked and stacked magnetostrictive rod configurations for homogeneous magnetization.

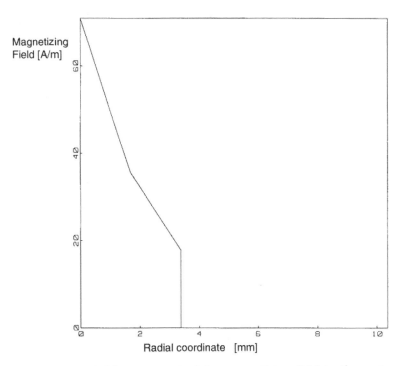

**FIGURE 3.12.** Axial component of the magnetizing field in the center of the TERFENOL-D rod along a line from the center of the rod out to the magnetic flux path according to the configuration in Fig. 3.11a.

The series magnetization is straightforward because it involves a MMF along some section in the flux path. The MMF can be located.

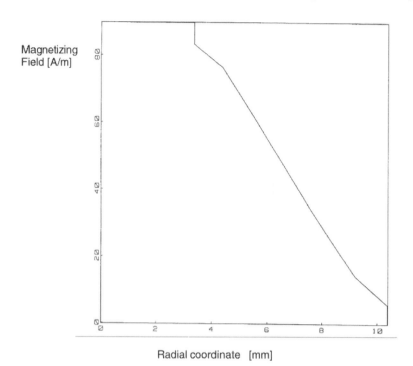

Magnetizing
Field [A/m]

Radial coordinate   [mm]

**FIGURE 3.13.** Axial component of the magnetizing field in the center of the TERFENOL-D rod along a line from the center of the rod out to the magnetic flux path according to the configuration in Fig. 3.11b.

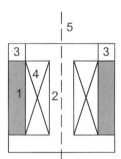

**FIGURE 3.14.** An inverted magnetic circuit of a magnetostrictive actuator comprising a hollow cylinder of magnetostrictive material or discrete rods (1), a high-permeability flux path (2), a high-permeability and mechanically stiff force outlet (3), and a driving coil (4) with no permanent magnets and symmetry axis (5).

- in the magnetostrictive material as permanent magnet discs inserted along the rod axis (stacked magnets)
- at the rod ends as disc magnets
- in the flux return path as hollow cylindrical magnets (barrel magnets)
- in the flux return path as hollow disc magnets

The principal configurations are illustrated in Fig. 3.15. The last two configurations can be interpreted as a continuous and a discrete variation of the same basic principle.

Irrespective of whether or not the geometry is inverted, there is no unambiguous solution to the design problem. The reluctance $R_p$ can easily be estimated by adding all reluctances that correspond to the series-connected magnets to the total reluctance of the soft magnetic parts of the magnetic circuit, where all reluctances are calculated by using Eq. (2.193) and (2.198).

The stacked magnet magnetic bias technique intersperses magnets between the magnetostrictive parts so as to form a "stack" consisting of magnet/TERFENOL-D/magnet/TERFENOL-D/magnet, etc. until the desired full length is achieved. This stack is inserted into the solenoid drive coil. A ferromagnetic material outside the coil establishes a magnetic circuit return path. This method of magnetic biasing is very beneficial for magnetic bias points >64 kA/m because the magnets are in line with the rod, relieving magnetic saturation concerns. High device coupling coefficients can be attained with this technique due to the magnetic circuit return path.

In the stack method, magnet to magnet spacing is critical because it affects magnetic bias field uniformity within the TERFENOL-D part. Some magnetic circuits can have as much as a 60% variation in magnetic intensity within the TERFENOL-D from end to end, in which case the magnetic domains will have very different performance. The result is lower efficiency and an increase in transducer harmonic distortion. Improper magnet spacing can result in magnetic flux leakage which tends to saturate the outer edge of each TERFENOL-D part. Saturated TERFENOL-D produces no work.

In stiffness- or size-constrained designs, the stacked-magnet method can be problematic. Magnets in the load path reduce overall transducer stiffness, remove available TERFENOL-D volume, and require tightly toleranced TERFENOL-D and magnet perpendicularity and parallelism. In addition, as much as 30% of the entire length of the stack is composed of magnets which are energized by the coil. This increases coil power

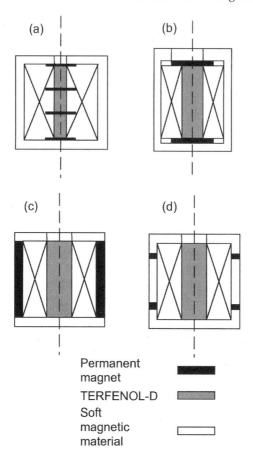

**FIGURE 3.15.**    Various magnetic circuits with permanent magnets.

consumption by $\approx 30\%$ and hence reduces efficiency. The stacked magnet biasing method should be applied to designs requiring long lengths of TERFENOL-D (>150 mm) and large diameters (>38 mm), when high magnetic bias points are required (>64 kA/m), or when magnetic leakage reduction is of great importance. This method should be avoided when tight volume constraints are imposed, when high-frequency operation (>15 kHz) is used, when high stiffness is required, when rod diameters are small (<13 mm), or when low harmonic distortion is required.

The barrel magnet magnetic biasing technique comprises hollow cylindrical (shaped like a barrel) magnets surrounding the TERFENOL-D rod and solenoid coil. Ferromagnetic pole pieces on the top and bottom of both the rod and the magnets direct magnetic flux from the magnet down through the TERFENOL-D rod. This design can give a field uniformity in the rod within ± 4%. The pole pieces combine with the solenoid drive coil to deliver a uniform field throughout the length of the rod, which increases magnetostriction and reduces harmonic distortion.

Barrel magnet design is simple and implementation does not require tight tolerances. However, this method has the following limitations:

- High magnetic bias levels for large TERFENOL-D rods are difficult to attain.
- Ferromagnetic pole pieces may be subject to magnetic saturation.
- The magnetic bias may be subject to conditions outside of the transducer. That is, external ferromagnetic materials in close proximity to the transducer can divert magnetic flux away from the TERFENOL-D rod, reducing magnetic bias.
- The magnet becomes the magnetic return path for the drive coil. Without a ferromagnetic return path, the coil must work harder to produce the desired field intensities within the rod. In general, magnetic return path is not important for magnetic designs in which the rod length to diameter ratio is 4 or more.

Barrel magnet designs should be avoided when long lengths of TERFENOL-D (>150 mm) or large diameters (>38 mm) are required, when high magnetic bias points are needed (>64 kA/m), or when magnetic leakage is of great importance. This method should be used when tight volume constraints are imposed, for high-frequency operation (>15 kHz), when high rod stiffness is required, when rod diameters are small (<13 mm), or when low harmonic distortion is required.

The parallel magnetization implies two magnetic circuits—one DC circuit for the bias magnetization and one AC for the dynamic excitation. Figures 3.16 and 3.17 show two patented examples of these circuits. The first circuit is suitable to stack with an appropriate number of cells to meet the effective rod length demand (8). One drawback with this concept is that the AC flux path comprises an air gap that increases the reluctance $R_p$. The second circuit, the bipolar design, has no such air gap, which gives a lower $R_p$ value. However, it is not possible to stack the second circuit in a meaningful way because it comprises two counteracting rods with the actuator force outlet in the center where the two rods meet. The length of these two rods can be extended if the flux leakage and flux nonhomogeneity are managed in a similar way as for nonpermanent magnet circuits. It is also possible to arrange the second circuit to obtain an actuator giving a mechanical torque (the bipolar actuator) (Fig. 3.18).

In all practical cases, the following additional constraints must be considered simultaneously:

- Weight demands
- Volume demands
- Material preferences
- Thermal aspects
- Mechanical aspects
- Manufacturing aspects

After some estimations of the flux path and leakage reluctances of some feasible magnetic designs, the normal procedure is to then perform finite element analyses. For the magnetic design, it is in most cases sufficient to perform 2-D analysis. The following are the most important questions that such an analysis can answer:

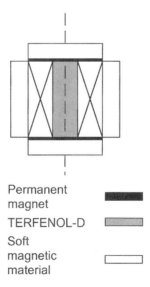

Permanent magnet

TERFENOL-D

Soft magnetic material

FIGURE 3.16.   A patented magnetic circuit with parallel circuits for AC and DC magnetization of the magnetostrictive material.

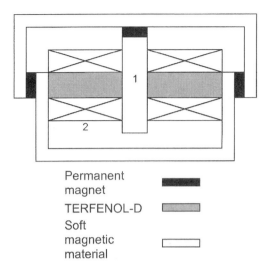

Permanent magnet

TERFENOL-D

Soft magnetic material

**FIGURE 3.17.** A patented magnetic circuit with parallel circuits for AC and DC magnetization of two magnetostrictive rods in which the force outlet (1) is located between the rods with separate drive coils (2).

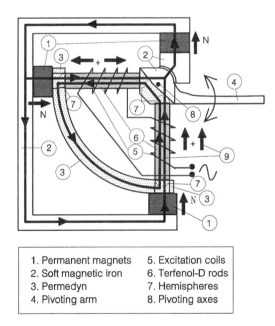

| | |
|---|---|
| 1. Permanent magnets | 5. Excitation coils |
| 2. Soft magnetic iron | 6. Terfenol-D rods |
| 3. Permedyn | 7. Hemispheres |
| 4. Pivoting arm | 8. Pivoting axes |

**FIGURE 3.18.** Application of the magnetic circuit in Fig. 3.17 in the form of a bipolar torque actuator.

- What is the expected magnitude of the bias magnetization $H_{bias}$?
- What is the expected magnitude of magnetizing field $H_{ex}$ for a given excitation current?
- How homogeneous is the magnetic flux in the magnetostrictive material?
- What is the flux density in the flux path materials?
- Are there any "hot spots" in the design?
- Is it possible and how can the design be improved?

In addition, there are other characteristics related to the magnetic circuit that can be studied with FE methods, such as eddy currents and high-frequency effects. Such studies are preferably performed by means of a 3-D analysis in order to identify and quantify hot spots in the design.

By means of a transient 3-D FE analysis and appropriate material models, it is in principle possible to estimate the magnetic efficiency $\eta_{mag}$ of the magnetic circuit. Such efficiency estimations are a subject of current research. One possible way to do this is to use a combination of a lumped element model and a FE analysis.

The important results from the magnetic design activity are that it defines the magnetic range of the actuator and its principle magnetic design and related design data. This forms a basis for the mechanical design of the actuator, which is often developed after the magnetic design.

## ELECTRICAL DESIGN

### Geometry of the Coil and Flux Leakage

The electrical design of a magnetostrictive actuator refers to the design of its coil for dynamic excitation by the input current. The function of the drive coil is crucial for obtaining a high total efficiency of the actuator because it generates the total available magnetic power $P_{mag}$ from which a fraction is used for the magnetoelastic transduction in the active material. The coil function can be described by the following parameters:

$H_{coil}$:  The $H$-field in the center of the coil

$k_c$:  The magnetic coupling factor of the coil

$R_{coil}$:  The coil resistance giving $P_{coil\ losses} = R_{coil}I_{coil}^2$

$\lambda$: The fill factor for the coil wire; circular and rectangular wire cross sections give $\lambda$ values $\pi/4$ and 1, respectively

$\rho$: Electrical resistivity of the coil wire.

$Q_{\text{coil}}$: The $Q$ value of the coil defined as $E_{\text{mag,max}}/E_{\text{coil losses}}$

$E_{\text{coil losses}}$: the dissipated energy in the coil during one cycle

$E_{\text{mag, max}}$: the maximum stored energy in the coil

$L_{\text{leak}}$: The leakage inductance of the coil

$K_{\text{coil}}$: Defined as $H_{\text{coil}} = K_{\text{coil}} I_{\text{coil}}$

$G_{\text{coil}}$: Defined as

$$H_{\text{coil}} = G_{\text{coil}} \sqrt{\left(\frac{P_{\text{coil losses}} \lambda}{\rho a_1}\right)} \tag{3.7}$$

To distinguish the coil performance, it is assumed that the magnetic circuit is ideal (i.e., that no losses occur in the magnetic circuit).

Typical coil geometry is shown in Fig. 3.19. It is common to define

$$\alpha = \frac{a_2}{a_1}$$

$$\beta = \frac{l_r}{2a_1} \tag{3.8}$$

$$\gamma = \frac{a_1}{r_r}$$

Expressions comprising $\alpha$ and $\beta$ for the $K$ and $G$ factors of the coil can be derived by an integration of the Biot and Savart's law around a single current loop and then an integration over all current loops that constitute the whole coil winding (9). The following are the resulting expressions:

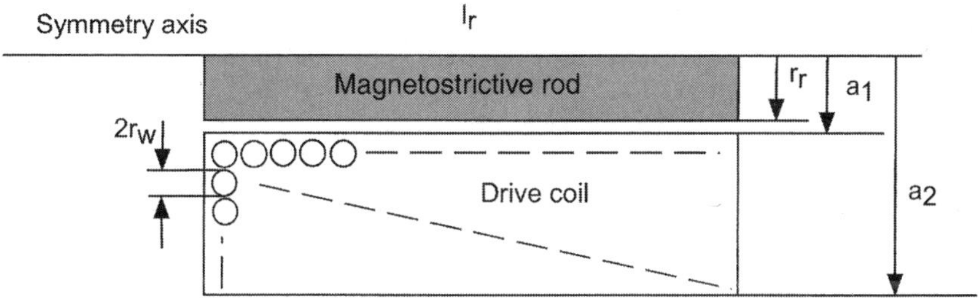

FIGURE 3.19. Drive coil geometry.

$$K(\alpha, \beta) = \frac{N\pi}{2a_1(\alpha - 1)} \ln\left(\frac{\alpha + (\alpha^2 + \beta^2)^{\frac{1}{2}}}{1 + (1 + \beta^2)^{\frac{1}{2}}}\right) = \frac{N\pi}{2a_1\beta(\alpha - 1)} F(\alpha, \beta) \qquad (3.9)$$

$$G(\alpha, \beta) = \frac{1}{5}\left(\frac{2\pi\beta}{(\alpha^2 - 1)}\right)^{\frac{1}{2}} \ln\left(\frac{\alpha + (\alpha^2 + \beta^2)^{\frac{1}{2}}}{1 + (1 + \beta^2)^{\frac{1}{2}}}\right) \qquad (3.10)$$

$$F(\alpha, \beta) = \beta \ln\left(\frac{\alpha + (\alpha^2 + \beta^2)^{\frac{1}{2}}}{1 + (1 + \beta^2)^{\frac{1}{2}}}\right) \qquad (3.11)$$

$F(\alpha,\beta)$ and $G(\alpha,\beta)$ are usually called the "field factor" and the geometry factor $G$ (also Fabry factor) of the coil, respectively. In Fig. 3.20, the $G$ factor is illustrated, in which contours of constant $G(\alpha,\beta)$ are plotted versus $\alpha$ and $\beta$. The $G$ factor exhibits a maximum $G = 0.179$ for $\alpha = 3$ and $\beta = 2$. This corresponds to the case in which the coil has an inside diameter equal to the coil thickness and a length equal to double its inside diameter. This optimum represents the geometry in which most magnetic fields are produced for the least dissipated resistive power in the coil.

It is easy to derive an expression for the coil resistivity for given $\alpha$, $\beta$, and $\gamma$. In most coil designs one strives to minimize the coil inner diameter. Therefore, $\gamma = a_1/r_r \approx 1$ in most cases. Thus, the expression for the coil resistance is

$$R_{coil} = \frac{N^2\rho\pi}{\lambda l_r}\frac{(\alpha + 1)}{(\alpha - 1)} \qquad (3.12)$$

The coil losses can now be expressed as

$$P_{coil\ losses} = \frac{N^2\rho\pi(\alpha + 1)}{\lambda l_r(\alpha - 1)} I_{coil}^2 \qquad (3.13)$$

If this expression is inserted into Eq. (3.7),

$$H_{coil} = G_{coil}NI_{coil}\sqrt{\frac{\pi}{l_r a_1}\frac{(\alpha + 1)}{(\alpha - 1)}} \qquad (3.14)$$

For $\alpha = 3$ and $\beta = 2$,

$$H_{coil} = G_{coil}\frac{NI_{coil}}{a_1}\sqrt{\frac{\pi}{2}} = 0.224\frac{NI_{coil}}{a_1} \qquad (3.15)$$

which represents the most efficient design with respect to dissipated resistive power in the coil.

It is interesting to compare this result with the estimation of the

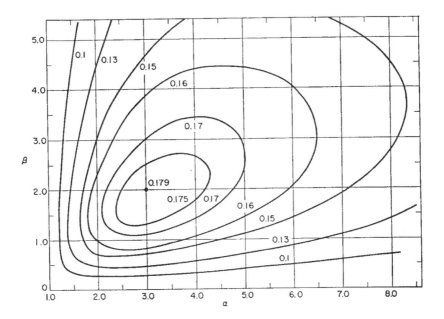

**FIGURE 3.20.** Illustration of the $G$ factor (9).

obtained H-field with a coil that is closed with an ideal magnetic circuit. This expression is the well-known equation

$$H_{\text{coil}} = \frac{NI_{\text{coil}}}{l_{\text{coil}}} \tag{3.16}$$

Recall that for $\beta = 2$ the length of the coil and the rod is $l_{\text{coil}} = l_r = 4a_1$, which gives $a_1 = l_{\text{coil}}/4$. Inserted into Eq. (3.16),

$$H_{\text{coil}} = G_{\text{coil}} \frac{NI_{\text{coil}}}{l_{\text{coil}}} \sqrt{8\pi} = 0.897 \frac{NI_{\text{coil}}}{l_{\text{coil}}} \tag{3.17}$$

The flux leakage of the coil in this design is at least 10% in this example because the H-field is considered along the central axis of the coil.

## $Q$ Value of the Driving Coil

By means of the $G$ factor, it is now possible to derive an estimation of the maximum stored magnetic energy in the magnetostrictive material divided by the dissipated energy in the coil resistance during one cycle (i.e., $Q_{\text{coil}}$). The maximum magnetic energy delivered by the coil to the magnetostrictive material can be estimated as

$$E_{\text{mag, max}} = \frac{1}{2}\mu_r H_{\text{coil}}^2 V_{\text{rod}} \tag{3.18}$$

where $\mu_r$ and $V_{\text{rod}}$ are the mean equivalent value of the relative magnetic permeability and the volume of the magnetostrictive rod, respectively.

From the definition of $G_{\text{coil}}$,

$$H_{\text{coil}}^2 = G_{\text{coil}}^2\left(\frac{P_{\text{coil losses}}\lambda}{\rho a_1}\right) \tag{3.19}$$

The coil losses during one cycle are

$$E_{\text{coil losses}} = P_{\text{coil losses}}\frac{2\pi}{\omega} = H_{\text{coil}}^2\frac{\rho a_1 2\pi}{\lambda G_{\text{coil}}^2\omega} \tag{3.20}$$

Therefore,

$$Q_{\text{coil}} = \frac{E_{\text{mag, max}}}{E_{\text{coil losses}}} = \frac{1}{4}\mu_r\frac{\lambda G_{\text{coil}}^2\omega l_r r_r}{\gamma\rho} \tag{3.21}$$

## The Magnetic Coupling Coefficient of the Coil

Another key parameter of the coil is its coupling coefficient, which defines the ratio of magnetic energy stored in the magnetostrictive rod to the total magnetic energy stored in the actuator. If we assume an ideal flux return path, this coupling coefficient refers entirely to the coil. With the geometry given in Fig. 3.19, the expression for the magnetic coupling coefficient of the coil can be written as

$$k_c^2 = \frac{\overbrace{\frac{1}{2}\mu_r H^2\pi r_r^2 l_r}^{\text{magnetic energy stored in the magnetostrictive rod}}}{\underbrace{\underbrace{\frac{1}{2}\mu_r H^2\pi r_r^2 l_r}_{\substack{E_{\text{rod}}\\ \text{magnetic energy stored in the magnetostrictive rod}}} + \underbrace{\frac{1}{2}\mu_0 H^2\pi(a_1^2 - r_r^2)l_r}_{\text{magnetic energy stored between the rod and coil winding}} + \underbrace{\frac{1}{2}\mu_0\int H^2(r)dV}_{\text{magnetic energy stored in the coil windings}}}_{E_{\text{mag, coil}}}$$

$$\tag{3.22}$$

It is assumed that the average effective permeability of the magnetostrictive rod is $\mu_r$.

For an ideal magnetic flux return path with no fringing fields at the ends of the rod, Amperes' law implies a linear variation of the H- and B-fields in the coil according to

$$H(r) = H_{ex} \frac{a_2 - r}{a_2 - a_1}$$
$$B(r) = \mu_0 H_{ex} \frac{a_2 - r}{a_2 - a_1}$$

$$(3.23)$$

where $H_{ex}$ is the H-field at the boundary of the magnetostriction rod (10).

Figure. 3.21 shows the geometry and field lines for the good and a "bad" magnetic circuits corresponding to the curves shown in Fig. 3.22. In Fig. 3.22 FE analysis illustrates such a linear variation of the magnetic field for a magnetostrictive rod with a reasonably "good" magnetic flux return path.

If the previous flux variation is inserted, we derive the following expression for $k_c^2$:

$$k_c^2 = \frac{1}{1 + \frac{1}{\mu_r}(\gamma^2 - 1) + \frac{\pi}{6\mu_r}\gamma^2(\alpha - 1)(\alpha + 3)}$$

$$(3.24)$$

From this expression, it is evident that there is a trade-off between a high coupling factor and low specific resistive coil losses. It is easy for the second term in the denominator to be kept low because it is relatively easy to obtain $\gamma = 1.05 - 1.1$. Regarding the third term, $\alpha = 1$ implies a maximal value of $k_c^2 \approx 1$, but this choice will imply infinite coil losses (see Eq. 3.13). A reasonable compromise between $\alpha = 3$ (minimal coil losses) and $\alpha = 1$ (maximal magnetic coupling factor of the coil) is $\alpha = 1.5$ that, with $\mu_r/\mu_0 = 5$, gives

$$k_c^2 = \frac{1}{1 + \frac{1}{5}0.1025 + \frac{\pi}{30}1.05^2 \cdot 0.5 \cdot 4.5} = \frac{1}{1 + 0.0205 + 0.25977} = 0.781$$

$$(3.25)$$

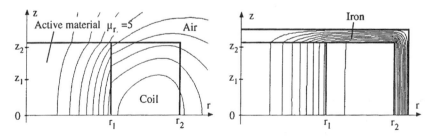

**FIGURE 3.21.** Magnetic flux lines in an actuator with and without a magnetic flux return path.

## Inductance of the Coil

Another important parameter is $L_{\text{coil leakage}}$, which can now be estimated. The expression for the magnetic energy $E_{\text{mag, coil}}$ in the coil winding and between the coil and the rod in Eq. (3.22) can be written as follows:

$$E_{\text{mag, coil}} = \frac{1}{2}\mu_0 H_{\text{ex}}^2 \pi r_r^2 l_r \left(\gamma^2 - 1\right) + \frac{1}{2}\mu_0 H_{\text{ex}}^2 2\pi l_r \int_{r_1}^{r_2}\left(\frac{a_2 - r}{a_2 - a_1}\right)^2 dr \qquad (3.26)$$

$$E_{\text{mag, coil}} = \frac{1}{2}\mu_0 H_{\text{ex}}^2 \pi r_r^2 l_r \left(\gamma^2 - 1\right) + \frac{\pi}{12}\mu_0 H_{\text{ex}}^2 a_1^2 l_r (\alpha - 1)(\alpha + 3) \qquad (3.27)$$

Now $L_{\text{coil leakage}}$ can be expressed as

$$L_{\text{coil leakage}} = \frac{2E_{\text{mag, coil}}}{I_{\text{coil}}^2} \qquad (3.28)$$

If we assume a homogeneous field inside the rod, we can set $H_{\text{ex}} = H_{\text{coil}}$, which gives

$$H_{\text{ex}} = H_{\text{coil}} = G_{\text{coil}} NI_{\text{coil}} \sqrt{\frac{\pi}{l_r a_1}\frac{(\alpha + 1)}{(\alpha - 1)}} \;\Rightarrow\; \frac{1}{I_{\text{coil}}^2} = \frac{G_{\text{coil}}^2 N^2 \pi}{l_r a_1 H_{\text{ex}}^2}\frac{(\alpha + 1)}{(\alpha - 1)} \qquad (3.29)$$

and

$$L_{\text{coil leakage}} = G_{\text{coil}}^2 N^2 \pi^2 \mu_0 \left( r_r \frac{(\gamma^2 - 1)}{\gamma}\frac{(\alpha + 1)}{(\alpha - 1)} + \frac{1}{6}a_1(\alpha + 1)(\alpha + 3) \right) \qquad (3.30)$$

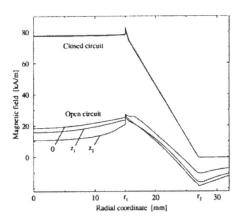

**FIGURE 3.22.**  The corresponding axial component of the magnetic field along the radius of the magnetostrictive rod for a couple of axial positions in the actuator in Fig. 3.21 without and with a magnetic flux return path.

In most cases, the first term in the parentheses can be neglected, which leads to

$$L_{\text{coil leakage}} = \mu_0 \frac{\pi^2}{6} G_{\text{coil}}^2 N^2 a_1 (\alpha + 1)(\alpha + 3) \qquad (3.31)$$

From the equation, one can see that the number of coil turns has the main impact on the coil leakage inductance. For example, $\alpha = 1.5$ and $\beta = 3$ gives $G = 0.11$ and

$$L_{\text{coil leakage}} = 2.81 \times 10^{-7} a_1 N^2 \qquad (3.32)$$

The $L_{\text{coil leakage}}$ must not be confused with the inductance $L_{\text{rod}}$ related to the magnetostrictive rod. If one omits the magnetic energy stored in the return flux path, an estimate of $L_{\text{rod}}$ can easily be done by instead considering the magnetic energy in the rod, $\frac{1}{2}\mu_r H_{\text{ex}} \pi r_r^2 l_r$, which with

$$\frac{1}{I_{\text{coil}}^2} = \frac{G_{\text{coil}}^2 N^2 \pi}{l_r a_1 H_{\text{ex}}^2} \frac{(\alpha + 1)}{(\alpha - 1)} \qquad (3.33)$$

gives

$$L_{\text{rod}} = \frac{2E_{\text{rod}}}{I_{\text{coil}}^2} = \frac{G_{\text{coil}}^2 N^2 \pi}{l_r a_1 H_{\text{ex}}^2} \frac{(\alpha + 1)}{(\alpha - 1)} 2 \frac{1}{2} \mu_r H_{\text{ex}}^2 \pi r_r^2 l_r = \mu_r G_{\text{coil}}^2 N^2 \pi^2 r_r \frac{(\alpha + 1)}{\gamma (\alpha - 1)} \qquad (3.34)$$

which, with the same parameter values, gives

$$L_{\text{rod}} = 3.57 \times 10^{-6} r_r N^2 \qquad (3.35)$$

For 100 turns and $r_r = 0.01$, $L_{\text{rod}} = 0.357\,\text{mH}$ and $L_{\text{coil leakage}} = 29.5\,\mu\text{H}$, which implies $\approx 8.25\%$ flux leakage. However, one must remember that we have assumed an ideal magnetic flux return path.

An important conclusion is that there is in principle no problem with flux leakages in the coil if one can use a good magnetic flux return path that is equivalent to a high value of $L_p$ which limits $I_p$ and consequently the stored magnetic energy in the return path.

For permanent magnetic circuits the magnetic energy in the the flux return path normally dominates. Therefore, in many cases one of the most important features of the coil is its capability to generate maximum magnetic field per dissipated losses. (i.e., a high $G$ factor). On the other hand, the magnetic coupling factor of the coil is also important; therefore, there will always be a trade-off between $k_c$ and $G_{\text{coil}}$.

## Number of Coil Turns and Ratings of the Power Supply

Regarding the choice of number of turns $N$, this is a matter of electrical adaptation to the available electrical power source. From the derived equations it is evident that coil losses are proportional to $N^2 I^2$ and the generated field to $NI$. Therefore, the MMF $NI$ determines the situation; if the power supply can give only low current, one can increase the number of turns at the cost of a corresponding higher voltage to obtain the same MMF. The coil losses are always proportional to $N^2 I^2$. What is optimal in a specific design situation depends on the importance of the earlier mentioned additional constraints regarding the construction.

## MECHANICAL DESIGN

The mechanical design of a magnetostrictive actuator concerns the physical configuration and therefore in principle comprises ordinary mechanical engineering aspects. However, there are two additional aspects that are specific for magnetostrictive actuators: actuator type (resonant or nonresonant) and prestress mechanism. Both are related to the principal design of the actuator and its intended use.

### Nonresonant and Resonant Designs

Nonresonant actuators are preferably used at low or medium-high frequencies ($<2\,\text{kHz}$). Often, the intent is to use the actuator as a broadband controllable motion/force source. In fact, all actuators always show resonant properties. The physical background is that the intrinsic actuator stiffness/compliance and the stiffness/compliance of the mechanical load resonate with the intrinsic actuator inertial mass and the mass of the load. Occurring frequencies are $f_0, f_s, f_{\text{res}}$:

$$f_0 = \frac{1}{2\pi}\sqrt{\frac{k_{\text{eff}}}{m_{\text{eff}}}} \qquad \text{(natural frequency)} \qquad (3.36a)$$

$$f_s = \frac{1}{2\pi}\sqrt{\frac{k_{\text{eff}}}{m_{\text{eff}}}\left(1 - \zeta^2\right)} \qquad \text{(damped natural frequency)} \qquad (3.36b)$$

$$f_{\text{res}} = \frac{1}{2\pi}\sqrt{\frac{k_{\text{eff}}}{m_{\text{eff}}}\left(1 - 2\zeta^2\right)} \qquad \text{(resonance frequency)} \qquad (3.36c)$$

where

$$\zeta = \frac{1}{2Q} \quad \text{and} \quad Q = \frac{\sqrt{k_{\text{eff}} m_{\text{eff}}}}{d_{\text{eff}}} \tag{3.37}$$

are the relative damping and the $Q$ value of the system. In resonant systems with high $Q$ values $f_0$, $f_s$, and $f_{\text{res}}$ can be regarded as equal.

The effective stiffness $k_{\text{eff}}$ and mass $m_{\text{eff}}$ depend on the structural buildup of the application. The effective spring constants and masses can be series and/or parallel connected. For higher frequencies one needs to perform a modal analysis of the structure to determine effective masses and spring constants. If the modal analysis also comprises the magnetostrictive material, for small excitation levels it is possible to estimate occurring resonances directly by a magnetoelastic FE program (i.e., ATILA).[1] For higher excitations, if one intends to account for nonlinearities the lumped element method can be used (see Chapter 2).

The damping coefficient $d_{\text{eff}}$ reflects the effective damping of the system; therefore, power dissipation and/or acoustic radiation processes are involved. An actuator used for generation of acoustic waves in, for example, a fluid medium can be subjected to substantial damping due to the emitted acoustic radiation. This will give a comparatively high $\zeta$ and therefore a low $Q$ value, which implies broadband features of the system.

A characteristic of a resonant system is that the strain and stress amplitudes can be considerable for frequencies in the vicinity of its resonance frequency. This dynamical strain can exceed $4000 \, \mu\text{m/m}^*$ for low effective damping coefficients. The mechanical strength (tensile strength) of the magnetostrictive material limits the possible excitation amplitude. The corresponding low effective damping, however, is in some cases synonymous with low efficiency if the damping represents the useful mechanical work done by the actuator. For actuators in resonant operation the strain is limited by

- the excitation H-field or the accessible current amplitude below resonance,
- the tensile strength of the active material around resonance,
- the accessible voltage or resulting B-field above resonance.

---

[1] ATILA is CAD software for the analysis of structures based on active materials. It was developed for CERDSM by ISEN, Lille, France.
$^*\mu\text{m/m} = $ micrometer per meter.

Nonresonant or "broadband" actuators are generally used in situations in which an accurate motion/force control is needed with respect to both amplitude and phase. This is physically feasible for relatively low frequencies.

Resonant actuators can be designed for low, medium, and high frequencies with feasible application preferably at medium and high frequencies. At frequencies <70 Hz, however, the actuator size increases considerably. Therefore, magnetostrictive actuators operating at these frequencies are seldom used.

Although all magnetostrictive actuators are more or less resonant, the designation resonant actuator normally refers to how the acoustic wavelength in the active material at resonance is related to its length. One therefore distinguishes between half- and quarter-wavelength actuators. Figures 3.23 and 3.24 show the nodal and antinodal planes of these actuators. A common feature of such resonant actuators is that the rod and the connected mechanical transmission element must be treated as waveguides. This constitutes a difference between resonant and nonresonant actuators, in which the rod and mechanical transmissions in many cases are regarded as comparatively stiff elements.

The quarter-wavelength actuator has a nodal plane at one end and an antinodal plane at the other. A half- wavelength actuator has antinodal planes at both rod ends and a nodal plane in the middle of the rod. The two designs differ in crucial aspects.

The half-wavelength actuator does not need inertial tail mass because the generated forces and motions in the two rod halves are directed opposite each other. The net force in the center of the rod does not imply any acceleration of the whole structure. Therefore, there is no need for a dedicated mass to keep the actuator at rest if one attaches the actuator fixture at this point. One drawback is that mechanical power is transmitted at both rod ends. In cases in which one needs mechanical power at only one rod end, there will be a significant power loss at the other end.

The quarter-wavelength actuator needs a substantial inertial tail mass because the generated forces and motions are not balanced. The balance is obtained by a tail mass, which is a mass of at least 10 times the effective mass of the active material and the mechanical load. In some cases this tail mass can be considerable, which increases the total weight and/or volume of the actuator. In other cases, an effective tail mass can be realized by the structure in which it operates. One advantage is that the actuator normally transmits power in only one direction. An interesting configuration is the arrangement of two quarter-wavelength actuators toward each other. If they are operated at a 180° mechanical phase shift

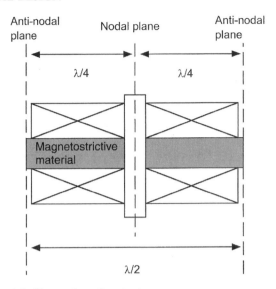

**FIGURE 3.23.**    A half-wavelength actuator.

one can obtain an electrically controlled resonator, in which the force/motion is taken out in the interfacing plane between the two actuators, (Fig. 3.25). The device, however, requires a quite rigid fixture that keeps the opposite rod ends at rest relative to each other.

**Prestress Mechanism**

The need for a magnetostrictive actuator prestress mechanism arises from the fact that there is a requirement of the active material to be mechanically compressed during operation. The reason is twofold: The tensile strength of the material is limited ($\approx$ 28 MPa) and the efficiency and coupling factors are considerably higher under compression.

In some applications, the required prestress can be supplied by the mechanical load (e. g., for the vibration control example in Fig. 3.26). In most cases, however, one must use a special prestress mechanism, which can be based on the following:

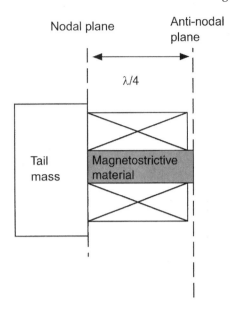

**FIGURE 3.24.**   A quarter-wavelength actuator.

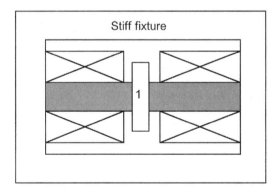

**FIGURE 3.25.**   Configuration with two opposite quarter-wavelength actuators operated at 180° mechanical phase shift against a central force outlet (1).

- Screw thread and spring
- Wedge mechanism
- Prestress wires
- Prestress rods

- Prestress tube
- Hydraulic mechanism
- Pneumatic mechanism
- Rigid fixture and shims

In Figs. 3.26–3.28, some of these mechanisms are shown. The different prestress mechanisms require a stiff exterior reaction path to more or less extent. For example, the prestress wire does not need a stiff path, but this is needed in a wedge mechanism. Key performance parameters of the prestress device are its effective spring constant $k_p$, adjustable force/stress range $\Delta F_p$ or $\Delta T_p$, and effective mass. Additional features are size, user-friendliness, and reliability.

The mechanisms have various advantages and drawbacks regarding parameters and features. Normally, a low spring constant is favorable because the prestress mechanism does not appreciably affect the total effective spring load that the magnetostrictive material "feels." A pneumatic or hydraulic mechanism can give a quite low effective spring constant. A prestress wire and a pneumatic mechanism imply a quite small effective mass. For high prestress, a hydraulic, wedge, or a stiff path with shims are preferable. The remaining mechanisms are multi-purpose and can be used in cases in which no special demand regarding the prestress device is given.

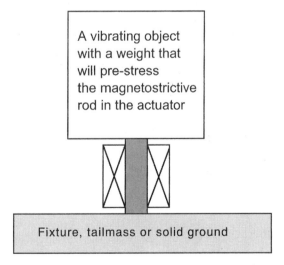

A vibrating object
with a weight that
will pre-stress
the magnetostrictive
rod in the actuator

Fixture, tailmass or solid ground

FIGURE 3.26.   A magnetostrictive actuator that is prestressed by the weight of a vibrating object.

## Transmission and Impedance Matching

An important feature of a magnetostrictive actuator is its ability to deliver mechanical power into a load or process in which the power is utilized. This task is performed by a mechanical transmission element that is often integrated in the actuator.

The transmission is different for low and high frequencies. At low frequencies an optimal transmission system functions as an infinitely stiff link between the active material and the load. Of course, it is not possible to obtain such a link. The resultant lumped stiffness of a cylindrical element can be estimated, however, as

$$K_{\text{trans}} = \frac{E_{\text{trans}} A_{\text{trans}}}{l_{\text{trans}}} \tag{3.38}$$

where $E_{\text{trans}}$, $A_{\text{trans}}$, and $l_{\text{trans}}$ are the Young's-modulus, cross-sectional area, and axial length of the transmission element, respectively. This value has

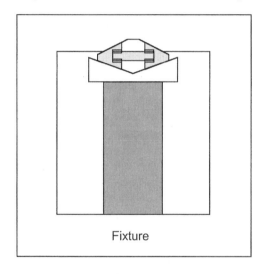

Fixture

WEDGES WITH
INTERMEDIATE SCREW
AND THREADS

MAGNETOSTRICTIVE ROD

FIGURE 3.27.  A magnetostrictive actuator that is prestressed by a wedge mechanism.

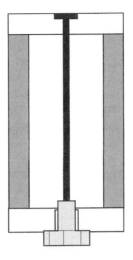

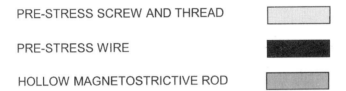

PRE-STRESS SCREW AND THREAD

PRE-STRESS WIRE

HOLLOW MAGNETOSTRICTIVE ROD

**FIGURE 3.28.** A magnetostrictive actuator that is prestressed by a wire mechanism.

to be corrected because the mechanical power in most actuators is transferred through the magnetic return flux path material, which can show a significantly lower E-modulus.

In fact, this conflict between high mechanical stiffness and high-performance magnetic properties is difficult to resolve. No ideal solution exists. Some patent applications[2] related to the problem are pending, and it is likely that additional patent applications will be made in the future.

The most common physical embodiment of the mechanical transmission element for low frequencies is a rod of a stiff material. If the magnetic flux is guided properly in the magnetic circuit there will be only a small leakage of the AC flux. The rod can be made of mechanically

[2]Swedish patent pending.

high-performance steel when it constitutes a part in the DC flux path in cases in which the AC and DC flux paths are separated. Ceramics can be used in cases in which it is important to use nonmagnetic materials.

Using the transmission device, it is possible to match the mechanical impedances of the actuator to the mechanical load because the transmission can also function as a lever mechanism with a mechanical advantage ratio $n_g$. Assume that the actuator can simultaneously deliver the force $F_{act}$ and the velocity $v_{act}$. The lever mechanism will then deliver the force and velocity, $F_{act}n_g$ and $v_{act}n_g$, to the load, respectively. The lever mechanism can be regarded as an analog to an electrical transformer, which implies that the mechanical impedance $Z_{load, \ actuator}$, that the actuator "sees" is

$$Z_{load, \ actuator} = n_g^2 Z_{load} \qquad (3.39)$$

Therefore, in principle, it is possible to adapt the actuator to the requirements of the load. One example of such a case is when one needs more strain and the generated force is substantially higher than needed.

Lever mechanisms should normally be used only at low frequencies because the transfer functions of the mechanisms have a limited bandwidth (a few hundred hertz). If there is a need for lever ratios higher than 5, one maybe needs to think over the whole design and perhaps select some other actuator technology, for example electromagnetic or electrodynamic actuators.

The most common low-frequency leverage principles are the lever mechanism and hydraulic mechanism (3). Figure 3.29 shows these principles, which have various advantages and drawbacks. A lever mechanism is often an integrated part of a design in which the actuator and load are built together. A hydraulic mechanism is a more stand-alone component which can replace an ordinary rod transmission (Fig. 3.29a). The hydraulic mechanism shows more damping but can manage very high ratios ($n_g > 10$). The bandwidth for higher ratios will be narrow. The lever mechanism is used for smaller ratios and shows less damping and broader bandwidths. A lever mechanism can save magnetostrictive material in a configuration, in which the strain demands imply a long effective rod length. The disadvantages are a more complicated design, reduced bandwidth, and transmission losses. A trade-off between material cost and performance features is evident. Performance simulations of the whole actuator application can support such an optimization.

For higher frequencies the speed of sound must be considered. One has to regard all mechanical transmissions as waveguides. Therefore, they all need to be tuned according to the operation frequency. For higher

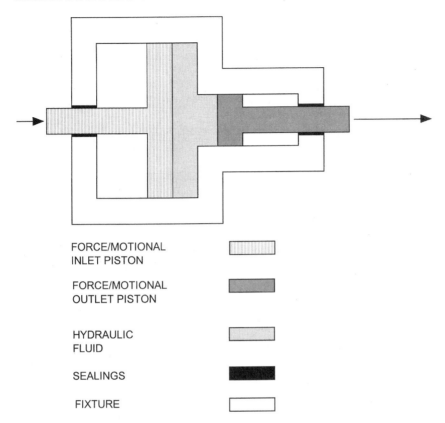

FORCE/MOTIONAL
INLET PISTON

FORCE/MOTIONAL
OUTLET PISTON

HYDRAULIC
FLUID

SEALINGS

FIXTURE

**FIGURE 3.29.** (a) Illustration of a hydraulic mechanism. (b) Illustrations of a
hydraulic mechanism. Observe that the lever mechanism also reverses the motion.

frequencies the actuator normally operates in its resonant mode at a
certain resonance frequency. The transmission lengths must be multiples
of $\lambda/2$, where $\lambda/2$ is a half-wavelength in the transmitting material, in
order to conserve the mechanical amplitudes. For transmission lengths of
multiples of $\lambda$, both amplitude and phase are conserved. The acoustical
signal, however, will also be attenuated because of the material damping
$\nu$ in the transmission element, which can be expressed as a percentage of
the critical damping $2A_r\sqrt{\rho_r/s}$, where $\rho_r$ is the rod density and $s$ is the
effective compliance of the rod material.

The transmission elements also have to adapt the impedance of the
actuator to the impedance of the load process. In this design work, the

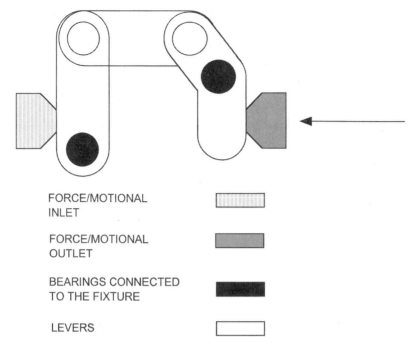

FORCE/MOTIONAL
INLET

FORCE/MOTIONAL
OUTLET

BEARINGS CONNECTED
TO THE FIXTURE

LEVERS

**FIGURE 3.29.** Continued.

following expression can be used for the wave impedance $Z_r$ of a rod with cross-sectional area $A_r$, speed of sound $c_r$, and density $\rho_r$:

$$Z_r = \rho_r c_r A_r \tag{3.40}$$

The previous expression for the wave impedance indicates that it is possible to obtain similar ratios as that for low frequencies. Changing the transmission rod cross-sectional area along the rod can alter the transmission impedance. Therefore, the local strain and mechanical stress in the transmission rod can reach very high values because they increase proportional to $n_g$, which can be expressed as

$$n_g = \frac{A_a}{A_l} \tag{3.41}$$

where $A_a$ is the cross-sectional area of the transmission of the rod attached to the magnetostrictive rod or the actuator force outlet, and $A_1$ the cross-sectional area of the transmission rod attached the mechanical load.

Regarding the mechanical transmission path, it is important that no abrupt changes occur. An immediate change of the impedance from $Z_1$ to $Z_2$ will result in a reflected wave with reflection and transmission coefficients:

$$\beta_{12} = \frac{Z_2 - Z_1}{Z_2 + Z_1} \quad \text{(reflection coefficient)} \quad (3.42)$$

$$\alpha_{12} = \frac{2Z_2}{Z_2 + Z_1} \quad \text{(transmission coefficient)} \quad (3.43)$$

From the previous equations, it is evident that impedance changes must be smooth, i.e., they must occur at least within a whole wavelength. If the operating frequency $f$ is 25 kHz and the corresponding waveguide is of aluminum, the wavelength is $\lambda = c_r/f \approx 5000/25000 = 0.2m$. Therefore, transmissions with large ratios preferably should be made of a material with comparatively low speed of sound. TERFENOL-D is such a material. It is then possible to integrate an impedance adapter in the actuator by a dedicated geometrical design of its magnetostrictive material (4) (see Fig. 5.17). A more common solution is to design the impedance adapter with a special cross-sectional area variation along its axial direction. Various shapes have been studied (5). In Fig. 3.30 some designs are illustrated in which the Fourier shape exhibits favorable features. Gains up to 100 have been reported.

Equation (3.40) reveals that, because of different speeds of sound in the magnetostrictive rod and the transmission element, it is difficult to achieve a proper impedance matching between the active material and the transmission if their corresponding cross-sectional areas are the same. Figure 3.31 shows an example of how this problem is solved in an ultrasonic surgical tool application.

## Detail Engineering Aspects

A characteristic feature of magnetostrictive devices is that the resulting strains are on the order hundredths to tenths of millimeters. Therefore, special attention must be paid to tolerances in the construction.

There is a need to manufacture or machine the magnetostrictive material and mechanical transmission parts with a tolerance level within a couple or tens of micrometers to achieve predictable performance. It is also important that all surfaces that transmit force and strain are flat and

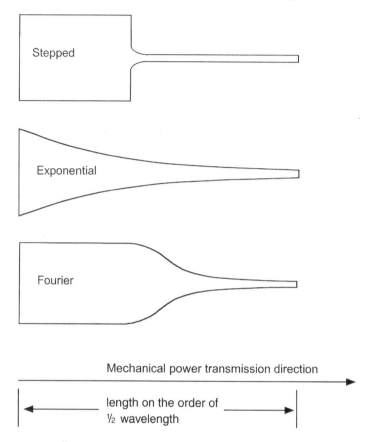

FIGURE 3.30.   Illustration of various impedance adapters.

smooth. The smoothness demand is normally within a couple of micrometers. The flatness demand can be difficult to achieve if the contacting surfaces are comparatively large.

An effective method to manage surface and flatness imperfections is

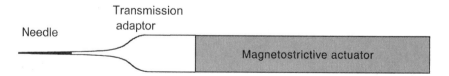

FIGURE 3.31.   Principal design of an ultrasonic surgical tool.

to put a thin shim of a soft metal such as copper in between interfacing surfaces. This method can also be used if there is a slight misalignment between the actuator and the mechanical load.

Another possibility is to use coated materials. In cases in which one needs permanent magnets in the mechanical transmission path, the physical hardness of the high-performance NdFeB material can be problematic. By coating the magnets with a thin layer of tin, it is possible to manage this problem and obtain a well-functioning mechanical transmission between the active material and the load. This coating can also supply an effective inclusion of small magnetic particles that normally are attached to magnetized high-performance magnets. Tolerance and surface management is crucial for a well-functioning actuator. A failure in this respect can substantially degrade its performance.

In an actuator there are always interfaces among surfaces of the active material, magnetic conductive materials, permanent magnets, etc. Actuators with long, efficient active material lengths can comprise many interfacing surfaces, especially if several actuator elements are stacked in series. In these situations, it is important to guide the resultant stack of parts comprising magnetostrictive, magnetic, and mechanical transmission materials (the actuator stack) in such way that an alignment is achieved. Critical situations occur especially during assembly when no or only a low force prestresses the actuator. Steering needles placed in the interface between two adjacent materials (Fig. 3.32) can in principle supply the necessary guidance. Another method is to use guiding flanges or slide bearings (Fig. 3.33).

A popular method is to glue all parts together to provide the necessary alignment and guidance. A problem related to this approach is that two adjacent materials with different E-moduli will be subjected to high stresses in the radial direction. A common material failure in such material interfaces is cracks around the circumference of the magnetostrictive rods.

An estimation of displacements and implied stresses can be made by finite element calculation. In Fig. 3.34, a FE analysis by ATILA of such a phenomenon is shown. In most applications the reliability of an actuator is crucial. For example, failure of the magnetostrictive actuators in a hydroacoustic transducer application is quite costly. Ensuring an acceptable mechanical reliability is synonymous with proper surface management. All surfaces of the active material must be smoothed by soft grinding or spark erosion methods and then be treated with glue that can wet all existing microcracks and pores (e.g., Loctite).

It is well-known that magnetostrictive, soft and hard magnet

FIGURE 3.32.   Illustration of steering needles.

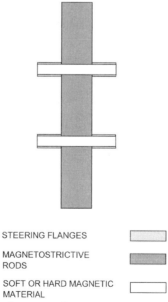

FIGURE 3.33.   Illustration of steering flanges.

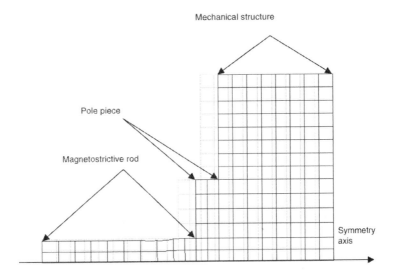

**FIGURE 3.34.** Illustration of the radial displacement of a magnetostrictive rod at its end that acts against a mechanical structure.

materials have a pronounced tendency to oxidize in air. The oxidation rate also increases with the air humidity. Moisture protection measures must be taken. For this purpose, it is common to use oil or grease. Another method is to coat all exposed surfaces with a thin layer of a plastic material. In some cases, it is sufficient to use an adhesive film. If the active material is contained in a limited cavity with no ventilation, it is highly recommended to use a moist protector such as silicagel.

## Assembly and Mechanical Verification

The assembly of a magnetostrictive actuator with permanent magnets is obstructed by the fact that the existing magnetic forces tend to distort the relative positions of the different parts in an unpredictable way. The parts also have a tendency to hit each other due to the magnetic forces in such a way that the brittle magnetostrictive materials crack into pieces. Therefore, in many cases it is worth investing in a dedicated assembly tool, which in principle can be built as a surrounding fixture in which the different magnetic and magnetostrictive parts of an actuator can be attached on a movable linear guide. This guide can be equipped with an electromagnet by which the actuator parts can be attached and released in a controllable way. In other cases, it may be possible to feed the magnetizing coil with current such that the permanent magnet forces are countered.

The assembly of the actuator is a critical step in the manufacturing process. Errors that are made in this step will in most cases cause reduced performance during the actuator lifetime. A common error is that adjacent surfaces are guided in a wrong way relative to each other. This can result in (i) misalignment of the whole actuator stack, (ii) poor mechanical contact between some of the adjacent surfaces in the actuator stack, and (iii) the actuator stacks coming in contact with surrounding fixture details.

The first result will cause an uneven stress distribution over the cross section of the magnetostrictive material. This leads to an unpredictable and probably nonoptimal magnetomechanical operation range of the magnetostrictive material. The second result will decrease effective stiffness of the actuator stack. This is very critical with regard to the performance. Assume, for example, that the stiffness is reduced by a factor of two. This immediately decreases the force capability of the actuator by a factor of two. This a very common malfunction of the actuator. The third result can reduce the strain and force capability to some extent because the actuator stack will be more or less locked by the fixture. Then there will also be an uneven stress distribution over the cross section of the magnetostrictive material.

To ensure that a magnetostrictive actuator is properly assembled, it is necessary to test its functionality. One simple way to do this is to use a mechanical position indicator. By first fixing the actuator and the mechanical indicator relative to each other, it is then easy to investigate the strain response when energizing the actuator with different DC currents. For an actuator equipped with permanent magnets, it is easy to estimate at what current level the magnetizing effect of the permanent magnets is canceled. The strain at this level will change sign. The orientation of the magnetization of the permanent magnets and the coil current can also be checked. In one direction, the field generated by coil adds to that of the permanent magnet and in the other direction, it subtracts. Actuator linearity is also checked by recording the strain versus coil current. A comparison between the strain in the different current directions reveals any imbalances and nonlinearities. A large strain difference indicates that the balance of the actuator can be improved. The recorded strains should approximately agree with the performance data given by the material supplier. It is normally possible to balance an actuator without taking it apart. However, if the actuator shows major malfunctions it must be disassembled and properly re-assembled. A task that normally requires specialized equipment.

## System Interaction

A magnetostrictive actuator is often designed for a dedicated application. The demands corresponding to such an application are at first given in terms of strain and force. Selecting an appropriate working range and possibly an impedance matching device is the normal way to fulfill such demands. In order to achieve a good total efficiency, it is necessary to examine additional performance data of the actuator. Mechanically and electrically, these data are the effective stiffness $K_{act}$ and the effective inductance $L_{act}$ of the actuator, respectively (see System Integration).

It has been reported (6) that actuators operating at resonance show their maximal efficiency when the mechanical load comprises an effective damping element of magnitude $K_{act}/\omega_0$, where $\omega_0$ is the resonance frequency of the system with no damping. To achieve optimal efficiency the equivalent stiffness of an actuator that is intended to operate in resonance has to be adjusted to the effective damping coefficient of the load process such that

$$D_{load} = \frac{K_{act}}{\omega_0} \tag{3.44}$$

One way to alter the efficient actuator stiffness is to connect series or parallel acting spring elements between the actuator and the load. This measure, however, implies a change of $\omega_0$, so an iterative technique can be used.

Regarding the effective inductance $L_{act}$ of the actuator, it is possible to increase the total efficiency of the system by resonant operation using a series capacitor $C_{series}$ such that electrical series resonance occurs at the actual operation frequency $\omega_{op}$:

$$\omega_{op} = \sqrt{\frac{1}{C_{series}L_{act}}} \tag{3.45}$$

The effective inductance of an actuator can be defined in several ways because magnetostrictive actuators behave like a fixed inductance. A useful definition that is appropriate if one wants to tune in an electrical resonance with a series capacitor is that one feeds the actuator with a constant voltage source with frequency $\omega$ and identifies the capacitance $C(\omega)$ where the magnitude of the actuator impedance is minimal. The effective actuator inductance is then given by

$$L_{\mathrm{act}}(\omega) = \frac{1}{\omega^2 C(\omega)} \qquad (3.46)$$

This definition is useful for both measurements and inductance estimations based on dynamic simulation.

For nonresonant actuators the selection of $K_{\mathrm{act}}$ and $L_{\mathrm{act}}$ is made to obtain a high degree of controllability, i.e., $K_{\mathrm{act}}$ should be as stiff as possible and $L_{\mathrm{act}}$ as small as possible. This is synonymous with a low flux path reluctance and a small magnetic flux leakage.

## ELECTROMECHANICAL DESIGN

Electromechanical design constitutes the phase in the development process in which all the design data are determined on the basis of the electrical, magnetic, magnetostrictive, and mechanical specifications. The main aid for this purpose is the use of linear and nonlinear actuator models.

In an actuator construction all the basic design data obtained in the electrical, magnetic, magnetostrictive, mechanical design, etc. can be used to evaluate the functional behavior of the actuator in its intended application. A very direct method is the use of available models in time simulations of the actuator function.

For resonant operation at low excitation levels, linear models can be sufficient. In principle, it is possible to implement the presented models in, for example, MATLAB or to use the FE program ATILA, in which it is possible to perform a transient analysis. At higher excitation levels, substantial nonlinear effects occur. In this case, it is possible to use the presented nonlinear models, which can be implemented using a dynamic simulation package (see Table 2.2). Using this latter approach, it is possible to simultaneously take into account nonlinearities, eddy currents, hysteresis effects, and mechanical and electrical system interaction.

The presented nonlinear model is implemented in the dynamic simulation package SANDYS.[3] The code has been used successfully in

[3]Simulation and Analysis of Dynamic Systems, a program developed by ABB Corporate Research (Västerås, Sweden) for dynamic simulation of physical systems.

many applications, including hydroacoustic transducers, high-pressure pumps, and vibration control. An example from a hydroacoustic transducer application will be given to illustrate the power of the dynamic simulation approach and how it can be used in electromechanical design.

The hydroacoustic transducer design problem can be divided into two parts: acoustic and actuator. A FE analysis of the structural mechanics and fluid dynamics of the transducer can be used for the acoustic problem. This analysis can give effective mechanical impedance values corresponding to the effective mechanical load for the actual frequency range. By using these frequency-dependent impedance values in a dynamic system simulation that includes the whole electrical source it is possible to obtain the radiated acoustic power, voltage, and current variation of the transducer.

In fact, the time variation of all system variables in the system description of the actuator can be investigated. Therefore, it is possible to study the functional behavior of the actuator in the $H - S$ and $S - T$ planes. Thus, one can check the actuator working range in detail. Estimation of transducer efficiencies based on the time variation of eddy currents, hysteresis, and ohmic losses can also be done easily. Figures 3.35 and 3.36 show examples of output from such simulations.

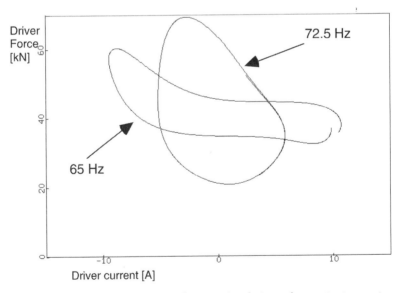

**FIGURE 3.35.** Example of output from a simulation of an actuator system off resonance with various prestress levels.

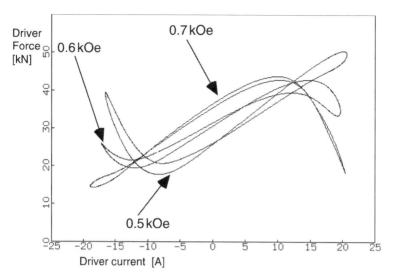

FIGURE 3.36. Example of output from a simulation of an actuator system operated near resonance.

In the electromechanical design activity, it is possible to study the influence of all basic design data on the actuator function by using a dynamic simulation tool. The nonlinear models can also be used to deliver appropriate magnetostrictive constants, such as $d$, $s^H$ and $\mu^T$, to a linear FE program such as ATILA for a given magnetic bias, mechanical prestress, and excitation level. An additional FE analysis of magnetoelastic structural phenomena inside the rod structure and surrounding mechanical structures can then be performed. On the basis of such analysis, a nonlinear dynamic simulation can give information regarding the mean influence of nonlinearities and eddy current and hysteresis effects in the active material. In this study the surrounding structure must be described in the form of equivalent lumped elements.

Finally, a close interaction between a magnetoelastic FE analysis and lumped element dynamic simulation has the potential to give good accuracy in the estimation of magnetostrictive actuator performance data in most applications.

## THERMAL DESIGN AND COOLING

The experience and knowledge regarding electric, magnetic, electrome-chanical, and mechanical design are relatively established compared to those of thermal design of magnetostrictive actuators because, for many applications, the main efforts are directed at primary functional features such as force and strain capabilities. Therefore, thermal design and cooling can be regarded as an optional possibility that can be used if one can afford to use it. Another motivation is that there is often a strong need for high efficiency so that no thermal design and cooling are considered. However, with an appropriate magnetic design it is possible in most applications to achieve bias magnetization levels that correspond to dynamical magnetization currents that are limited only by the tempera-ture increase of the actuator.

Evidently, there is a trade-off between high-efficiency actuators and high-performance actuators. A high-efficiency actuator has a lower power density because the power loss processes such as hysteresis and eddy currents increase with the excitation level. This leads to comparatively large and heavy high-efficiency devices. A very strong argument for using magnetostrictive technology is its high power density. In many cases, it is feasible to accept lower efficiency if considerable volume and weight are saved. In fact, it is possible that there is no space for a "bigger" more efficient actuator. This is a fact in hydroacoustic transducer and ultrasonic applications.

The cooling of the actuator, which will improve its overload capabilities, can be divided into natural and forced.

## Natural Cooling

The following are facts to consider:

- In steady state all the dissipated losses in the active material, coil, and magnetic circuit have to be removed.
- A relatively smooth temperature distribution can be obtained if the heat conduction coefficients of the actuator materials only slightly differ.
- The temperature level of the actuator depends on the heat transfer coefficient(s) at the interface(s) between the actuator exterior and its environment.

On the basis of the previous facts, some conclusions can be formulated. From the first statement it follows that very high temperatures can be reached even at moderate heat dissipation levels if there is no effective heat-removal system provided. Therefore, it is very important to achieve a thermal design with few heat flow obstacles in the path between the heat source and the actuator exterior. In practice, all kinds of air gaps should be avoided inside the actuator. In the gap between the active material and the driving coil it is recommended to use grease or mineral oil. Other potential cavities should be filled with heat-conductive glue (e.g., epoxy with a glass or $Al_2O_3$ filler).

The second statement indicates that failure in preventing heat flux obstacles can cause locally high temperatures in the actuator even when the heat removal at the interface(s) between the actuator exterior and its environment is effective. The third statement means that one must always ensure high heat removal efficiency at the interface(s) between the actuator exterior and its environment.

Studies of the thermal properties of a design can be done by a FE analysis, in which appropriate heat-conductivity data are defined in all parts of the geometry. In addition, the boundary conditions must reflect the actual thermal interaction with the surrounding environment. The latter is intriguing because, as stated previously, the heat removal efficiency is implicitly determined by the boundary conditions.

The boundary conditions are described in terms of heat transfer coefficients that quantify the rate of heat transfer per cross-sectional interface area and celsius ($W/m^2K$). Heat transfer coefficients depend on the interfacing materials, their surface properties such as roughness, possible material speeds, physical orientation (horizontal and vertical), etc. and can be determined experimentally.

In the case in which the surrounding medium is immobile air, it is possible to use the empirical formula

$$P = \alpha_k A(t_1 - t_2) \; (W) \tag{3.47}$$

where $\alpha_k$ is the free convection heat transfer coefficient for a vertical surface $A$. The temperatures $t_1$ and $t_2$ are those of the actuator exterior surface and the surrounding air. The heat transfer coefficient for a vertical plane of height $h < 0.3\,m$ or a horizontal pipe of diameter $h$ can be expressed as

$$\alpha_k = K \sqrt[4]{\frac{t_1 - t_2}{h}} \tag{3.48}$$

where $K$ for $t_m = (t_1 - t_2)/2$ are as follows:

<Air>

| $t_m$(°C) | 0 | 50 | 100 | 200 | 300 | 400 | 500 |
|---|---|---|---|---|---|---|---|
| $K$ | 1.38 | 1.31 | 1.28 | 1.17 | 1.10 | 1.05 | 0.99 |

<Water>

| $t_m$(°C) | 20 | 40 | 60 | 80 | 100 | 150 | 200 |
|---|---|---|---|---|---|---|---|
| $K$ | 107 | 150 | 181 | 208 | 229 | 278 | 312 |

This simple formula can be used to estimate the actuator case temperature on the basis of the total dissipated losses. For small temperature differences it is also possible to account for radiation losses by addition of a correction $\alpha_r$. For a shiny metal surface, $\alpha_r$ is given by

$$\alpha_r = 0.23\varepsilon \left(\frac{t_m}{100}\right)^4 \tag{3.49}$$

where $\varepsilon$ is the emissivity coefficient of the metal.

An example of an actuator of length 0.1 m and diameter of 0.05 m with case and ambient air temperatures of 60 and 20°C can be provided by using the previous equations. The influence of radiation can be neglected:

$$P = 1.3 \cdot \sqrt[4]{\frac{40}{0.1}} \cdot 2\pi \cdot 0.05 \cdot 0.1 \cdot 40 = 7.31\text{W} \tag{3.50}$$

From the previous example, it is evident that even for a relatively small power dissipation some thermal design has to be performed. The natural way to do this in this example is to supply the actuator with cooling flanges, which can easily yield a 10-fold increase in effective area of the actuator case.

The temperature distribution inside a cylindrical actuator can be easily estimated if the actuator is regarded as long and axially symmetric by using the expression for the radial heat flow in a pipe with length $l$ and an inner diameter $d$ and an outer diameter $D$:

$$P = \frac{2\pi l (t_1 - t_2)\lambda}{\ln\frac{D}{d}} \tag{3.51}$$

where the thermal conductivity of the pipe material is $\lambda$, and $t_1$ and $t_2$ are the temperatures of the inner and outer wall of the pipe, respectively.

First, the given ambient and case temperatures provide a total dissipated power loss of the actuator by means of Eq. (3.47). The actuator can then be modeled as many concentric pipes of different materials, with different effective heat conductivity coefficients. The previous equation then gives the temperature at the interfacing surface between the outer material and the material inside. This calculation procedure can be repeated to give the temperature at the next interfacing surface inside the previous surface. In such a way, the whole temperature distribution of the actuator can be estimated. If the losses in the different layers are known the stated power on the left side of the equation should adjusted. It is assumed that only radial heat flow occurs. Therefore, the losses in an inner material can be transferred to its outside layer. If there are also losses in this layer the sum of these losses is transferred to the next outside layer, etc.

In many cases no actuator temperatures and only the dissipated powers in different parts of the actuators are available by, for example, lumped element dynamic simulation. In these cases, it is possible to solve the equation system that can be set up by using Eqs. (3.47) and (3.51) for all interfacing surfaces in the actuator. The solution of this system gives the actuator temperatures. The temperature-dependent expressions for $\alpha_k$ should then be used.

The previous estimation approach should be replaced by a more rigorous FE analysis if one also wants to study the axial heat flow. A 2-D model could be used.

## Forced Cooling

The cooling of a magnetostrictive actuator can be improved considerably if it is forced, which means that a flow of a cooling medium removes the heat. The heat transfer coefficient $\alpha_k$ can then be expressed as

$$\alpha_k = B \cdot \frac{\lambda}{d} \cdot \mathrm{Re}^m \mathrm{Pr}^n \tag{3.52}$$

where $\mathrm{Re} = vd/\nu$ is the Reynolds number, $\mathrm{Pr} = \nu\rho c/\lambda$ is the Prandtls number, $\nu$ is the kinematic viscosity of the fluid, $\rho$ is the density of the fluid, and $c$ is the specific heat of the fluid.

For a fluid flow perpendicular against a cylinder, $B$ is given as

| Re | B | m | n |
|---|---|---|---|
| 0.1–50 | 3810 | 0.385 | 0.310 |
| 50–10000 | 2512 | 0.500 | 0.310 |

If Pr is constant, the equation reduces to

$$\alpha_k = B \cdot \frac{\lambda}{d} \cdot \text{Re}^m \qquad (3.53)$$

with

| Re | B | $m$ |
|---|---|---|
| 1–4 | 3730 | 0.330 |
| 4–40 | 3437 | 0.385 |
| 40–4000 | 2575 | 0.466 |
| 4,000–40,000 | 729 | 0.618 |
| 40,000–250,000 | 100 | 0.805 |

The previous formulae can be used to estimate the cooling due to forced fluid flows against the exterior of the actuator.

If cooling tubes with diameters of 10–30 mm are used, the corresponding $\alpha_k$ can be calculated as

$$\alpha_k = C \cdot v^{0.87} \qquad (3.54)$$

The values of $C$ depend on the type and temperature of the cooling medium. The following are examples of these values:

| | |
|---|---|
| Water (10°C) | 3920 |
| Water (20°C) | 4470 |
| Water (30°C) | 4980 |
| $NH_3$ (20°C) | 5420 |
| F-12 (20°C) | 1690 |

Cooling pipes can be located in a casing that surrounds the magnetic circuit of an actuator and/or centrally in actuators with an inverted magnetic circuit (Figs. 3.37 and 3.38).

In principle, it is possible to perform similar estimations of temperatures and cooling capacities by means of the approach presented regarding natural cooling. The practical arrangements for forced cooling, however, in most cases imply significant deviations from axial symmetric geometry. Therefore, the detail design of a forced cooling system would benefit from a 3-D FE program such as FLUX3D,[4] which can handle thermal analysis. In such an analysis the presented $\alpha_k$ values can be used.

As previously mentioned, thermal magnetostrictive actuator design is a topic under development. It has been hindered by the fact that there has been no reliable way to determine the power dissipation rates in all

---

[4]FLUX3 is a CAD program for electromagnetic and thermal analysis in 3-D developed by Laboratoire Electrotechnique (Grenoble, France).

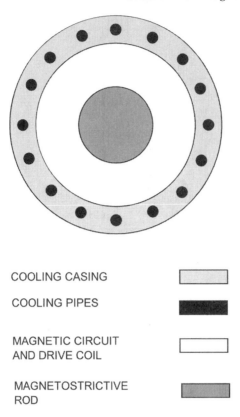

COOLING CASING

COOLING PIPES

MAGNETIC CIRCUIT
AND DRIVE COIL

MAGNETOSTRICTIVE
ROD

**FIGURE 3.37.** Cooling pipes in a surrounding casing of a magnetostrictive actuator.

actuator parts. Recent progress regarding estimation of hysteresis and eddy current losses in the active, soft, and hard magnetic materials of magnetostrictive actuators has provided the potential for better thermal design tools, which will enhance future development of effective high-power actuators.

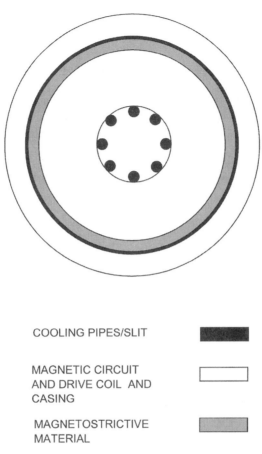

COOLING PIPES/SLIT

MAGNETIC CIRCUIT
AND DRIVE COIL  AND
CASING

MAGNETOSTRICTIVE
MATERIAL

**FIGURE 3.38.** Tentative cooling slit and pipes in an actuator with a so-called inverted magnetic circuit.

## REFERENCES

1. A. E. Clark, J. P. Teter, and O. D. McMasters, Magnetostriction "jumps" in twinned $Tb_{0.3}Dy_{0.7}Fe_{1.9}$, *J. Appl. Phys.* **63**(8), 3910–3912.

2. F. Stillesjö and G. Engdahl, Performance studies of a magnetostrictive actuator with varying mechanical and magnetic bias levels by dynamic simulations, *J. Phys. IV France*, 8 (1998).

3. D. J. Jendritza, H. Janocha, and H. Schmidt, Displacement amplifies for solid-state actuators, *in* "Conference Proceedings on Actuator '96 the Institute Supérieur d'Elecronique du Nord" Bremen, Germany, pp. 300–303.

4.  T. Cedell, Magnetostrictive materials and selected applications. Licentiate thesis, Deptartment of Production and Materials Engineering, Lund University, Sweden, 1995.

5.  A. E. Crawford, Large amplitude resonant actuators using TERFENOL-D actuators, *in* "Conference Proceedings on Actuator '96," Bremen, Germany, pp. 268–271. Published by MESSE BREMEN GERMANY.

6.  G. Engdahl, F. Stillesjö, and A. Bergqvist, Performance simulations of magnetostrictive actuators with resonant mechanical loads, *in* "Conference Proceedings on Actuator 96," Bremen, Germany.

7.  M. E. H. Benbouzin, G. Reyne, G. Meunier, "Dynamic Magnetic Circuit Design for Magnetostrictive Actuators" Conference Proceedings on Actuator '94, Bremen, Germany

8.  U. S. patent No. 4914412

9.  D. B. Montgomery, "Solenoid Magnet Design." Wiley-Interscience John Wiley & Sons Inc. 1969

10. H. Tiberg, Design in Applied Magnetostriction, Licentiate Thesis, Royal Institute of Technology, Sweden, 1994

11. T. T. Hansem, "Magnetostrictive materials and Ultrasonics," CHEMTECH, August 1996, pp. 56-59

## CHAPTER 4

# Magnetostrictive Material and Actuator Characterization

Göran Engdahl
*Royal Institute of Technology*
*Stockholm, Sweden*

## GENERAL

The high strains that can be achieved with magnetostrictive materials compared to piezoelectric materials make them more fit for low-frequency applications, in which higher strains are needed, for example, in hydroacoustic transducers. The described inductive behavior also implies lower required feeding voltages due to a low reactance for lower frequencies.

In recent years the usable frequency range of highly magnetostrictive materials has extended toward higher frequencies because of improved technology regarding eddy current prevention measures such as lamination and powder technology combined with an increased demand for high-power, high-frequency actuators. Therefore, magnetostrictive material and actuator characterization both comprise the traditional methods for characterizing high-frequency piezoelectric actuators and low-frequency characterization methods (5, 6, 7, 8, 9, 10).

As previously mentioned, low-frequency magnetostrictive actuators are often used in nonresonant applications for control purposes. The excitation levels are normally comparatively high which, together with the off-resonance operation mode, implies a nonlinear and hysteretic material behavior. Low-frequency magnetostrictive material and actuator characterization have been shown to be best for time domain methods.

## TIME DOMAIN METHODS

A commonly used low-frequency method is the so-called quasi-static characterization of magnetostrictive materials. It is performed by measuring the mechanical strain and magnetic induction as a function of the magnetizing field from a negative saturating value to a positive saturating value and then back to the negative value. This is done for a constant stress. The field is then decreased to zero, a larger stress is applied, and a new major loop is taken. To decrease effects of a nonsaturating maximum field, an extra major loop is traced for each stress level before the measured major loop is taken.

Figure 4.1 shows an example of an experimental sample holder for quasi-static measurements that is placed in the pole gap of an electromagnet. The mechanical stress on the sample rod is applied by a hydraulic system in which the pressure of the oil is transduced to stress in the rod by a movable plunger. This sample holder design is based on magnetic and mechanical finite element (FE) calculations with the

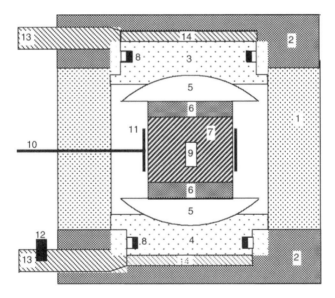

FIGURE 4.1.   Sample holder for quasi-static characterization of magnetostrictive materials. 1, Housing; 2, lid; 3, movable plunger; 4, fixed plunger; 5, spherical bearing; 5, spherical bearing; 6, pole piece; 7, TERFENOL-D rod; 8, mounting fixture; 9, sealing; 10, noncontact displacement sensor; 11, strain gauge; 12, Hall sensor; 13, induction coil; 14, pressure sensor; 15, hydraulic oil connection; 16, oil volume.

purpose of obtaining a homogeneous magnetic field and mechanical stress distribution in the rod.

Measured quantities are the magnetizing field $H$, the magnetic induction $B$, the mechanical strain $S$, and the mechanical stress $T$. $H$ is measured by a Hall probe connected to a Gaussmeter, $B$ is measured by a fluxmeter (integrating voltmeter) from the induced voltage of a pick-up coil, $S$ is measured by strain gauges, and $T$ is obtained by measuring the hydraulic oil pressure with a piezoresistive pressure transducer. All quantities are measured in the axial (longitudinal) direction of the rod.

A computer program including control and data acquisition and processing can manage the entire setup. Figure 4.2 shows a configuration in which the variations of the field can be completely arbitrary and applied independently of each other. The Gaussmeter signal must be feed-backed to the power supply to get full control of the imposed magnetizing field. This suppresses the influence of the magnetic yoke of the electromagnet.

In Fig. 4.3 examples of experimental curves $S_{exp}(H, T)$ and $B_{exp}(H, T)$ obtained by a quasi-static setup are shown for two different material qualities . The curves clearly show the combined influence of mechanical stress and magnetizing field on the resulting strain and magnetization. It is instructive to note that the strain is larger for a prestressed rod. As discussed in Chapter 1, the prestress has the effect of orienting the domains perpendicular to the rod axis. A perfect perpendicular orientation implies that the magnetization $M = M_s \cos \theta$ (along the rod axis only is due to domain rotation, where $\theta$ is the angle between the measuring direction (the rod axis) and the domain orientation and $M_s$ is the saturation magnetization of a domain.

When assuming that the magnetostriction, and thus the strain, has a quadratic dependence of $\cos \theta$, then we have $S \propto B^2$. This is confirmed by the parabolic-shaped $S_{exp}(B, T)$ curves in Fig 4.4 at high prestresses, in

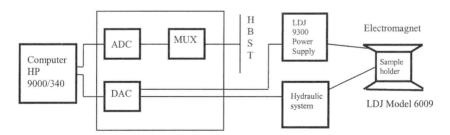

**FIGURE 4.2.** A schematic of the system for static studies of highly magneto-strictive materials.

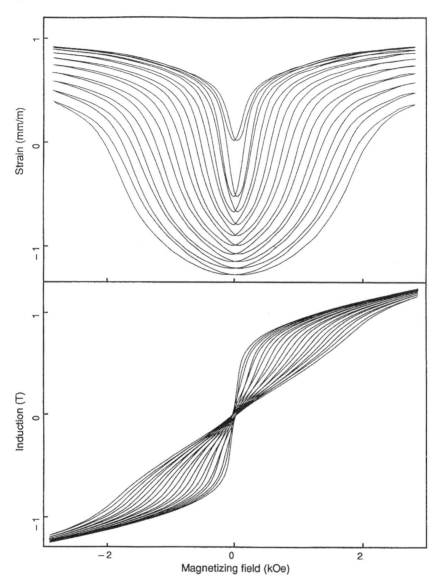

**FIGURE 4.3.** (a) Typical magnetostriction and magnetization curves of a commercial magnetostrictive material I as a function of the magnetizing field. The prestress values are 1, 6.5, 12, 19, 26.5, 34.5, 42.5, 50, 57.5, and 65 MPa. (b) Typical magnetostriction and magnetization curves of a commercial magnetostrictive material II as a function of the magnetizing field. The prestress values are 1, 6.5, 12, 19, 26.5, 34.5, 42.5, 50, 57.5, and 65 MPa (7).

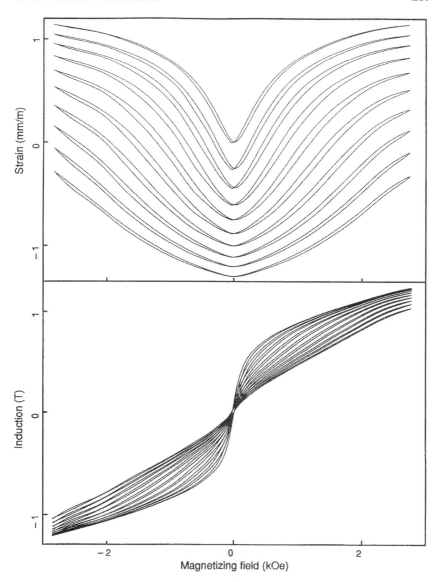

**FIGURE 4.3.** Continued

which a good alignment of domains is obtained.

Obtained $S_{\exp}(H, T)$ and $B_{\exp}(H, T)$ curves can be used with the previously described lumped element dynamic simulation model. Such

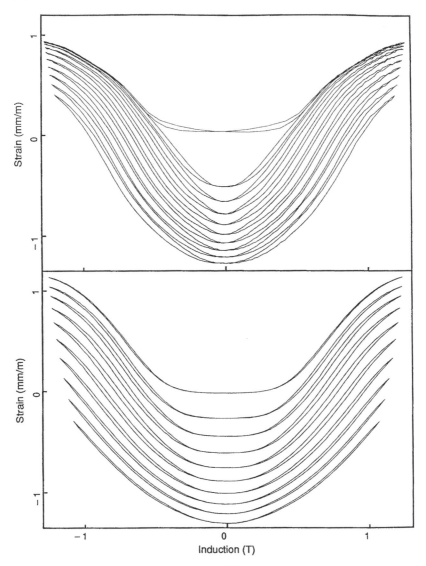

**FIGURE 4.4.** Typical magnetostriction and magnetization curves of a commercial magnetostrictive material I as a function of the flux density. The prestress values are 1, 6.5, 12, 19, 26.5, 34.5, 42.5, 50, 57.5, and 65 MPa (7).

models can comprise eddy currents and magnetomechanical hysteresis phenomena. An estimation of magnetostriction parameters $d$, $s^H$, $s^B$, $\mu^T$, $\mu^S$ and $k$ around a given operation point can be done by using the linearized constitutive magnetostriction equations and simulations under small signal excitation.

The magnetostriction parameters can also be determined by direct inspection of obtained strain and magnetization curves from the setup. The following methods can be used:

1. Incremental measurements: superposition of small oscillatory signals on the static signals, i.e., measurements based on minor loop excitations

2. Differential measurements: measurements based on major loop excitations

In Fig. 4.5 both methods are illustrated. The estimated values of $d$, $\mu^T$, and $s^H$ are identified as the slopes of the minor and major loops according to the definition of the magnetostriction parameters in the linear model.

$$s^H = \left.\frac{\partial S}{\partial T}\right|_H \tag{4.1a}$$

$$\mu^T = \left.\frac{\partial B}{\partial H}\right|_T \tag{4.1b}$$

$$d = \left.\frac{\partial S}{\partial H}\right|_T = \left.\frac{\partial B}{\partial T}\right|_H \tag{4.1c}$$

After $d$, $\mu^T$, and $s^H$ are determined, it is possible to obtain the coupling factor $k$ from the expression

$$k^2 = \frac{d^2}{\mu^T s^H} \tag{4.2}$$

Table 4.1 presents numerical results corresponding to the curves in Fig 4.5. It is evident that calculated $d$, $\mu^T$, and $s^H$ differ between the incremental and differential methods. It is also evident that the coupling factor $k$ is nearly the same for both methods, thus confirming that the coupling factor is an intrinsic material parameter that is invariant with respect to the shape of excitation.

The presented quasi-static characterization can in principle also be performed on an entire magnetostrictive actuator. Such measurements, however, will differ in some significant aspects. The actual magnetic field and mechanical stress in the active material cannot be controlled directly and independent from each other in an easy way. No guarantee can be given regarding homogeneous magnetic and stress fields. Therefore, the quasi-static characterization method is primarily suitable for character-

**Table 4.1**  Incremental and differential values of $d$, $\mu^T$, $s^H$, and $k$

| Measurement method | Load case | $d\ (m/A)$ | $\mu^T/\mu_0$ | $s^H(m^2/N)$ | $k$ |
|---|---|---|---|---|---|
| Differential | 1 | $2.18 \times 10^{-8}$ | 9.07 | $6.26 \times 10^{-11}$ | 0.816 |
| Incremental | 1 | $1.43 \times 10^{-8}$ | 7.15 | $3.46 \times 10^{-11}$ | 0.811 |
| Differential | 2 | $1.67 \times 10^{-8}$ | 6.50 | $5.21 \times 10^{-11}$ | 0.812 |
| Incremental | 2 | $1.14 \times 10^{-8}$ | 5.20 | $2.69 \times 10^{-11}$ | 0.826 |

*Note.* Load case 1: $H = 0.5\,kOe$, $T = 25\,MPa$ (corresponds to the curves in Fig. 4.5); load case 2: $H = 1\,kOe$, $T = 43\,MPa$.

ization of magnetostrictive materials that are considered to be used in magnetostrictive actuators.

Characterization of magnetostrictive actuators in the time domain is preferably performed in experimental setups for dynamic studies. There are many demands that such a setup must meet, including the following:

- The fixture should be extremely stiff in order to withstand the high forces generated by the evaluated magnetostrictive actuators.
- The resonance frequency of the setup should be higher than the working frequency range of the evaluated actuators.
- The evaluated actuators should be easy to mount and dismount.
- Prestress and operating forces should be high.
- The mechanical load should be easy to define.
- The measurement procedure should be computerized.
- The environmental conditions should be easy to control.

Figure 4.6 shows an example of such a setup (*1, 2*). The evaluated actuator is clamped against a heavy backmass (900 kg) with a controllable prestressing force (0–70 kN) by means of a hydraulic cylinder. The actuators can be placed in a test window with a width of 200 mm and a height controllable from 25 to 500 mm using the same cylinder.

Two yokes and cylindrical legs form the fixture that can be clamped to the heavy backmass by hydraulically operated self-locking wedges. This clamping is for most loading conditions equivalent to an ideal clamping with a well-defined length between the backmass and the yoke. The yoke is designed for a shear and bending of <5 $\mu$m at 70 kN and a resonance frequency of 1.3 kHz. The leg diameter is 100 mm and the total length is 1600 mm; these give an estimated resonance frequency of approximately 1 kHz. The fixture rests on four rubber wheels and can consequently move

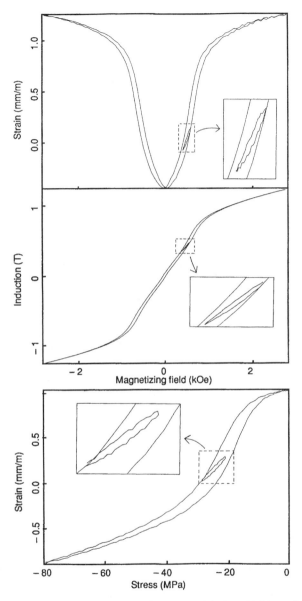

**FIGURE 4.5.** Major and minor loops for $S$ vs $H$ (a), $B$ vs $H$ (b), and $S$ vs $T$ (c). In a and b, the stress is 25 MPa, and in c the field is 0.5 kOe. In c, compressive stress is defined as negative according to the stress definition in Eq. (2.25) (7).

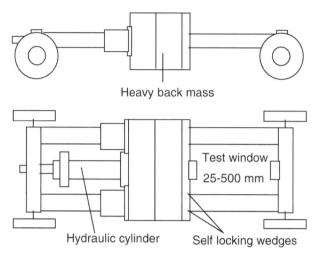

Heavy back mass

Test window
25-500 mm

Hydraulic cylinder          Self locking wedges

FIGURE 4.6.   Setup for dynamic characterization of actuators.

through the backmass. The air pressure in the wheels can be controlled to fine-tune the height of the fixture. Finally, the rigid fixture rests on a platform that functions as a guide or "rail" for the rubber wheels.

When the test object is placed and locked in the rigid fixture, many state variables can be monitored by appropriate sensors, including

- Actuator current $(I_{act})$
- Actuator voltage $(U_{act})$
- The magnetic flux density $(B_{act})$
- The magnetizing field $(H_{act})$
- The actuator force $(F_{act})$
- The actuator displacement $(x_{act})$
- The actuator acceleration $(a_{act})$

Furthermore, an electronically operated mechanical load can be mounted in the test rig. This load can also be a magnetostrictive actuator. This approach admits resistive, spring, and mass loads and combinations of those within wide ranges.

A schematic of the test rig with the actuators and sensors is shown in Fig. 4.7. The actuator load currents and voltages are supplied from Techron 7700 power amplifiers. The magnetic flux density $B_{act}$ and magnetizing field $H_{act}$ are measured with a pickup coil and a Hall probe,

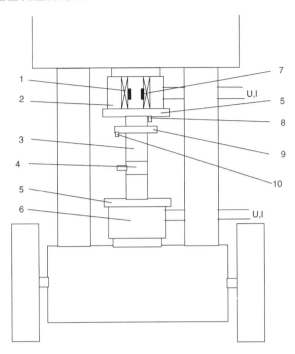

**FIGURE 4.7.** Schematic showing the test rig and the measurement setup and appropriate sensors. 1, Driving coil; 2, test actuator; 3, transmission rod; 4, piezoelectric force transducer; 5, circular plate; 6, load actuator; 7, pickup coil; 8, eddy current displacement sensor; 9, circular plate; 10, accelerometer.

respectively. Regarding the $H_{act}$ and $B_{act}$ measurements, the same difficulty occurs here as that encountered for quasi-static measurements in obtaining representative signals because in general the H-and B-fields are nonhomogeneous in the active material of the actuator. The obtained values, however, can be used as boundary conditions in an additional magnetic FE analysis and/or in a lumped element dynamical simulation of the actuator.

The actuator force $F_{act}$ is measured with a piezoelectric force transducer. The displacement $x_{act}$ and acceleration $a_{act}$ are measured with an eddy current displacement sensor and an accelerometer, respectively. The sensor signals are processed by a multiplexed digital signal processor (DSP) via A/D conversion. Also, analog output signals from the DSP are supplied to the amplifiers via D/A converters. The DSP signals can in real time be presented on a PC and transferred for postprocessing in, for example, MATLAB for presentation, comparisons with dynamic simulations, etc. The measurement system is schematically depicted in Fig. 4.8.

Regarding magnetostrictive actuators with low $Q$ values for non-resonant operation, there is no common standard to quantify their performance features. Some attempts have been made (3), but much remains to be done in this field. Important features, however, are the actuator total efficiency and its transition time or phase lag between, e.g., the fed current and the corresponding displacement coordinate of the actuator force outlet. The following are important load parameters:

- The impedance level of the load
- Nearness to the main resonance of the actuator-load system
- Degree of damping of the system

Regarding the efficiency of magnetostrictive actuators, a recent study has shown that it is maximized at resonance if it is loaded with a viscous damper $D$ equal to $D \approx K_{act}/\omega_0$, where $\omega_0$ is equal to the natural frequency or the resonance frequency of the system with no damping, and $K_{act}$ is the efficient stiffness of the actuator.

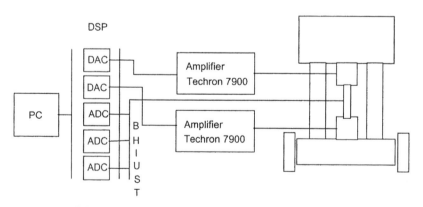

FIGURE 4.8.   Schematic showing the test rig and the data acquisition system. The actuators are driven by two Techron power amplifiers via DACs and the sensor signals are monitored via ADCs on a DSP. DAC = Digital/Analogue Converter ADC = Analogue/Digital Converter.

## FREQUENCY DOMAIN METHODS

Frequency domain methods for characterization of electromechanical actuators have been used in electroacoustics since the beginning of the century (4). The common features of such methods and the previously described procedures for determining magnetostrictive coupling factors are that they assume linear material properties. The $Q$ values of such systems are also usually comparatively high, which implies that the frequencies $f_o$, $f_s$, and $f_{res}$ can be regarded as equal. Therefore, such methods can be used for evaluation of magnetostrictive actuators in resonant operation.

A brief review of the commonly used electric-impedance analysis approach is provided. A schematic representation of a studied actuator is shown in Fig. 4.9.

The relation between the electrical and mechanical quantities can be described in its so-called canonical form:

$$U = Z_e I + T_{em} v \tag{4.3a}$$

$$F = T_{me} I + Z_m v \tag{4.3b}$$

where $U$, $I$, $v$, and $F$ are defined in Fig. 4.9.

The transduction coefficient $T_{em}$ is defined as the electromotive force appearing in the electrical branch per unit velocity. Similarly, $T_{me}$ is defined as the force acting in the mechanical branch per unit current. $Z_e = R_e + jx_e$ and $Z_m$ denote electrical and mechanical impedances. The effective electric operational impedance $Z_{ee}$ can be achieved by setting $F = 0$ in Eq. (4.3b) and substituting the result in Eq. (4.3a), which gives

$$Z_{ee} = R_{ee} + jx_{ee} = Z_e + \frac{-T_{em}T_{me}}{Z_m} = Z_e + Z_{mot} \tag{4.4}$$

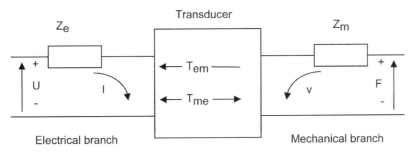

**FIGURE 4.9.**   Schematic representation of a magnetostrictive actuator.

The second term $Z_{mot}$ is identified as the motional impedance, which vanishes when the actuator is clamped, i.e., it is prevented from moving by some external constraint. The mechanical impedance $Z_m$ can, in the frequency domain, be described as

$$Z_m = D + j\omega M + \frac{K}{j\omega} = D + jx_m \tag{4.5}$$

The locus of $Z_m$ is evidently a line parallel to the imaginary axis crossing the real axis at $D$ for the mechanical resonance frequency $\omega_0$:

$$\omega_0^2 = \frac{K}{M} \tag{4.6}$$

It is easy to show be means of conform imaging that the corresponding admittance locus $Y_m = 1/Z_m$ is a circle on the complex plane described by the equation

$$|Y_m| = \frac{1}{D}\cos\theta \tag{4.7}$$

and with a diameter from origin to $1/D$.

The tip of the admittance vector traverses this circle clockwise when $\omega$ increases. The frequencies $\omega'$ and $\omega''$, where the magnitudes of the real and imaginary parts of the mechanical admittance or impedance are the same, are designated "quadrantal frequencies."

At these frequencies the phase angles of the mechanical admittance or impedance are $\pm 45°$, respectively, and the dissipated power is half that at the mechanical resonance frequency if the voltage across the motional impedance is maintained constant. By inserting $\omega'$ and $\omega''$ in Eq. (4.5) and subtracting the resulting equations from each other, it can be shown with some manipulation that $\omega'\omega'' = \omega_0^2$ and

$$Q_m = \frac{\omega_0 M}{D} = \frac{\omega_0}{(\omega'' - \omega')} \quad \text{or} \quad \omega'' - \omega' = \frac{D}{M} \tag{4.8}$$

The frequencies $\omega'$ and $\omega''$ evidently coincide with the previously defined half power points in Chapter 2.

Experimental determination of $\omega_0$, $\omega'$, and $\omega''$ will uniquely define the relative values of $D$, $M$, and $K$. These constants characterize the mechanical system and could in principle also be measured directly by mechanical means.

What is observed from the electrical terminals, however, is the admittance locus multiplied by the scaling factor $-T_{em}T_{me}$. For a linear system this factor can be regarded as a constant complex operator that changes the scale and dimension of the admittance circle into a locus of electric impedance. It will also change the diameter of this circle and rotate it about the origin by an angle $2\beta$, where $\beta$ is designated as the loss angle of the actuator.

From Eq. (4.4), it is evident that the effective operational electric impedance $Z_{ee}$ also includes the pure electric impedance $Z_e$. The tip of $Z_e$ will traverse its own impedance locus as the frequency changes. Thus, when $\omega$ varies from low to high values the tip of $Z_e$ moves along its dashed locus while the $Z_{mot}$, drawn from a moving origin at the tip of $Z_e$, simultaneously traverses its own circular locus. The situation is illustrated in Fig. 4.10. For systems with several vibration modes there will be additional loops in the impedance locus. The impedance locus also reveals parasitic modes or reactions from the mechanical load occurring as, for example, reflecting waves.

The motional impedance $Z_{mot}$ can easily be obtained by simply subtracting $Z_e$ from $Z_{ee}$. $Z_e$ is obtained by measuring the electric impedance under clamped conditions. However, it is sometimes difficult to block the motion of the mechanical system. In such cases, $Z_e$ can be

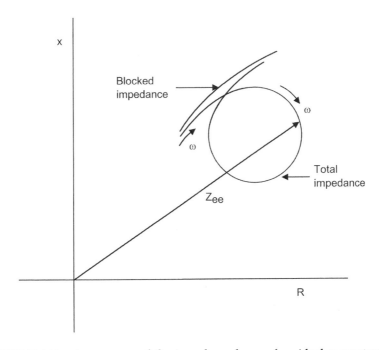

**FIGURE 4.10.** Appearance of the impedance locus of an ideal magnetostrictive actuator.

obtained by graphic interpolation when $Z_{ee}$ is plotted in rectangular coordinates as shown in Fig. 4.11.

Now, $R_{mot} = \text{Re}\{Z_{ee} - Z_e\}$ and $X_{mot} = \text{Im}\ \{Z_{ee} - Z_e\}$ form a circular locus in the complex impedance plane, where its diameter of length $d$ and dip angel $2\beta$ can be obtained directly or deduced by interpolation (Fig. 4.12).

When the locus coincides with this diameter the frequency will be $\omega_0$, i.e., there is resonance and the mechanical impedance $Z_m$ reduces to $D$. At resonance Eq. (4.4) yields

$$d = |Z_{mot}| = \left|\frac{-T_{em}T_{me}}{D}\right| = \frac{A^2}{D} \tag{4.9}$$

If $D$ is known, Eq. (4.9) gives $A$, which is designated the magnitude of the force factor. $D$, however, is normally not known. Therefore, Eqs. (4.6), (4.8), and (4.9) need to be completed using an additional relation to obtain $D, M, K$ and $A$ from analysis of the electric impedance of the actuator. Two methods are feasible.

The added-mass method requires an attachment of an additional mass $m$ to the moving system and a measurement of the new resonance frequency $\omega_{0m}$. Elimination of $K$ in the resulting equation system

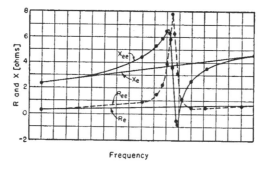

FIGURE 4.11. Equivalent resistance and reactance data for a magnetostrictive actuator operating with no mechanical and interpolated straight lines corresponding to blocked operation conditions.

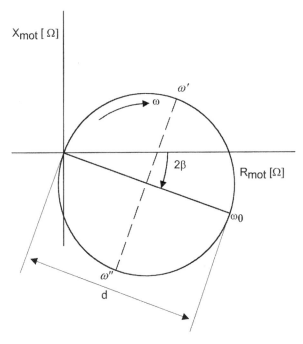

**FIGURE 4.12.** Principal appearance of a motional impedance circle of a magnetostrictive actuator.

$$\omega_0 = \sqrt{\frac{K}{M}} \tag{4.10a}$$

$$\omega_{0m} = \sqrt{\frac{K}{M+m}} \tag{4.10b}$$

then gives

$$M = \frac{m\omega_{0m}^2}{\left(\omega_0^2 - \omega_{0m}^2\right)} \tag{4.11}$$

$D$ from Eq. (4.8) can now be used in Eq. (4.9) to determine the magnitude of the force factor $A$.

The alternative method involves a measurement of the amplitude of the mechanical vibration at the resonance frequency for a measured value of the exciting current. In principle, it is more complicated because of difficulties in determining the appropriate resonance peak and in measuring it correctly. The following expression can be achieved after algebraic manipulation:

$$A = |T_{em}| = |T_{me}| = \frac{d \cdot I_0}{\omega_0 u_0} \tag{4.12}$$

where $I_0$ and $u_0$ are the current and displacement amplitudes at resonance.

The electric impedance method can also give estimations of actuator efficiencies by using studies of the impedance locus with and without external mechanical resistive loads. Some useful results are presented here.

If the diameters of the motional impedance circles for unloaded and loaded conditions are designated $d_V$ and $d_L$, it can be shown that the efficiency at resonance under loaded condition is

$$\eta_{res} = \frac{(d_V - d_L)}{d_V} \frac{d_L}{(R_e + d_L \cos 2\beta)} \tag{4.13}$$

The indices V and L stand for vacuum (i.e., unloaded) and loaded conditions, respectively.

The mechanical impedance $Z_m$ can be divided into one component related to the intrinsic mechanical properties of the actuator $Z_{mi}$ and one component related to the mechanical load $Z_L$:

$$Z_m = Z_{mi} + Z_L = D_i + D_L + j(x_{mi} + x_L) = (D_i + D_L)(1 + j2Qp) = |Z_{mi} + Z_L|e^{j\theta} \tag{4.14}$$

where

$$2Qp = \frac{x_{mi} + x_L}{D_i + D_L} = \tan\theta \tag{4.15}$$

$$p \equiv \frac{1}{2}\left(\frac{\omega}{\omega_0} - \frac{\omega_0}{\omega}\right), \quad \omega_0^2 = \frac{K}{M}, \quad Q = \frac{\sqrt{MK}}{(D_i + D_L)} \tag{4.16}$$

where the indices i and L stand for intrinsic actuator and load-related quantities, respectively.

At off-resonance when $p \neq 0$, the efficiency can be estimated as

$$\eta = \frac{D_L}{(D_i - D_L)} \frac{R_{em}^2 + X_{em}^2}{R_e(D_i - D_L)(1 + 4Q^2p^2) + R_{em}^2 - X_{em}^2 + 4R_{em}X_{em}Qp} \tag{4.17}$$

where $R_{em}$, $X_{em}$, $D_i$, and $D_L$ are calculated from the already derived $A$, $d_L$, $d_V$, and $\beta$ and the following equations:

$$d_L = \frac{R_{em}^2 + X_{em}^2}{D_i + D_L} \tag{4.18a}$$

$$d_V = \frac{R_{em}^2 + X_{em}^2}{D_i} \tag{4.18b}$$

$$R_{em} + jX_{em} = Ae^{-j\beta} \tag{4.18c}$$

It can be shown that maximal efficiency, $\eta_{max}$,

$$\eta_{max} = \left(\frac{d_V - d_L}{d_V}\right)\left[\frac{d\cos^2\theta_m}{R_e + d_L\cos\theta_m\cos(2\beta + \theta_m)}\right], \quad \tan\theta_m = \frac{d_L\sin 2\beta}{2R_e} \tag{4.19}$$

in fact is obtained off-resonance, i.e., for

$$p = \frac{-R_{em}X_{em}}{2R_e\sqrt{MK}} \tag{4.20}$$

Finally, it can be proven that this maximal efficiency can be maximized when the load is purely resistive and equal to

$$D_{L,opt} = \left(D_i + \frac{R_{em}^2}{R_e}\right)^{\frac{1}{2}}\left(D_i - \frac{X_{em}^2}{R_e}\right)^{\frac{1}{2}} \tag{4.21}$$

This maximized efficiency is called the potential efficiency and is the highest attainable efficiency of the system. For this reason, this is very important in the characterization of an actuator. If Eq. (4.21) is inserted into Eq. (4.19), one derives

$$\eta_{max,opt} = \frac{\left(D_i + \frac{R_{em}^2}{R_e}\right)^{\frac{1}{2}} - \left(D_i - \frac{X_{em}^2}{R_e}\right)^{\frac{1}{2}}}{\left(D_i + \frac{R_{em}^2}{R_e}\right)^{\frac{1}{2}} + \left(D_i - \frac{X_{em}^2}{R_e}\right)^{\frac{1}{2}}} = \frac{\sqrt{R_{max}} - \sqrt{R_{min}}}{\sqrt{R_{max}} + \sqrt{R_{min}}} \tag{4.22}$$

where

$$R_{max} = R_e + d_V\cos^2\beta = R_e + \frac{R_{em}^2}{D_i}$$

$$R_{min} = R_e - d_V\sin^2\beta = R_e - \frac{X_{em}^2}{D_i} \tag{4.23}$$

The expressions of $R_{max}$ and $R_{min}$ can also be interpreted graphically in the unloaded total impedance circle as the maximum and minimum values of $R_{ee}$. In Fig. 4.13, these values can be identified as the real coordinates of the crossing points between this circle and its horizontal

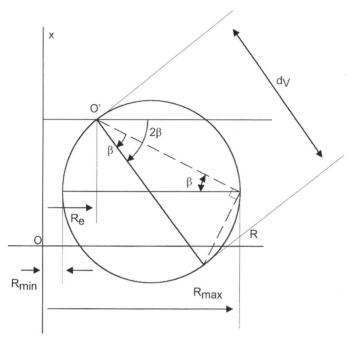

**FIGURE 4.13.** Construction lines showing geometrical properties of the total impedance circle of a magnetostrictive actuator and how $R_{min}$ and $R_{max}$ can be estimated geometrically.

diameter. It is interesting to note that this efficiency estimation requires only the electric impedance when the actuator is operated with no load.

However, there is one restriction regarding the validity of the previous estimation. It is assumed that the clamped electric resistance $R_e$ is constant while the total impedance traces its motional-impedance circle. This occurs when the mechanical $Q_m$ at no load is high enough for there to be relatively little change in the clamped impedance $Z_e$. For $Q$ values of approximately 30, the estimation is accurate, but the expression can also be used for significantly lower $Q$'s and also constitutes in these cases a valuable tool regarding magnetostrictive actuator evaluation.

From obtained impedance circles it is also easy to identify the defined electric resonance and antiresonance frequencies $f_r^E$ and $f_a^E$, (Fig. 4.14). On the other hand, this estimation is also valid for low $Q$ values, where $f_0, f_s,$ and $f_{res}$ differ. Therefore, corresponding coupling factors can be achieved for various operational conditions.

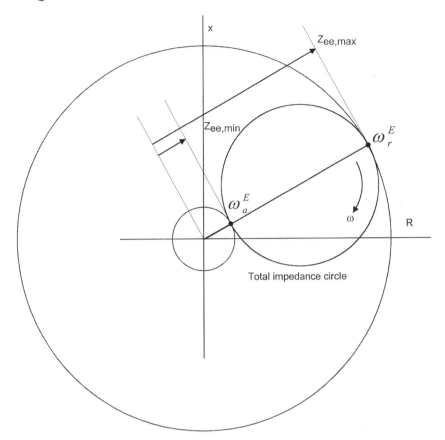

**FIGURE 4.14.**   The electrical resonance and antiresonance frequencies in the total impedance circle of a magnetostrictive actuator.

It can be concluded that all dissipative phenomena, such as eddy currents, hysteresis, and material damping, for a studied resonance are included in the efficiency estimations. The effect of various actuator designs can be studied with the electric-impedance method by examining the input actuator voltage and current. This makes the method very popular among magnetostrictive actuator designers, even though the developed actuators show significant nonlinear and/or hysteretic behavior.

# REFERENCES

1. H. Tiberg and G. Engdahl, A facility for evaluation of actuators based on giant magnetostrictive materials, Paper presented at the MRS fall meeting, November 28–December 2, 1994 , Boston.

2. F. Stillesjö and G. Engdahl, An experimental evaluation system for characterization and assesssment of magnetostrictive actuators, paper will be presented at INTERMAG 99 and published in *IEEE Trans. Magn.*

3. G. Engdahl and F. Stillesjö, Estimation of the intrinsic magnetomechanical response of magnetostrictive actuators by dynamic simulation, *J. Alloys Compounds* **258**, pp. 79–82 (1997).

4. F. V. Hunt, "Electroacoustics—The Analysis of Transduction, and Its Historical Background." Acoustical Society of America, 1954. Melville, NY.

5. F. Claeyssen, Design and construction of low-fequency sonar transducers based on rare-earth alloys. Doctoral Thesis, INSA, Lyon, France, 1989.

6. F. Claeyssen, Comparative study of TERFENOL-D Piezomagnetic Constants, in "Proceedings of the Second International Conference on Giant Magnetostrictive and Amorphous Alloys for Sensors and Actuators," Marbella, Spain, October 12-14, 1988.

7. L. Kvarnsjö, On Characterization, Modelling and Application of Highly Magnetostrictive Materials, Royal Institute of Technology, Sweden, 1993.

8. T. Cedell, Magnetostrictive materials and selective applications, Doctoral Thesis, Lund University, Sweden 1996.

9. M. Wun-Fogle, J. B. Restorff, A. E. Clark, J. F. Lindberg, "Magnetic Properties of Giant Magnetostrictive $Tb_xDY_yHo_zFe_{1.95}(x + y + z = 1)$ Rods under Compressive Stress" Conference Proceedings on Actuator '94, Bremen, Germany.

10. P. Bouchilloux, N. Lhermet, F. Claeyssen, "Dynamic shear characterization in a magnetostrictive rare earth alloy" *M. R. S. Symp. Proc.* **360**, pp. 265-272 (1994).

# CHAPTER 5

# Device Application Examples

Göran Engdahl
*Royal Institute of Technology*
*Stockholm, Sweden*

Charles B. Bright
*ETREMA Products, Inc.*
*Ames, Iowa*

## BACKGROUND

One of the first studied applications of highly magnetostrictive materials was as a generator of force and motion for underwater sound sources. Starting in the 1960s, the exceptional force and strain capabilities of these materials motivated the U.S. Navy to carry out extensive research in this field.

In the mid-1980s, a particular formulation of highly magnetostrictive materials dubbed TERFENOL-D became commercially available (1). It opened the way to developing totally new electromechanical devices with higher energy density, faster response, and better precision than previously possible. Many potential applications were then suggested:

- Sound and vibration sources
- Sonar systems
- Underwater information exchange
- Mechanical impact actuators
- Experimental acoustics
- Structure mechanics
- Active vibration control
- Micromotional control
- Magnetostrictive motors
- Hydraulics

**287**

- Ultrasonics
- Mechanical treatment
- Chemical processing
- Sensors
- Electric generators

In this chapter, the background, functional principles, design aspects, and technical and commercial potential of some of the listed applications are discussed. Because the field is expanding, it is not possible to discuss it completely, and the listed applications can therefore only be regarded as representative of the actual situation. Design methods, active materials, soft and hard magnetic materials, manufacturing techniques, etc. are undergoing continuous improvement. This book should be regarded as one source of basic knowledge and inspiration.

The previously listed applications can be further grouped as follows:

- Sound and vibration sources
- Vibration control
- Direct and nondirect motional control
- Material processing
- Electromechanical converters

The following application examples were chosen from these categories with preferences originating from the experiences of the authors.


## SOUND AND VIBRATION SOURCES

### Acoustic Underwater System

One of the earliest applications of highly magnetostrictive materials was as underwater sound sources, also known as hydroacoustic transducers (2). This is due to the fact that their force and strain capabilities are superior to other force and strain sources, such as piezoelectric, electromagnetic, and electrodynamic. Furthermore, these capabilities improve the mechanical impedance matching to water.

One objective in the search for highly magnetostrictive materials was the need for small, high-power, low-frequency transducers. Underwater sound transducers based on TERFENOL-D were, quite naturally, first developed by the U.S. Navy, from which TERFENOL-D also originates.

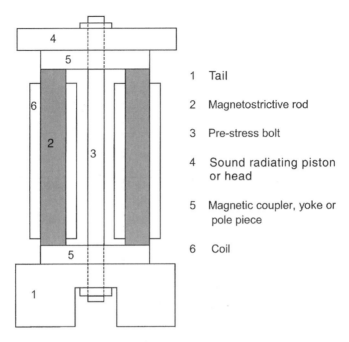

1   Tail

2   Magnetostrictive rod

3   Pre-stress bolt

4   Sound radiating piston
    or head

5   Magnetic coupler, yoke or
    pole piece

6   Coil

**FIGURE 5.1.**   Cutaway of a Tonpilz transducer.

An example of a unidirectional piston-type transducer, commonly called
Tonpilz transducer (2), is shown in Fig. 5.1. Another early design is the
ring type-transducer (3) shown in Fig. 5.2

A more effective concept (4), the flextensional transducer further
developed by the electrotechnical company ABB (Asea Brown Boveri of
Sweden) (5), represents the state of the art regarding high-performance,
low-frequency magnetostrictive underwater sources. The flextensional
shell serves to match the mechanical impedances between the magnetostrictive drive unit and the water (Fig. 5.3).

When compared to competing technologies, the main advantages of
TERFENOL-D are superior controllability, reliability, and efficiency.
Recently, CelsiusTech of Sweden produced flextensional transducers in
series production (6). The following are examples of emerging applications of the transducer technology:

- Geophysical surveying and exploration (7)
- Ocean tomography (8,9)
- Mine clearance (7)

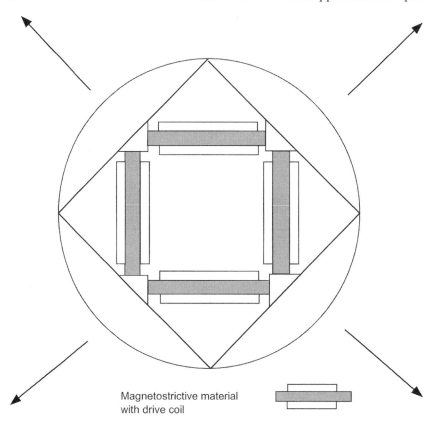

Magnetostrictive material
with drive coil

FIGURE 5.2.   Cutaway of a ring type-transducer or a rare earth ring.

- Underwater information exchange (*10*)
- Underwater sonar systems (*11*)

Obtaining broadband seismic energy sources for seafloor and underground imaging (*12*) using magnetostrictive transducers has been shown
to be technically and economically feasible for a variety of applications,
including seismic imaging to detect and identify hazardous waste
containers, spills, and underground obstacles in cleanup activities. Oil
companies have shown interest in using the technology for location of
petroleum reservoirs and seismic stimulation of known reservoirs.

Seismic shear waves (S-waves) are generated by clamping a reaction
mass driven by a TERFENOL-D magnetostrictive transducer to the inside
of a well pipe. The transducer drives the reaction mass back and forth.

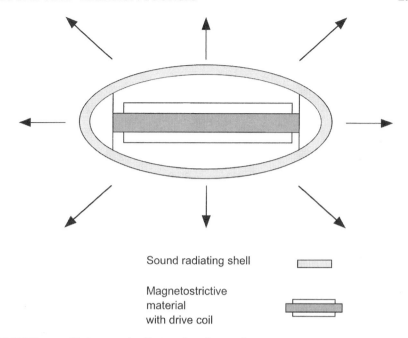

FIGURE 5.3.    Cutaway of a flextensional transducer.

Through the clamping mechanism the forces created by the accelerated reaction mass are countered by the well pipe, thus introducing shear waves from the pipe into the surrounding geological formations. Figure 5.4 shows a magnetostrictive-based device for a seismic application.

Other technologies cannot combine the broad bandwidth with uniform dynamic force output over the desired frequency band of 50–2 000 Hz in this application. The high frequency is necessary to detect small objects not previously detectable. The available power improves imaging capability at low frequency and reduces the number of bore holes necessary to map a volume since transmission distance increases.

## Sound Sources

TERFENOL-D has some basic characteristics that make it suitable for a wide variety of sound and vibration sources. High force capabilities and high strain at off-resonant frequencies with the consequential broad resonant peak provide flexibility in using it to drive a vibration source. When operated using a constant current source, most magnetostrictive

**FIGURE 5.4.**   A TERFENOL-D-based device for seismic applications.

transducers exhibit a reasonably constant displacement response below the first system resonance. Thus, using careful design of the system and its resonances, a magnetostrictive transducer can be tailored to have the desired frequency characteristics. Figure 5.5 shows components of a generic TERFENOL-D transducer.

An example of a broadband device is a solid-state speaker (Fig. 5.6). The speaker consists of a TERFENOL-D transducer driving the center of a solid circular polymer plate. The first resonance of the speaker system, when operated with the polymer edges free, is well above the frequencies required for voice reproduction. Therefore, the response of the speaker is reasonably uniform in the frequency range in which it will be used. The speaker is intended for use in clean rooms or other harsh environments in which traditional cone speakers are not practical. The speaker may be

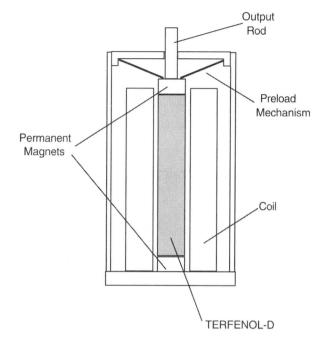

**FIGURE 5.5.** Components of a generic TERFENOL-D transducer.

**FIGURE 5.6.** Appearance of a speaker for clean rooms.

painted, covered with wallpaper, or otherwise disguised. Cleaning the speaker face and the surrounding surfaces presents no problems if the speaker is properly mounted.

One drawback of the speaker design is its audio quality. Voice reproduction is quite adequate. However, music reproduction is lacking in the low and high-frequency ranges.

Other broadband devices intended for general use are shown in Fig. 5.7. The transducers are generally operated below or at the first axial resonance of the TERFENOL-D rod. Rod length greatly influences how much transducer displacement can be expected and the resonant frequency. Uses of these broadband vibration sources include laboratory shakers, industrial shakers, and sound sources.

## VIBRATION CONTROL

The high controllability of the highly magnetostrictive materials makes them suitable for vibration control. Several studies have been performed

FIGURE 5.7. General-purpose actuators available from ETREMA.

using various approaches (*13–16*). In principle, there are two fundamentally different modes of vibration control: the infinitely soft and the infinitely stiff modes.

The principle of the soft mode is to control the force transmitted in mechanical transmission so that it is constant. The principle is illustrated in Fig. 5.8, in which it is applied on a vibrating electrical machine. The

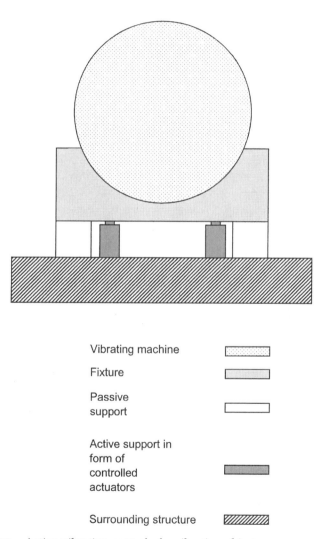

Vibrating machine

Fixture

Passive
support

Active support in
form of
controlled
actuators

Surrounding structure

**FIGURE 5.8.** Active vibration control of a vibrating object.

required control can be achieved by means of, for example, a PID (Proportional, Integrating and Deriving) controller. In a recent RTD (Research and Technical Development) project, MADAVIC (17), supported by the European Community, the technology of vibration control was studied extensively using dedicated actuators and controllers developed for this purpose.

The principle of the stiff mode is to control the position of an object so that it is kept at a fixed position. This can be done by displacement signal feedback to the controller, which then compares it with a position reference value. Figure 5.9 shows an example of infinitely soft and infinitely stiff vibration control, where force and position are damped 20 dB, respectively.

A promising application is adaptive control of tool vibrations in turning processes (18). Successful results have been achieved by using two bipolar actuators discussed in Chapter 3. One actuator compensates tangential movements and the other compensates axial movements. The resonant frequency of the whole system is approximately 2 kHz. This could be doubled if all moving parts were made of fiber-reinforced aluminum instead of iron. Figure 5.10 shows the turning tool integrated with the two bipolar magnetostrictive actuators.

Vibration control by means of magnetostrictive actuators is an emerging technology. It is enhanced by progress in other technologies (19, 20) such as better control algorithms, transputer technology, and switched amplifiers with use in, for example, aircraft and vehicles.

## DIRECT MOTIONAL CONTROL

The comparatively high strain and force capability of magnetostrictive materials makes them ideal for micromotional control in situations (21–23) in which the performance features of piezoelectric materials are insufficient. The following application examples illustrate the diverse potential of magnetostrictive technology. In each application, the magnetostrictive actuator works directly against its load.

### Diesel Engine Fuel Injector

It can be expensive to comply with environmental regulations regarding combustion engine emissions of particulate and other noxious pollutants. Meeting stricter regulations without redesigning or scrapping engines

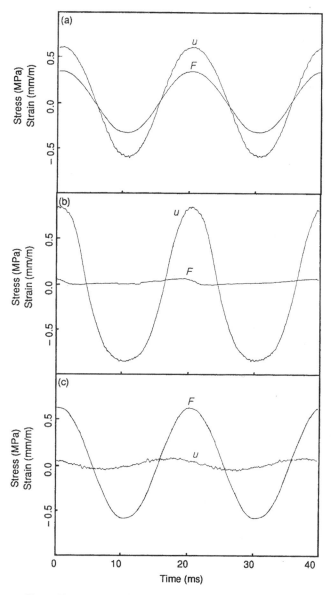

**FIGURE 5.9.** Force $F$ in terms of stress and displacement $u$ in terms of strain obtained with and without vibration control. (a) Amplitudes without energizing the magnetostrictive actuator; (b) amplitudes when the actuator is controlled to keep a constant force (soft system); (c) amplitudes when the actuator is controlled to keep a fixed position (stiff system).

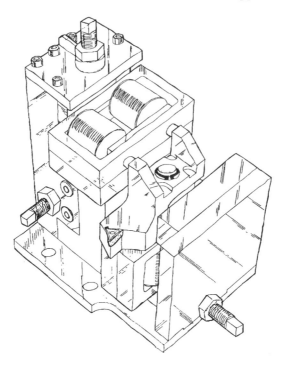

**FIGURE 5.10.**    Test rig used for active vibration control.

that are not yet worn out creates a new retrofit market. This allows immediate usage and immediate air pollution reduction.

The specific advantages of TERFENOL-D compared to other transducing (or "smart") materials in this application are the fast response, high stroke, and high force. They combine to make feasible a fuel injector driven by a magnetostrictive transducer for better control of the combustion process. In addition, TERFENOL-D does not exhibit any known significant deterioration characteristics that cause it to rapidly "age." That is, it does not lose its fundamental magnetostrictive properties when subjected to high temperature, high cyclic magnetic field intensity, and high cyclic strain. Its inherent physical mechanism of magnetostriction allows its performance to fully recover after overtemperature exposure, overfield exposure, and overstrain exposure, provided a catastrophic overexposure such as melting temperature does not occur.

TERFENOL-D's chemical composition of terbium, iron, and dysprosium allows direct contact with nonionic liquid hydrocarbon fuel, which can then be circulated as a coolant in direct contact with the heat-dissipating portions of the transducer. In one example, an existing diesel fuel injector was operated by a solenoid, which limited its "bang-bang" operation capability. In place of the bang-bang solenoid, a magnetostrictive transducer was required to drive a 0.565-kg mass from 0 to 30 $\mu$m displacement with a response time of < 500 $\mu$s. The transducer was required to use < 375 VA and < 10 A and work against a 2 600 N retraction force within a certain allowable length and diameter and under the condition that the diesel engine environment runs from cold to hot.

## Laser Optical Scanning System

Magnetostrictive transducers have been used in a new confocal laser scanning microscope developed by the Dutch research organization TNO. Existing scanners were slow, expensive, and bulky. Improvements were sought to make the instrument cheaper and smaller. These improvements resulted in the development of a new optical scanner applicable to other markets such as barcode scanners. The improved scanner required a small mirror to be driven at high scan rates. Because of its strain and force capability for a given size, a magnetostrictive transducer was developed and delivered to "flutter" the mirror at 8 kHz.

## Astronomical Image Stabilization Platform

Certain optics magnify apparent pointing errors. These errors can be reduced or eliminated by tilting or tipping a mirror. In one ground-based application, the piezoceramic drivers used to point these mirrors displayed low-stroke, high-mechanical quality factor $Q$ that limited bandwidth and required high voltage. These factors were unacceptable for use in a balloon-borne observatory package.

It was desired to fly a balloon-borne observatory near 40-km altitude, at which height the atmospheric pressure drops to about 1 kPa. High voltage needed to drive the piezoceramic tends to arc under these conditions. In contrast, developing a cylindrical magnetic field requires a certain number of amperes per meter in a current sheet around the cylinder. The voltage necessary to drive the electrons can be kept low enough to avoid arcing.

This application used a standard off-the-shelf TERFENOL-D magnetostrictive transducer model for both the ground-based and balloon-borne observatories. Mirror-positioning bandwidth increased by an order of magnitude, stroke increased by six times, and voltage was reduced to that available from a $\pm$ 15-V operational amplifier (op-amp) circuit.

## Movie Film Pin Registration

There is a demand to convert movies that were originally made using film into a format usable by television. Older technology conversion processes introduced jitter and blur onto the video. The next generation of converters used registration pins to accurately position each film frame, but these machines were extraordinarily slow and expensive to operate.

A vendor to the entertainment industry then developed a much faster way to operate the registration pin(s). TERFENOL-D was the optimum choice for speed, number of cycles of life, and low-cost, low-voltage power supply. To keep cost low, a standard transducer model was slightly modified to accept compressed air cooling between the magnetostrictive rod and its solenoid drive coil. This restrained rod thermal growth during continuous operation. To quickly establish current flow and thereby a magnetic field, a short pulse of relatively high voltage was used followed by the voltage required to maintain steady-state current flow. Quickly building the magnetic field reduces mechanical response time, thus improving speed. To completely retract the TERFENOL-D, a small reverse current is applied to overcome hysteresis.

## 1/4-Inch Magnetic Tape Head Micropositioner

A data-storage tape drive manufacturer found the low-frequency stiffness of TERFENOL-D attractive, which allowed construction of a tape drive head positioner that could increase data density on a common tape size. Specifically, the number of tracks increased eleven-fold in the designed version, with an additional increase in the next version. The competing technology was a stepper motor driving a lead screw without closed-loop position feedback. An innovative magnetic bias circuit prevented the transducer magnetic field from contaminating the information moving between the read/write head and the magnetic tape.

## NONDIRECT MOTIONAL CONTROL

### Linear Motors

Examples of linear motors based on magnetostrictive materials are the freewheel concept, the "inchworm" motor (24), and the Kiesewetter motor (25, 26) illustrated in Fig. 5.11. The inchworm motor consists of one clamper at each end of a plunger driven by a magnetostrictive rod. When

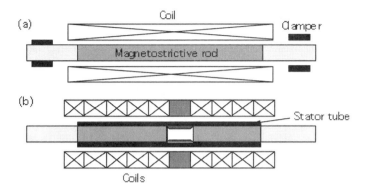

**FIGURE 5.11.** Linear motors based on magnetostrictive rods: the inchworm motor (a) and the Kiesewetter motor (b).

not actively powered, the clampers exert force to prevent motion. When the magnetostrictive rod is magnetized, one of the clampers is active so the plunger is bound to move in one direction. By activating the clampers and the magnetostrictive rod in sequence, a linear motion of the plunger will be obtained.

The freewheel concept (Fig. 5.12) is a self-locking mechanism. The Kiesewetter motor, named after its inventor, comprises a magnetostrictive

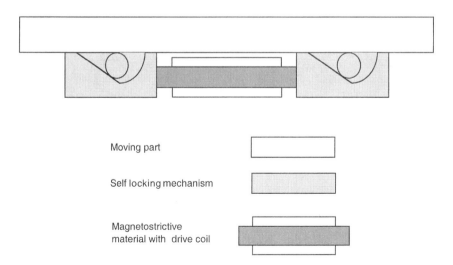

**FIGURE 5.12.** Principal function of the freewheel concept.

rod placed inside a stator tube which is magnetized sequentially along its length by its coils. When not magnetized, the rod is clamped inside the stator tube along its whole length. During operation, the magnetized section increases in length and, because magnetostriction is a constant volume process in the material, at the same time it loses its contact with the stator tube. A sequential magnetization to the right (see Fig. 5.11) would cause translational motion to the left. The Kiesewetter motor also provides holding capability without input electrical power.

Throughout the years, several magnetostrictive linear motors have been suggested and studied. One common feature is that they use friction phenomena that are controlled externally or internally by the actuator. The Kiesewetter motor represents a concept that from a pure mechanical point of view is very attractive. With respect to its electromagnetic function, however, it is very ineffective. This implies a poor total electromechanical efficiency. Further development of the Kiesewetter therefore is an interesting challenge for future designers of magnetostrictive devices (i.e. hopefully readers of this book).

The U.S. company ETREMA Products, Inc. offers a linear motor based on a variant of the inchworm concept for commercial and military applications in which high-force, extended-stroke, high-precision, and fail-safe characteristics are needed. Commercial applications extend over a wide variety of industries and include paper production, medical dispensing, and automotive accessories. Military applications range from positioning of trim tabs on aircraft, helicopters, and submarines to cargo latches.

In 1998, ETREMA began development of a magnetostrictive-based electrohydraulic actuator (EHA), which consists of a magnetostrictively powered pump and flow-control valves, an electronic controller, pump-control algorithms, and a displacement amplification unit integrated in a single block.

The system benefits in a modern aircraft are significant. EHAs and/or electromagnetic actuators (EMAs) can eliminate the need for a central hydraulic system, which reduces maintenance complexity and increases reliability. Maintenance then becomes a "remove and replace" procedure and is accomplished with a minimum amount of downtime. Figure 5.13 illustrates the EHA actuator.

## Fast-Response Valve Actuator

A military service required a valve plunger to be driven by a high-performance motional actuator. A magnetostrictive transducer was

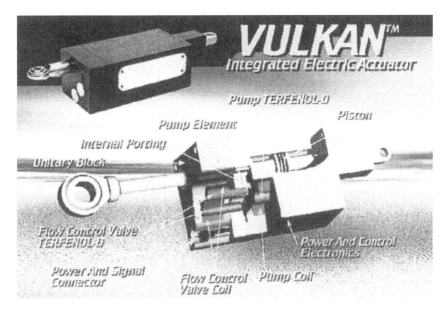

FIGURE 5.13.   A magnetostrictive-based EHA actuator available from ETREMA.

designed and delivered to fulfill this need. It replaced an existing solenoid-operated ball valve to control the flow of pressurized helium to the fin actuator of an extremely high-speed missile. The missile environment required the transducer to operate in a temperature range of − 40 to + 177°C after being subjected to a lateral acceleration of 1 000 g. The transducer was required to displace a mass full stroke in less than 0.5 ms.

The tight size envelope necessitated developing an internal stroke amplifier. Due to the short operational life of several seconds, high power dissipation was acceptable, allowing maximum stroke to be achieved in a minimum amount of time.

## Fast Servo Valves

Studies have been performed to consider using the controllability, force, and strain properties of highly magnetostrictive materials to drive the pilot stage of a servo valve (27). A prototype of the patented design has been evaluated experimentally at KTH (Sweden). In this prototype, standard ETREMA off-the-shelf actuators were used. The technical potential is to produce high-flow (1001/min) servo valves with a bandwidth exceeding 1 kHz.

## Pumps

It is technically feasible to use the favorable force capability of magnetostrictive materials to achieve low-volume, high-pressure pumps (*28*). In this patented design (*29*) the main piston is driven by two opposed actuators attached to a rigid fixture fed with an electrical phase shift of 180° and provided with biasing magnets. A prototype that can achieve 25 MPa has been built by ABB for offshore oil production. A cutaway of this is shown in Fig. 5.14. Another application is low-power highly controllable solid-state pumps for various process applications (*12, 30*).

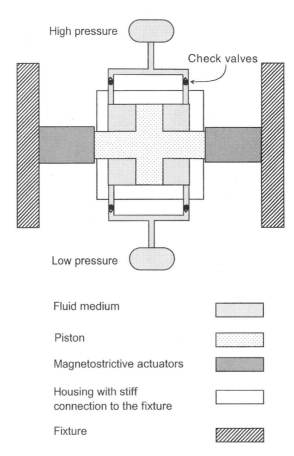

**FIGURE 5.14.**   Cutaway of the ABB high-pressure pump.

It is also possible to use magnetostrictive films for pump applications (*31*). This technology is in its first development phase and is discussed in Chapter 6 with regard to magnetostrictive films.

## Rotary Motors

Several types of rotating motors based on magnetostrictive materials have been presented (*32, 33*). Their main advantages compared to traditional electromachines are higher controllability and comparatively high torque at low speeds. The principle of the motor concept developed at Toyota Research Institute (*34*) is illustrated in Fig. 5.15. It has been used successfully in motors based on piezoelectric ceramics. With the use of TERFENOL-D, an improved force capability is obtained. Two perpendicular magnetostrictive rods, A and B, that are excited sinusoidally push a driving piece. The phases of the magnetizing fields of the two exciting rods are shifted 90°. Hence, an elliptical motion of the driving piece will be obtained. The tip of the driving piece transfers power to a rotating shaft by frictional forces. A different principle based on a roller-locking principle has been developed in the United States and used for an aerospace application (*34*).

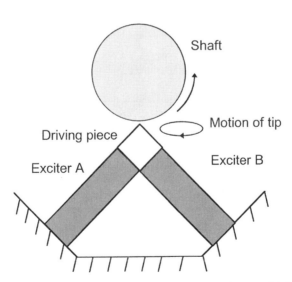

**FIGURE 5.15.** Principle of a rotational motor based on TERFENOL-D.

The elliptical motion concept has been further developed by Cedrat (35) (France) which uses one translational and one flexure mechanical resonance mode of a stator ring, which are excited by two linear actuators acting perpendicularly and diametrically against the stator ring. The actuators are fed with the same frequency but with a 90° phase shift. The resonant operation implies a significant increase in efficiency.

The motivation for development of magnetostrictive rotary motors is their higher forces compared to piezoelectric motors and their increased reliability and reduced maintenance demands compared to corresponding hydraulics and electromachines. Micromechanical prototype rotary motors based on highly magnetostrictive films have also been studied (36). This magnetostrictive thin film technology is described more thoroughly in Chapter 6.

## MATERIAL PROCESSING

### General

Affecting material processing by means of magnetostrictive technology can in principle be performed by direct interaction or indirectly by means of sound penetration. One application example of direct interaction is gravure printing engraving. An engraving company developed and patented a faster method of engraving a gravure cylinder for presses used in packaging industry printing. In this method, a magnetostrictive transducer replaces a pair of opposing electromagnets to drive a diamond-tipped stylus arm engraving a copper-plated surface, thus reducing processing time and increasing throughput. The specific TERFENOL-D properties that are taken advantage of are the unique speed, displacement, and force combination. The maximum frequency range of the electromagnets is 3–5 kHz, whereas the magnetostrictive transducer operates at 12 kHz.

A similar application is used for fretting machines for studies of interconnection failure of electrical contacts and connectors (37). A new computer-based fretting test machine is being developed to study such failures and to develop solutions. These magnetostrictively driven machines have been successfully used to test commercial electronic devices and military electronic equipment.

The majority of magnetostrictive actuator applications to material processing, however, are based on high-intensity sound or vibration penetration into the treated material. The frequency used depends on the process.

One such application is wave-actuated infiltration in the production of metal matrix composites (MMCs) (38). The method makes it possible to produce MMC materials that combine the properties of metals and ceramics. The process includes a magnetostrictive actuator operating in the frequency range of 0.5–7.0 kHz.

A similar application is vibration-assisted production of plastic composites for higher quality and improved material properties. Promising results have been achieved in an ongoing project at Lund University, Sweden.

## Ultrasonics and Actuators in General

Magnetostrictive actuator technology has made it possible to achieve high-frequency, high-power actuators. These are now penetrating medical, dental, petrochemical, and sonochemical applications markets. This relatively new technology is pushing the state of the art in the ultrasonics industry. Traditional technologies such as PZT® and the nickel class of magnetostrictive alloys (permendur, magnetostrictive iron alloys, etc.) have fully exploited their respective material limits. TERFENOL-D, due to its much higher energy density, high thermal conductivity, giant strain capabilities and very high fatigue life, is providing new possibilities for production applications requiring ultrasonic power.

TERFENOL-D has increased the state of the art of ultrasonic power available from a single device up to 6 kW. (Because transducers are reactive loads, the power input rating should be measured in KVA rather than kW. However, we adopt industry jargon in the interests of consistency.) A project with the U.S. government's Advanced Technology Program is under way in which a 25-kW ultrasonic transducer is being developed. The 25-kW transducer will be continuously operable at its rated power, an order of magnitude higher than that of the competing technologies in continuously available ultrasonic power.

In general, nickel, permendur, and other highly elastic materials with small magnetostrictive strains employ a half-wavelength design. The high strengths and low displacement potential of these materials are the main reasons for this configuration. High-power half-wavelength designs are plagued with long transducer lengths and large masses.

Nearly all PZT® ultrasonic transducers employ a quarter-wavelength configuration. PZT® is a piezoceramic material requiring voltage as opposed to magnetic field to produce strain and hence work to its environment. To reduce high-voltage requirements, PZT® wafers are stacked on top of each other, with thin electrodes placed between each wafer. For quarter-wavelength designs, a bolt is inserted through the inside diameter of the stack and screwed into the passive section. The bolt serves to apply mechanical compressive bias to the wafer stack to prevent damage to the PZT® during dynamic operation. By nature, PZT® ceramic is very brittle and will not survive a tensile environment. Material compression also ensures that each member of the stack is in direct contact, preventing appreciable acoustic energy losses. This design methodology has resulted in the entry into the marketplace of efficient, compact, and lightweight ultrasonic transducer designs.

PZT® and the nickel-class technologies are very mature in the ultrasonic marketplace and have their distinct market niche applications. The prospect of exciting new devices from either technology is unlikely. Both have low energy densities, effectively limiting new innovations. Nickel energy density values are so low that devices of any appreciable power higher than 3 kW are exceedingly large and cumbersome. Acoustic physics dictate how large a device can be without gross inefficiency from transverse modes of vibration and harmonics.

PZT® possesses an energy density about an order of magnitude higher than that possessed by the nickel-class materials. During the past 30 years, this has led to smaller, more powerful, and compact designs despite its very narrow bandwidth or high mechanical quality factor $Q$. However, the Achilles heel of PZT® is its very low thermal conductivity. Low thermal conductivity prevents heat caused by hysteresis, Joule heating, friction, and harmonic distortion to escape the PZT® stack. Cooling the stack is possible, but the cooling medium dampens stack vibration amplitude and hence reduces power output. Even with cooling present, the low thermal conductivity causes very large temperature gradients within the material. High temperatures lead to rapid performance degradation of the material as it depolarizes under its Curie temperature (as well as under conditions of high stress, strain, or voltage). This may be in part why the most powerful PZT®-based ultrasonic systems are limited to 4.5 kW. PZT® ultrasonic systems may be rated to 3 kW, but this occurs for only a short period of operation. In practice, the continuously operated rating can be as low as 60% of the short-duration rating. This is where TERFENOL-D shows great promise in ultrasonic applications.

TERFENOL-D ultrasonic transducers are designed in a similar manner as PZT® transducers (Fig. 5.16). They comprise a mechanical compressive bias mechanism, in which a disc spring or a bolt provides the static compression stress on the TERFENOL-D drive rod.

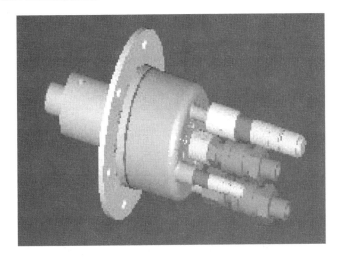

**FIGURE 5.16.** A TERFENOL-D ultrasonic transducer.

## Physical Treatment

### General

Previously mentioned material processing examples can be regarded as physical material treatment by direct interaction or indirectly by means of sound penetration. In the following examples, the specific physical effects of the high-intensity ultrasonic power are used.

There are three processes in widespread commercial use that benefit from the high-frequency, high-power features of magnetostrictive actuators:

Ultrasonic cleaning

Sonic cell disruption for analytical purposes

Material bonding (welding)

Ultrasonic cleaning and sonic cell disruption prior to chemical analysis of cell constituents are in very widespread use, and many manufacturers produce commercial equipment for these processes. Variations of the commercial equipment for the three ultrasonic processes are often used in laboratory studies of other processes.

Ultrasonic cleaning and sonic cell disruption depend on cavitation. Most commercial equipment for these processes operate at 20 kHz.

Ultrasonic bonding of materials, either plastic or metal-to-metal welding, operates by frictional heating of the parts. Operation can be from <15 kHz to well over 125 kHz. Generally, frequency is inversely proportional to size.

All the given examples, including emulsification/deemulsification, foaming, mixing, vibration, and boundary layer control, can be regarded as physical material treatment applications in the sense that the ultrasonic power primarily affects the physical state of the treated materials.

## Ultrasonic Cleaning

An example of ultrasonic cleaning is periodic removal of undesirable accumulation of solid materials in chemical reaction pipes to prevent process shutdown. In a specific case, the pipe was excited in its hoop-mode resonance. Because of the pipe material and dimensions and the solid material accumulation, the resonant frequency was in the range of 10–15 kHz. To maximize output displacement it was necessary to drive the pipe into resonance. The first resonance of the transducer was designed to be just above this range at about 16 kHz. Two transducers, each with sufficient tail mass, were mounted in diametric opposition across the pipe.

In this particular application, the ability to sweep across the frequency range was important because the internal material accumulations affected pipe resonance frequency. The choice of magnetostrictive transducers was driven by the desire to maximize output displacement to maximize the vibrational power within the whole resonant frequency range. Sealing against low-pressure water immersion was required. The water acted as a heat sink and thereby helped to maintain output for 100% duty cycle. The transducers were impedance matched to readily available electrical power supplies.

## Cell Disruption and Sterilization

There are many processes in which application of ultrasonic power results in a useful physical effect; that is, an effect in which no chemical bonds are broken (43). Ultrasonic sterilization is thought to result when some vital function is irrevocably disrupted in a living organism. Commonly, this is breakage of the cell wall in any single-cell organism. Cavitation is the likely mechanism for cell wall disruption.

Until recently, this was a common process, but it was limited to laboratories, chiefly medical. With the introduction of high-power commercial ultrasonic equipment, companies are developing techniques for sterilization of food, sterilization of bilge water discharged from ships at sea (without chemicals which might be harmful to marine life), and several other useful processes (40).

There have also been reports that changing the frequency of applied energy can selectively break cell walls. This technique has the potential to allow sterilization without harm to food products or to make medical analyses more sensitive by breaking only the cell of interest. Virtually all non-TERFENOL-D ultrasonic equipment for cell wall disruption operates at 20 kHz so that adequate power can be produced, at resonance, to produce the effect yet keep the equipment compact and efficient. Because of the broad bandwidth of TERFENOL-D, it is possible to construct equipment that operates at nearly any frequency under 35 kHz. Recently, a pilot production lot size of TERFENOL-D transducers operating at 50 kHz was delivered to a customer. Research into ribbons of 20–50 $\mu$m thickness permits even higher frequencies and composite materials capable of megahertz operation is now under way.

## Ultrasonic Friction Welding

One way of joining two sheets of polyethylene-coated paper is to vibrate the material in a direction perpendicular to the surfaces (41). The velocity of the deformation in the polyethylene layer generates heat due to internal friction. As a result of the high $d$ coefficient and low anisotropy energy of TERFENOL-D, it can be used directly without a booster or a wave concentrator.

By using a wedge shape rather than a rod with a uniform cross-sectional area, the following advantages are gained: higher resonance frequency and lower amplitude of stress at the attached end. This shape also has a drawback: It is difficult to produce a uniform flux density throughout the length of the rod.

The enhancement of the resonance frequency means that, at a given frequency, a longer rod can be used, thus increasing vibration amplitude. Because a longer rod means a greater volume of the magnetostrictive material, the internal power available to drive the rod also increases. The high stress levels in a uniformly shaped GMPC rod, when driven in

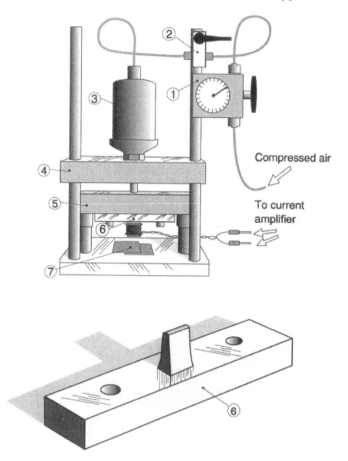

**FIGURE 5.17.** The friction-welding test setup. 1, Pressure regulator; 2, on/off valve for air; 3, pressure cylinder; 4, fixture; 5, vertically adjustable press table; 6, magnetostrictive actuator; 7, two samples of polyethylene-coated paper sheets.

resonance, cause rupture at the attached end. By using an appropriate wedge shape, the amplitude of the stress at the same magnetostriction is reduced by more than one-third.

The test setup used in the experiments is shown in Fig. 5.17. This test setup can apply an arbitrary pressure against the fixed press table, thus holding the samples between the actuator and the fixed press table. Feeding the coil with a high-frequency current tuned to the natural frequency of the loaded actuator causes the polymer coating of the papers to melt.

## Emulsification, Deemulsification, and Foaming

All ultrasonic processes involving emulsification, deemulsification, and foaming are believed to require cavitation. Emulsification processes are thoroughly discussed in the literature (39). Sonic power is an excellent means of causing a fluid to foam. Carbonated beverages often require foaming, either to measure the amount of dissolved gas as a production control technique or to drive air out of the beverage container before capping. In the beer industry, foaming is accomplished by injecting a very narrow stream of high-pressure water into the container just before it is sealed. This technique limits the filling machine speed because of the time it takes the foam to form after injection of the high-pressure water stream. It was found that stimulating the beer container with a TERFENOL-D-driven ultrasonic source immediately after filling caused almost instantaneous foaming (39). Controlling the power level of the transducer can easily control the foam level. This technique can allow significant speed up of existing bottling lines.

## Mixing, Vibration, and Boundary Layer Control

In general, techniques for mixing and similar processes are best performed at relatively low frequencies (often less than 10 kHz) because mixing is improved by application of relatively high amplitudes. TERFENOL-D is well suited to such applications because it works well at lower frequencies and provides significantly higher displacements per unit length than other solid-state transducer drive materials. In one such chemical production process, fluids are introduced into a pipe in which they react exothermically to produce a very valuable solid material. The process pipe must be run in a water bath to control the heat produced. Over time, the solid material coats the inner pipe walls, reducing process efficiency. Periodically, the plant must be shut down and the pipe walls manually cleaned. Adding TERFENOL-D transducers at selected points down the length of the pipe reduced time between cleanings and improved heat transfer. This was also described earlier as "Ultrasonic Cleaning." The ideal frequency for the process changes with time as coating builds on the walls. The electronic drive system periodically sweeps through a frequency range and determines the frequency at which maximum amplitude is achieved. The relatively broad bandwidth, high displacement, and temperature tolerance of TERFENOL-D made this upgrade of an existing chemical plant possible, significantly increasing plant capacity and reducing down time.

## Chemical Treatment

### Background

A remarkable number of chemical and biological processes have been reported to be affected by sonic energy (42). In this discussion sonic energy is considered to be any oscillatory motion imparted to a material over the range from a few hertz to at least 30 kHz. The promise of ultrasonic processing for new and improved products is now being realized due to the introduction of TERFENOL-D, which has overcome the power and reliability limitations of older solid-state transducer drive materials.

The following are some of the more important and better studied sonochemical processes:

Chemical Synthesis

Emulsification/emulsion breaking

Sterilization

Catalysis

Crystallization

Food processing

Polymer curing

Seed mutagenesis

Mixing

Some of these processes coincide with those mentioned in the physical treatment examples because those can also be regarded as sonochemical in a broader sense.

Some sonochemical processes are accelerated or improved by sonic energy, whereas others are possible only by application of sonic energy. It should be noted that few of the sonochemical processes studied in the laboratory have reached commercial application for reasons discussed later.

There have been very few studies of the mechanisms of sonochemical reactions. However, the vast majority of mechanisms are believed to involve cavitation (*40*). Cavitation can produce very large local variation in temperature and pressure at a point or surface at which a cavitation bubble forms (*43*).

Most of the published investigations of sonic energy have been performed at 20 kHz simply because the majority of commercial equipment available operates at this frequency. For some sonochemical mechanisms, the frequency band at which the process will operate is critical, but where maximum cavitation is needed to drive the process it is unlikely that frequency is critical.

For some processes there is evidence that frequency variation would be valuable. For example, several studies have shown that cells can be selectively broken by selection of the cell critical frequency (private communications). Little work has been done to determine the optimum frequency in most processes because frequency variation is not an option on available equipment.

## Terfenol-D in Sonochemistry

The following are barriers to successful scale-up of sonochemical processes to commercial production:

1. Production of sufficient energy, particularly for the continuous processes required for commercial application

2. Transducers with wide bandwidths

Both of these barriers are being successfully overcome by using equipment driven by giant magnetostrictive materials. The following are advantages that TERFENOL-D offers to sonochemistry:

High power, in a compact package

Broad bandwidth

Continuous operation

Long life/high reliability

High temperature tolerance

High efficiency

The key to taking sonochemical processes beyond the laboratory is higher continuous power production from the transducer. Nickel transducers are limited by relatively low inherent power and efficiency.

Piezoceramic transducers cannot be scaled-up because of internal heating, which if unchecked will destroy the ceramic. Generally, nickel transducers are limited to less than 1 kW output. Piezoceramic transducers can be operated continuously at a maximum of 1 or 2 kW, with short bursts to 4.5 kW, followed by a cooling period.

Commercial TERFENOL-D transducers are available to 6 kW, with continuous operation, and 25 kW systems are under development (39). The National Institute of Science and Technology of the United States has estimated that introduction of commercial 25-kW systems will result in the development of 46 new ultrasonic-based industries.

Some selected examples of TERFENOL-D sonochemistry are discussed in the following sections. These examples have been selected to provide insights into the possibilities for new commercial industrial processes.

## Petroleum Production and Processing

There have been many reports of beneficial effects of sonic energy in petroleum production and processing. Low-frequency (<1 kHz) energy has been reported to increase the flow of crude petroleum into producing wells (44). Energy in the 20-kHz frequency range has been reported to quickly and substantially reduce petroleum viscosity. Samples which are solid tar at room temperature (>30 Ns/m$^2$) can be quickly broken down into fluid with a viscosity about three times that of water (<0.01 Ns/m$^2$) (43, 45).

With the advent of equipment driven by TERFENOL-D, several petroleum production and benefaction processes are being commercialized. In each of these, the ability to continuously apply higher power is the key to success. However, many of the processes occur at the bottom of the well, where the ability of TERFENOL-D equipment to withstand the ambient temperature and pressure overcomes a previously insurmountable barrier. Lower viscosity crude petroleum is easier to pump and more valuable when it reaches the refinery. Increasing flow rate into the wells allows higher production rates and longer production life.

Flow rates through porous media have been shown to be frequency dependent (46). Most useful effects have been reported to occur between 100 and 2 000 Hz (47). TERFENOL-D transducers are particularly useful in these processes because the technology was specifically developed to provide higher power at lower frequencies compared to older transducer drivers.

## *Chemical Reactions*

The chemical literature contains many papers on chemical reactions that are initiated or accelerated by ultrasonic power. Here, it is assumed that acceleration of these chemical reactions is a form of catalysis. Several examples are discussed in which ultrasonic power initiates the formation or breaking of chemical bonds.

Several types of epoxies, both one-part and two-part formulations, have been shown to rapidly polymerize when high ultrasonic power is applied. Figure 5.18 shows an example of curing of one such epoxy commonly used to bond components to circuit boards. In the curing technique for this epoxy, a band of the uncured cement is applied between the components to be bonded, and the assembly is baked in an oven for 4 hr.

Figure 5.18 shows the results of application of ultrasonic power to the epoxy from an ETREMA Products, Inc., 6-kW ultrasonic system. Power was applied to an epoxy sample through a non-amplifying waveguide,

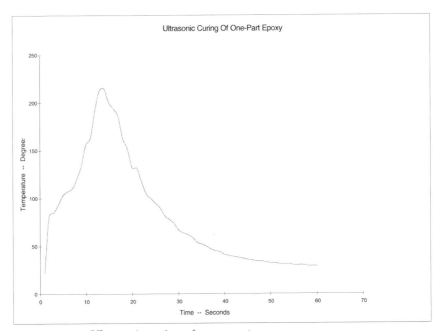

**FIGURE 5.18.**   Ultrasonic curing of a one-part epoxy.

such that the polymer sample was at an antinode. A thermistor on the back of the polymer "coupon" monitored change in temperature at the epoxy face opposite the waveguide face. The waveguide and the coupon were approximately 50 mm in diameter.

Note that the curing reaction begins within seconds after the power is applied. However, when power is left on, the coupon rapidly returns to ambient temperature. This suggests that the curing mechanism is not due to internal frictional heating of the epoxy. Most chemical reactions promoted by ultrasonic power are thought to depend on internal cavitation in the material.

## ELECTROMECHANICAL CONVERTERS

All magnetostrictive materials have the ability to convert magnetic energy to mechanical energy. In principle, this feature can be used in displacement or force sensors. Despite the huge strain of highly magnetostrictive materials, they are seldom used in sensor applications because a high coupling factor rather than absolute strain is important. Amorphous magnetostrictive alloys are ideal for this purpose. A high coupling factor of approximately 0.95 and a strain on the order of 30 $\mu$m/m with a figure of merit >200 000 make these alloys feasible for magnetostrictive sensor applications. It is possible to achieve a dynamic range of 104 dB between the maximal and minimal output signals with low dependence on temperature and good low-frequency response down to 0.01 Hz.

However, magnetostrictive amorphous alloys are beyond the scope of this book. Nevertheless, there are some activities regarding TERFENOL-D sensors that have been reported in Japan. A ring of magnetostrictive material made by powder technology with attached strain gauges was used for mechanical torque measurements (48). In another application, a magnetostrictive rod was used for a 3-D vibrational sensor, which is integrated with a neural network (49).

An odd application has been studied concerning the possibility of using magnetostrictive materials to generate electricity (50). Electric power would be generated in isolated and dangerous environments such as in those with high voltage potential and in radioactive locations.

# REFERENCES

1.  O. D. McMasters, J. D. Verhoeven, and E. D. Gibson, Preparation of TERFENOL-D by float zone solidification, *J. Magn. Magn. Mat.* **54–57**, 849–850, (1986).

2.  S. W. Meeks and R. W. Timme, Rare earth iron magnetostrictive underwater transducer, *J. Acoust. Soc. Am.* **62**(5), 1158–1164, (1977).

3.  J. L. Butler and S. J. Closek, Rare earth iron octagonal transducer, *J. Acoust. Soc. Am.* **67**(5), (1980).

4.  L. H. Royster, The flextensional concept: A new approach to the design of underwater acoustic transducers, *Appl. Acoust.* **3**, 117, (1970).

5.  R. Tenghamn and G. Engdahl, Flextensional transducers based on TERFENOL-D, *in* "International Symposium on Giant Magnetostrictive Materials and Their Applications," November 5–6, 1992, Tokyo pp. 83–89.

6.  L. Kvarnsjö, Underwater acoustic transducers based on TERFENOL-D, *J. Alloys Compounds* **258**, 123–125, (1997).

7.  K. R. Dhilsa, G. Markandeyulu, B. V. P. Subrahmanyeswara Rao, and K. V. S. Rao, Design and fabrication of low frequency giant magnetostrictive transducer, *J. Alloys Compounds* **258**, 53–55, (1997).

8.  I. Nakano, T. Tsuchiya, Y. Amitani, and T. Nakanishi, Giant magnetostrictive transducer and its application to acoustic monitoring of oceans, *in* "International Symposium on Giant Magnetostrictive Materials and Their Applications," November 5–6, 1992, Tokyo.

9.  H. Wakiwaka, K. Aoki, T. Yoshikawa, H. Kamata, M. Igarashi, and H. Yamada, Maximum output of a low frequency sound source using giant magnetostrictive material, *J. Alloys Compounds* **258**, 87–92, (1997).

10. K. Å. Magnusson and R. Tenghamn, Flextensional transducers for underwater communication, *ABB Rev.* **8–9**, 23–28, (1990).

11. H. Zhu, J. Liu, X. Wang, Y. Xing, and H. Zhang, Applications of TERFENOL-D in China, *J. Alloys Compounds* **258**, 49–52, (1997).

12. M. J. Goodfriend, K. M. Shoop, and T. Hansen, Applications of magnetostrictive TERFENOL-D, *in* "Conference Proceedings on Actuator '94," Bremen, Germany, pp. 214–217.

13. M. W. Hiller, M. D. Bryant, and J. Umegaki, Attenuation and transformation of vibration through active control of magnetostrictive terfenol, *J. Sound Vibration* **134**(3), 507–519, (1989).

14. K. Ohmata and Y. Nakahara, Semiactive damper using a magnetostrictive actuator, *in* "Nonlinear Phenomena in Electromagnetic Fields," Elsevier, Amsterdam, 1992.

15. K. Ohmata, M. Zaike, and T. Koh, A three-link arm type vibration control device using magnetostrictive actuators, *J. Alloys Compounds* **258**, 74–78, (1997).

16. M. G. Salloker, Active vibration damping with magnetostrictive actuators, *in* "Conference Proceedings on Actuator '98," Messe Bremen GmbH, Bremen, Germany, pp. 363–366.

17. L. Lecce and F. Franco, Results and prospects for the project MADAVIC, *in* "Conference Proceedings on Actuator '98," Messe Bremen GmbH, Bremen, Germany, pp. 389–394.

18. T. Cedell, Magnetostrictive materials and selected applications, Doctoral thesis, Lund University, Sweden, 1996.

19. M. Trimboli and R. Wimmel, Adaptive vibration isolation using a multifunctional actuator, *in* "Conference Proceedings on Actuator '94," Axon Technologie Consult GmbH, Bremen, Germany, pp. 261–265.

20. R. J. E. Smith, A. G. I. Jenner, A. J. Wilkingson, and R. D. Greenough, Magnetostrictive actuation performance under digital variable structure control, *J. Alloys Compounds* **258**, 101–106, (1997).

21. H. Eda and E. Ohmura, Development of TdDy(FeMn) alloys and ulra-precision actuators, *in* "International Symposium on Giant Magnetostrictive Materials and Their Applications," AMTDA, Tokyo, Japan, 97–101 (1992).

22. J. A. Dooley, C. A. Lindensmith, R. G. Chave, B. Fulz, and J. Greatz, "Cryogenic magnetostrictive actuators: Materials and applications," *in* "Conference Proceedings on Actuator '98," Messe Bremen GmbH, Bremen, Germany, pp. 407–410.

23. S. Okamoti and T. Mori, An application of giant magnetostrictive material to clutch with quick response, *in* "Electromagnetic Forces and Applications" (J. Tani and T. Takagi (Eds.) Elsevier, Amsterdam, 1992.

24. L. Clavier, O. Cugat, J. Delamare, and G. Reyne, Magnetostrictive inchworm linear actuator with two degree of freedom, *in* "Conference Proceedings on Actuator '98," Messe Bremen GmbH, Bremen, Germany, pp. 359–362.

25. L. Kiesewetter, TERFENOL in linear motors, *in* "Proceedings of Second International Conference on Giant Magnetostrictive and Amorphous Alloys for Sensors and Actuators," Marbella, Spain, October 12–14, 1988.

26. R. C. Roth, The elastic wave motor—A versatile terfenol driven, linear actuator with high force and great precision, *in* "International Symposium on Giant Magnetostrictive Materials and Their Applications", November 5–6, 1992, AMTDA, Tokyo, Japan pp. 109–116.

27. H. Tiberg, L. Kvarnsjö, and G. Engdahl, Magnetostrictive actuators in hydraulics, *in* "Conference Proceedings on Actuator '94," Axon Technologie Consult GmbH, Bremen, Germany, pp. 256–260.

28. G. Engdahl, Design of magnetostrictive devices using dynamic simulation, *in* "Proceedings of the Second International Conference on Giant Magnetostrictive and Amorphous Alloys for Sensors and Actuators," Marbella, Spain, October 12–14, 1988.

29. U.S. patent No. 4927334.

30.  M. Naoe and S. Nakamura, A magnetostrictive rod driven pump, *in* "International Symposium on Giant Magnetostrictive Materials and Their Applications," November 5–6, 1992, Tokyo, pp. 103–108.

31.  A. Ludwig, W. Pleging, and E. Quandt, Fabrication of a magnetostrictive membrane-type micropump by means of magnetron sputtering and laser assisted silicon micromaching, *in* "Conference Proceedings on Actuator '98," Messe Bremen GmbH, Bremen, Germany, pp. 376–379.

32.  J. M. Vranish, D. P. Naik, J. B. Restorff, and J. P. Teter, Magnetostrictive direct drive rotary motor development, *IEEE Trans. Magn.* **27**(6), 5355–5357, (1991).

33.  T. Akuta and Y. Ikegaya, Improved rotational-type actuators with TERFENOL rods, *in* "Conference Proceedings on Actuator '94," Axon Technologie Consult GmbH, Bremen, Germany, pp. 251–255.

34.  T. Akuta, Rotational-type actuators with TERFENOL-D rods, *in* "Conference Proceedings on Actuator '92," VDI/VDE, Bremen, Germany, pp. 244–248.

35.  F. Claeyssen, N. Lhermet, R. Le Letty, and L. Chouteau, A new resonant magnetostrictive rotating motor, *in* "Conference Proceedings on Actuator '96," Bremen, Germany, pp. 272–274.

36.  F. Clayssen, N. Lhermet, J. Betz, K. MacKay, D. Givord, E. Quandt, and H. Kronmuller, Linear rotating magnetostrictive micro-motors, *in* "Conference Proceedings on Actuator '98," Messe Bremen GmbH, Bremen, Germany, pp. 372–375.

37.  H. C. Hardee and N. L. Hardee, Magnetostrictive-driven fretting machines for interconnection research, *J. Alloys Compounds* **258**, 83–86, (1997).

38.  L. O. Pennander, "Wave-actuated Preform Infiltration Routines in MMC Production," Lund University, Sweden, 1998.

39.  Goodfriend, Weisensel, and Hansen, Development & Applications of very high power (>6 kW) ultrasonic transducers, *in* "Proceedings of the 28th Annual Ultrasonic Industry Association Symposium," UIA Valley Forge, PA.

40.  K. S. Suslik, Applications of ultrasound to material chemistry, *MRS Bull.*, 29–34 (1995; April).

41.  L. Sandlund, M. Fahlander, T. Cedell, A. E. Clark, J. B. Restorff, and M. Wun-Fogle, Properties and applications of low conductivity composite TERFENOL-D, *in* "Conference Proceedings on Actuator '94," Axon Technologie Consult, GmbH, Bremen, Germany, pp. 210–213.

42.  B. Jones, "Sonochemical Processing Using High Power Ultrasonics," Processing Technology, 1997.

43.  A. E. Crawford, "Ultrasonic Engineering, with Particular Reference to High Power Applications," Butterworth, London, 1960.

44.  Society of Petroleum Engineers of AIME, The effect of ultrasonic energy on the flow of fluids in porous media, Paper No. SPE 1316, 2nd Regional Eastern Regional Meeting, 1965.

45.  Hunter and Bolt, "Sonics," p. 95. Wiley, New York, 1955.

46.  Sadeghi, Lin, and Yen, Sonochemical treatment of fossil fuels, *in* "Energy Sources," Vol. 16, pp. 438–449, Taylor & Francis, London, 1992.

47.  M. S. Wong, Influence of ultrasonic energy upon the rate of flow of liquids through porous media, PhD thesis (H.V. Fairbanks advisor), 1969.

48.  A. R. Rathore and T. Mori, "A study on torque induction in the giant magnetostrictive material", *J. Alloys Compounds* **258**, 93–96, (1997).

49.  Y. Yamamoto, H. Eda, T. Mori, and A. Rathore, Three-dimensional magnetostrictive vibration sensor: development, analysis, and applications, *J. Alloys Compounds* **258**, 107–113, (1997).

50.  A. Lundgren, H. Tiberg, L. Kvarnsjö, G. Engdahl, and A. Bergqvist, A magnetostrictive electric generator, *IEEE Trans. Magn.* **29**, 3150, (1993).

# CHAPTER 6

# Giant Magnetostrictive Thin Film Technologies

Eckhard Quandt
*Center of Advanced European Studies and Research*
*Bonn, Germany*

For the realization of advanced microsystems—systems that combine electronics with sensorial and actuator functions employing miniaturization to derive highly integrated, complex, and intelligent devices—the development of suitable microactuators is a key issue (1). Because classical actuator concepts such as electromotors are limited in size reduction mostly due to friction losses, transducer materials which directly convert electrical into mechanical energy (magnetostrictive, piezoelectric, and shape memory materials) present a promising way of realizing new actuators in microsystems because they can easily be scaled down to small lateral dimensions. For these applications thin film deposition techniques of the transducer materials (2) are of special interest because they are capable of cost-effective mass production compatible to microsystem process technologies and they avoid packaging and interconnecting technologies. These thin film transducer materials can be used to give the appropriate actuation to existing micropatterns (cantilevers or membranes) which result in deflection or bending of these elements. In comparing the different transducer materials, giant magnetostrictive thin films are very promising for these microactuator devices because they combine high-energy output, high-frequency and remote-control operation, simple actuator layout, and comparably low process temperatures during fabrication. This potential of giant magnetostrictive thin films has resulted in a rapidly increasing interest in such films for microsystems applications during the past few years (3–6). Due to the interest in microsystem applications, research has focused on materials in thin film form exhibiting low-field magnetostriction and soft magnetic properties.

The development of low-field giant magnetostrictive thin films is based on the rare earth iron (RE-Fe$_2$) Laves phases, in particular (Tb$_{0.3}$Dy$_{0.7}$)Fe$_2$, which is, under the name of TERFENOL-D, well-known to exhibit giant magnetostriction in combination with reduced magneto-crystalline anisotropy and which was described in detail in previous chapters.

In order to exploit giant magnetostriction at reasonably low fields, which are obtainable in microsystem applications, it is essential to further lower the macroscopic anisotropy. Successful attempts include the use of amorphous, nanocrystalline, or multilayered Tb-based positive mag-netos-trictive or Sm-based negative magnetostrictive materials. Beneath the RE Tb or Sm the magnetic transition metals (TMs) Fe, Co, or FeCo alloys have been used with different RE/TM ratios to form the RE–TM exchange coupling which allows giant magnetostrictions at room temperature or higher.

In comparison to bulk giant magnetostrictive alloys, a large variety of different thin film materials, which are predominantly influenced in their magnetic properties by the choice of the substrate, indicates the need to review the most promising amorphous, nanocrystalline, and multilayered giant magnetostrictive thin film materials, their typical fabrication routes, and some of the specific characterization techniques, especially the magnetostriction measurements. This chapter also provides some examples of applications of giant magnetostrictive thin films in the field of microsystem technology.

## GIANT MAGNETOSTRICTIVE THIN FILM MATERIALS

According to the previous chapter, this review is restricted to low-field giant magnetostrictive materials in thin film form, especially amorphous, nanocrystalline, and multilayered RE–TM materials. For most applications magnetostrictive materials in thin film form must incorporate the following main features:

- Giant magnetostriction to be exploited in considerably low magnetic fields
- Sufficiently high magnetic ordering temperatures well above the maximum ambient temperature of use

Special attention must be paid to the combination of these two features because, in general, successful approaches to reduce the magnetic saturation field also decrease the magnetic ordering temperature. Therefore, the development routes to achieve these materials' properties only represent compromises between low-field giant magnetostriction and high Curie temperature, and the actual choice of the best suited material will generally depend on the specifications of the application.

As a result, the main development goal of giant magnetostrictive thin films is to obtain materials that exhibit low magnetic saturation fields while almost preserving magnetic ordering. Most important for achieving low magnetic saturation fields is the orientation of the magnetic easy axis with respect to the direction of the driving magnetic field. Because most applications are restricted to in-plane driving field directions in order to avoid large demagnetization losses, in general in-plane magnetic easy axes are required for magnetostrictive thin film materials. However, in this case the substantial in-plane saturation magnetostriction $\lambda_{s,\|}$ is reduced due to the magnetization processes involved. Although magnetization caused by $180°$ domain wall motions, which are important processes in the case of in-plane magnetic easy axes, do not contribute to the magnetostriction, the $90°$ domain rotations predominantly present for materials exhibiting perpendicular anisotropy induce maximum magnetostrictive strains (7). As an optimum, a material having a nonisotropic domain contribution at its demagnetized state, with directions being all inplane but perpendicular to the driving field direction, will show the maximum possible magnetostrictive strain at low fields.

The direction of the magnetic easy axis of a magnetostrictive film is directly related to the mechanical stress of the film. For an isotropic ferromagnet with minimum magnetoelastic energy (8)

$$E_{\mathrm{me}} = -\frac{3}{2}\sigma\lambda_{\mathrm{s}}\left(\cos^2\alpha - \frac{1}{3}\right) \tag{6.1}$$

the orientation of the magnetic easy axis ($\alpha$; angle between the directions of the magnetization and application of the stress) depends on the sign of the film stress $\sigma$ for a material with a given saturation magnetostriction $\lambda_{\mathrm{s}}$. This equation implies an in-plane magnetic easy axis for a positive product $\sigma\lambda_{\mathrm{s}}$, whereas a negative product results in a perpendicular anisotropy. Therefore, in the case of positive magnetostrictive materials tensile stress is required for in-plane anisotropy, whereas negative magnetostrictive films should be deposited with compressive stress to result in the same anisotropy. Because the sign and the magnitude of the film's stress can be controlled by the fabrication conditions (9), the thermal expansion coefficient of the substrate (11), or by stress annealing (10), an in-plane magnetic easy axis can be adjusted, in general, for these

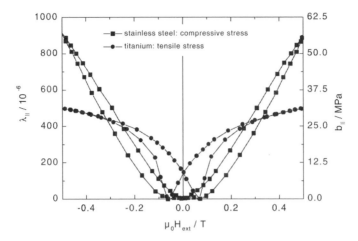

**FIGURE 6.1.** In-plane magnetostriction and magnetoelastic coupling coefficient ($E_f = 80$ GPa, $\nu_f = 0.3$) of nanocrystalline TbDyFe films (mean grain size 9 nm) deposited under the same conditions onto Ti and stainless steel.

materials. Figure 6.1 verifies the possibility of adjusting the magnetic easy axis by the choice of the substrate in the case of nanocrystalline $(Tb_{0.3}Dy_{0.7})Fe_2$ films. The results reveal that the sample with in-plane magnetic easy axis (Ti substrate) exhibits lower saturation magnetostriction obtained at lower saturation fields. Furthermore, the magnetostriction curves show that despite the small grain size of about 9 nm, which should be smaller than the ferromagnetic exchange length, the hysteresis of these nanocrystalline TbDyFe films is too large for most thin film applications. Additional reduction of the magnetic saturation field, which is defined by (11)

$$H_s = \frac{2K}{M_s} \tag{6.2}$$

is related to routes reducing the magnetic anisotropy constant $K$ and/or by increasing the saturation magnetization $M_s$.

For the most common approach to reduce the macroscopic anisotropy, amorphous RE–TM films (TM = Fe, Co, and $Fe_xCo_{1-x}$) are used. These amorphous materials show a saturation polarization of about $J_s = 0.4$ T, which is about half the magnitude of the crystalline polarization, and a Curie temperature which is, compared to the crystalline compounds, lower for the amorphous RE-Fe but higher for the amorphous RE-Co materials (10). Although some data on Tb–Co (10) and the negative magnetostrictive system Sm–Fe (12) are available, most of the investigations on these binary materials were concentrated on amorphous Tb–Fe thin films.

The magnetic and magnetostrictive properties of amorphous Tb–Fe films as a function of the Tb content have been the subject of many investigations (13–17). For these investigations, amorphous, single-phase Tb–Fe films were prepared in the composition range between 10 and 50 atomic percent (at.%) Tb, whereas films with higher Tb contents resulted in two-phase (Tb–Fe and Tb) amorphous films (18). Figure 6.2 shows the results of different investigations comparing the dependence of the magnetization on Tb–Fe films versus their Tb contents. Due to the antiferromagnetic coupling of Tb and Fe, the amorphous film exhibits the room-temperature magnetization compensation at about 22 at.% Tb. For composition higher than approximately 42 at.% Tb, the reduction of the magnetization, which decreases to zero at approximately 48 at.% Tb, corresponds to the decrease of the Curie temperature approaching room temperature at approximately 48 at.% Tb. Coercitivity exhibits its maximum at the room-temperature magnetization compensation composition and decreases with the Tb content (Fig. 6.3) resulting in minimum coercivities of 5 mT (16).

The in-plane magnetostriction values measured in the same compositional range (Fig. 6.4) reveal that giant magnetostriction is mainly obtained at rare earth-rich compositions having a Tb content that is

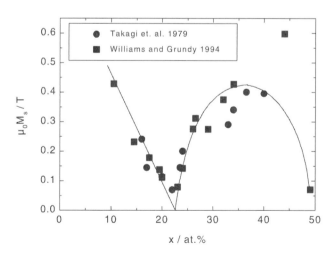

**FIGURE 6.2.** Compositional variation of the saturation magnetic polarization of amorphous $Tb_xFe_{1-x}$ films (14, 16, 17).

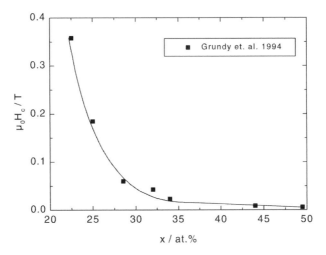

**FIGURE 6.3.** Compositional variation of the magnetic hysteresis of amorphous $Tb_xFe_{1-x}$ films (16).

significantly higher than those of the crystalline magnetostrictive Tb–Fe compounds $Tb_2Fe_{17}$, $Tb_5Fe_{23}$, and $TbFe_3$ and, most important, the Laves phase $TbFe_2$, although it must be noted that the applied magnetic fields of the different investigations were usually not sufficient to saturate the Tb–

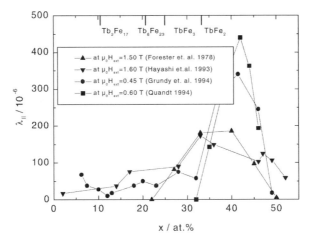

**FIGURE 6.4.** Compositional variation of the in-plane magnetostriction of amorphous $Tb_xFe_{1-x}$ films at different external applied fields (13, 15, 16, 19).

Fe films. Nevertheless, the obtained data clearly indicate that magnetostriction in amorphous Tb–Fe films increases dramatically for compositions higher than 30 at.% Tb to a maximum between 40 and 43 at.% Tb and decreases with higher Tb contents toward the composition having the Curie point at room temperature (48 at.% Tb).

The magnetization data revealed that under normal conditions, perpendicular or, for higher Tb contents (> 37 at.%), unclear orientations of the magnetic easy axis are common (15, 16), which can be related, considering the minimum of the magnetoelastic energy (Eq. 6.1), to the usually compressive stress state of the films. By applying a radio frequency (RF) bias voltage during magnetron sputtering in combination with most commonly used Si substrates, tensile film stress can be obtained which leads to an in-plane magnetic easy axis and to a dramatically improved magnetostriction at low fields (Fig. 6.5).

Assuming the same local environment in the amorphous and the crystalline states, another approach to lower the remaining anisotropy is to eliminate the fourth-order anisotropy by Tb/Dy substitution (10, 21). This approach leads to an additional reduction in the magnetic saturation field, but the lower Curie temperature of Dy-based amorphous RE–TM alloys leads to a significant reduction in the saturation magnetostriction resulting in no or only a very small gain in low-field magnetostriction (Fig. 6.6) of these amorphous films.

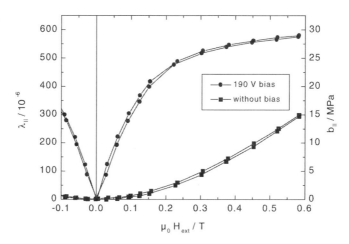

**FIGURE 6.5.** In-plane magnetostriction and magnetoelastic coupling coefficient loops ($E_f = 50\,\text{GPa}$, $\nu_f = 0$) of amorphous TbFe films as a function of the applied RF bias voltage during film deposition.

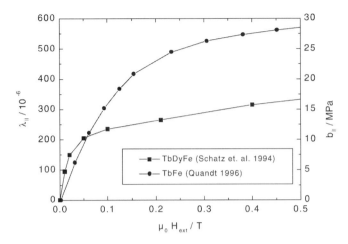

**FIGURE 6.6.** In-plane magnetostriction and magnetoelastic coupling coefficient $(E_f = 50\,\text{GPa}, \nu_f = 0)$ of an amorphous TbFe film (20) in comparison to an amorphous TbDyFe film (7).

Compared to Tb-based amorphous films, there is much less information available on SmFe thin films. The results of Honda $et\ al.$ (12) reveal that giant magnetostriction can be obtained in a wider composition range (between 25 and 50 at.%), although the saturation magnetostriction shows its maximum at 38.6 at% Sm, which is in accordance with both the TbFe results and our measurements (22). The easy axis was found to be in-plane for the magnetron sputtered films, which is related to the usually compressive stress state of these films. Applying a bias voltage, it was possible to alter the stress state to tensile stress, which resulted in a perpendicular anisotropy due to the negative magnetostriction but an increased saturation magnetostriction (Fig. 6.7). In the case of in-plane magnetic easy axis, the saturation magnetostriction of approximately $-300 \times 10^{-6}$ is significantly lower than that of the best Tb-based positive magnetostrictive materials $(\lambda_s \approx 700 \times 10^{-6})$ but is in accordance to published data of other Sm–Fe thin films (12). In contrast to Tb–Fe films, crystallization of Sm–Fe films does not result in a higher saturation magnetostriction; only the hysteresis was found to be significantly increased (Fig. 6.8).

The second way to lower the macroscopic anisotropy is to increase the saturation magnetization. Giant magnetostrictive thin films discussed so far have low magnetizations due to their ferrimagnetic nature. For the RE–TM compositions of interest (30–45 at.% total RE content), the RE moments dominate and so an increase in the TM content will only further

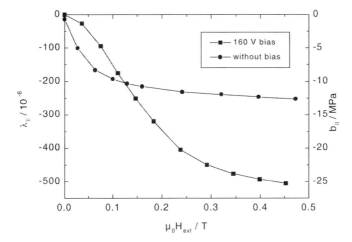

**FIGURE 6.7.** In-plane magnetostriction and magnetoelastic coupling coefficient $(E_f = 50\,\text{GPa}, \nu_f = 0)$ of amorphous Sm–Fe films as a function of the applied RF bias voltage during film deposition.

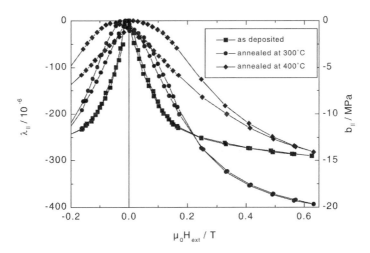

**FIGURE 6.8.** In-plane magnetostriction and magnetoelastic coupling coefficient loops $(E_f = 50\,\text{GPa}, \nu_f = 0)$ of Sm Fe films as a function of the annealing temperature.

reduce the magnetization, whereas an additional increase in the RE content results in a lowering of the Curie temperature, which is inappropriate for most applications. Therefore, the saturation magnetization cannot be notably increased using homogeneous amorphous RE–TM alloys.

However, using multilayers it is possible to engineer new composite materials which have properties that overcome these limitations. To create such a material, the layers must be magnetically coupled, which means that the layer thicknesses must be thinner than their magnetic exchange length to avoid domain wall formation at their interfaces. In this example, two materials—a giant magnetostrictive amorphous RE–TM alloy and a magnetically very soft material with a very high magnetization (e.g. Fe and Fe–Co)—are combined in a multilayer arrangement. As a consequence, the magnetic properties of such a multilayer system will be defined by the average of those of each individual layer, which leads to an important increase in saturation magnetization resulting in a considerable decrease in the saturation field while retaining relatively large values of the magnetostriction (23). This multilayer material is a result of a typical individual layer thickness in the range of 3–10 nm restricted to thin film applications but it offers approximately a 100% gain of saturation magnetization of the total system; magnetostriction and anisotropy are reduced by a factor of two assuming equal thickness of both layers and antiferromagnetic coupling of the layers. This leads to a reduction of the saturation field by a factor of four and an improvement in $\lambda/H_s$ by a factor of two. This change in magnetic properties is demonstrated in the case of a TbFe/Fe multilayer film (Figs. 6.9 and 6.10). Furthermore, by varying the thickness ratio the saturation field can be engineered in a wide range according to the specifications of the specific application. Figure 6.11 summarizes the theoretical change in the magnetic properties of a TbFe/Fe multilayer system in comparison to the experimental data.

Another important achievement in the saturation magnetostriction can be reached by using soft magnetic, high-polarization materials which exhibit positive magnetostriction. In the case of $Fe_{0.5}Co_{0.5}$ saturation magnetostriction exceeding $100 \times 10^{-6}$ leads to an improvement of the in-plane magnetostriction by a factor of two (Fig. 6.12), whereas the other magnetic properties are almost unchanged (24). Furthermore, these multilayers show an interesting large $\Delta E$ effect which is developed in a field of <20 mT (26), exemplarily shown for a TbFe/Fe multilayer in Fig. 6.13 (page 335).

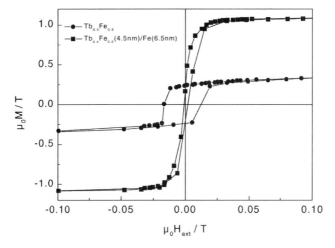

**FIGURE 6.9.** In-plane magnetization loops of a TbFe/Fe multilayer film in comparison to an amorphous TbFe film.

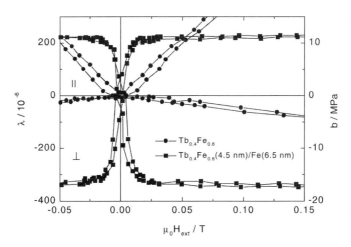

**FIGURE 6.10.** Magnetostriction and magnetoelastic coupling coefficient loops ($E_f = 50\,\text{GPa}, \nu_f = 0$) of a TbFe/Fe multilayer film in comparison to an amorphous TbFe film.

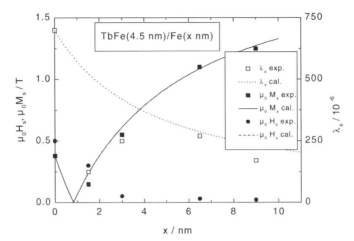

**FIGURE 6.11.** Magnetic properties of TbFe (4.5 nm)/Fe ($x$ nm) multilayer films as a function of the Fe layer thickness $x$ in comparison to a simple model (straight lines) that assumes ferrimagnetic exchange coupling of the layers.

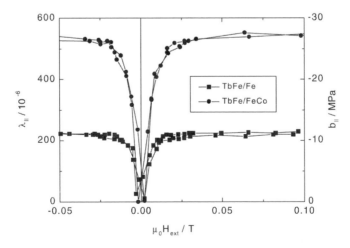

**FIGURE 6.12.** Magnetostriction and magnetoelastic coupling coefficient loops ($E_f = 50\,\text{GPa}, \nu_f = 0$) of a TbFe/Fe in comparison to a TbFe/FeCo multilayer film.

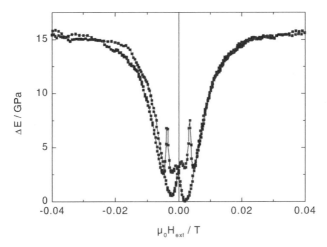

**FIGURE 6.13.** Change of Young's modulus ($\Delta E$ effect) of a TbFe/Fe multilayer film that is applied to an in-plane external magnetic field (25).

## FABRICATION

Giant magnetostrictive thin films are made using several physical vapor deposition techniques, including electron beam evaporation (13), laser ablation (27), and, most commonly used, sputtering (28, 29), especially magnetron sputtering (9, 10, 19). Important advantages of sputtering processes include

- The high energy of the coated atoms leading to good film adhesion
- The capacity to pass materials into the vapor phase while largely preserving the stochiometry of the target
- The high deposition rates especially achieved with sputtered metals
- The possibility of forming compounds by reactive sputtering, in which one or more species are introduced in the gas phase
- The well-developed technology to fabricate large area coatings with high homogeneity

Thus, in this chapter considerable reference is made to sputtered giant magnetostrictive thin films.

Sputtering techniques for the preparation of these films include DC or RF magnetron, diode, triode, or ion beam sputtering using either multitarget arrangements with pure element targets or single-target techniques using cast- or hot-pressed composite targets or mosaic targets. For magnetron sputtering the typical power per target area during sputtering is $0.1\,W/cm^2$, the Ar pressure of the plasma discharge is typically 0.2–10 Pa, and the target–substrate distance is 5–10 cm, leading to maximum deposition rates up to 20 $\mu$m/hr.

Typically, the substrates used are subject to special cleaning processes before the film deposition, for example, cleaning in an ultrasonic acetone bath and/or by sputter etching using an $Ar^+$ bombardment. The targets are normally cleaned by presputtering in case they are exposed to air in order to avoid high oxygen contents in the films due to oxide layers of the targets.

Multilayers can be fabricated by placing the substrates on a turntable operated in a stop-and-go mode, whereas the individual layer thickness is determined by the holdup time of the turntable at the corresponding target positions. Other techniques for multilayer fabrication include arrangements in which the targets are placed on a carousel or, by the use of different sources, which are directed onto a fixed substrate position. In this case the sources are sequentially selected either by switching of the sputtering power or by a shutter system. Depending on the required crystallographic state and stress of the films, the substrate temperatures are adjusted by cooling or heating devices up to 900 K (30). Stress annealing and crystallization as desired under magnetic field up to 2.2 T (10) are performed in HV furnaces, whereas in some cases the films are protected against oxidation by different protection layers such as W or Cr.

Protection against oxidation is needed not only during annealing processes but also for actuators being exposed to higher ambient temperatures or corrosive media. For example, the oxidation velocity of amorphous $Tb_{0.26}Fe_{0.74}$ films are dramatically increased ambient temperatures of 180°C (31), whereas for room-temperature applications the immediately formed oxide layer of about 30-nm thickness almost prevents further oxidation in-depth (9).

Additional materials needed for the fabrication of magnetostrictive thin film actuators include special antidiffusion, adhesive, and protection layers for lateral patterning of magnetostrictive films by lithography and etching processes. Because the choice of suitable materials very much depends on the involved substrate materials and micromachining processes, no universally valid selection can be given.

## CHARACTERIZATION OF MAGNETOSTRICTIVE THIN FILMS

The microstructure and the composition of the magnetostrictive materials in thin film form are determined by thin film characterization techniques. Wavelength or energy-dispersive X-ray, Rutherford backscattering, and Auger electron spectrometry (AES) analyses are used for the determination of the composition, whereas the crystallographic structure is investigated by transmission electron microscopy or X-ray diffraction. AES depth profiles are a powerful tool to investigate oxidation, interdiffusion, or the multilayered structure (Fig. 6.14). Elastic properties of the films, especially the Young's modulus, are determined by different techniques such as nanoindentation, beam bending (32), and resonance (33) methods. Some of these techniques are also applied to a magnetic field in order to measure the $\Delta E$ effect of the different magnetostrictive thin film materials.

Magnetic properties of magnetostrictive films are measured by vibrating sample magnetometry or by SQUID magnetometers in fields of up to 8 T between cryogenic and temperatures as high as the Curie temperature.

Special attention must be paid to the magnetostriction measurement in thin films; two formulas (34, 35) for the evaluation of the most

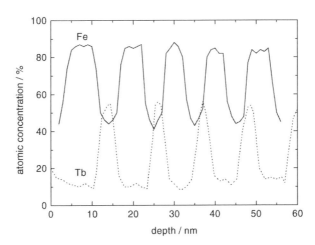

FIGURE 6.14. AES depth profile of an as-deposited TbFe (4.5 nm)/Fe(9 nm) multilayer film.

commonly used cantilevered-substrate technique (36) are used. This technique allows the in-plane magnetoelastic coupling coefficient of the films, $b$, to be directly determined (23) using

$$b = \frac{\alpha\, t_s^2}{L\, t_f} \frac{E_s}{6(1 + \nu_s)} \tag{6.3}$$

where $\alpha$ is the deflection angle of the beam coated with a magnetostrictive film as a function of the applied magnetic field, $L$ denotes the sample length, $t_f$ and $t_s$ are the thicknesses of film and substrate, respectively, and $E_s$ and $\nu_s$ are the Young's modulus and the Poisson ratio for the substrate.

In comparison to magnetostriction, the magnetoelastic coupling energy is independent of Young's modulus $E_f$ and Poisson ratio $\nu_f$ of the film, which are difficult to determine reliably. However, most published data provide values only for the magnetostriction, which is proportional to the magnetoelastic coupling energy by (34)

$$\lambda = \frac{-b(1 + v_f)}{E_f} \tag{6.4}$$

The different beam-bending techniques can be classified into two main techniques that are applied for measuring the deflection angle: capacitive (37) and different optical methods (38). Additional reported techniques include the incorporation of strain gages (39), the use of special piezoelectric substrates to compensate the beam bending (40), or evaluation of the initial susceptibility of the film, which allows the saturation magnetostriction to be determined (17). Nevertheless, the cantilevered-substrate technique is a standard for the determination of magnetostriction in thin films, especially because the obtained magnetoelastic coupling coefficient $b$ can be directly used for the modeling of magnetostrictive thin film actuators.

## APPLICATIONS OF MAGNETOSTRICTIVE THIN FILMS

Most of the applications of giant magnetostrictive thin films use the bending of coated cantilevers, beams, plates, or membranes. Applications include switches, valves, fluid-jet deflectors (4), optical beam deflectors, and magnetometers (41) in the case of cantilevers, (ultrasonic) motors in the case of beams and plates (42, 43), and microvalves (4) or pumps (44) in the case of membranes. All these applications, in comparison to other applications, have the main advantage that the magnetostrictive thin film solution can possibly be incorporated with remote-control operation.

For the practical realization of this kind of actuator the availability of both positive Tb-based and negative magnetostrictive Sm-based films is of importance. Because they are deposited on opposite sides of thin substrates, the combination of positive and negative magnetostrictive thin films forms bimorphs (19) enhances the total effect, reduces initial curvature of the cantilevers or membranes due to film stress, and almost eliminates temperature coefficients due to bimetallic effects. Furthermore, if the fabrication of the microsystems allows film deposition only on one side of micromachined substrates, the availability of both positive and negative magnetostrictive materials will be essential for structures which require convex and concave bending such as ultrasonic motors.

Because sophisticated giant magnetostrictive materials in thin film form have only recently become available on a laboratory scale, fabrication and standardization are non existent; most of the possible applications are either just demonstrations of approaches or have not been realized by now.

As demonstrators, two types of linear standing-wave ultrasonic motors have been fabricated and characterized: a bimorphous type with a polyimide substrate (45) and a silicon beam with teeth driven by magnetostrictive stripes (47) (Fig. 6.15). The maximum speed in the range of several millimeters per second is in both cases obtained at the resonance frequency of a few hundred Hertz and a driving field of 15–30 mT.

Special importance for the realization of advanced microsystems is related to the availability of fluid handling systems comprising micro-valves, micropumps, and flow rate detectors. Applications for fluid handling range from chemical process technology regarding fluid analysis to medical applications, in which the remote-control feature of magnetostrictive thin film actuators might offer important advantages compared to other mechanisms. Based on FEM calculations for the optimization of magnetostrictive bimorphous membranes (22), micro-pumps have been developed having either inlet and outlet check valves (46) and or specially formed fluid channels which result in a preferential flow direction (47). Using the latter micropump, a flow rate exceeding $100\,\mu l/min$ was achieved in an alternating radial magnetic field of approximately 20 mT operated at a resonance frequency of 200 Hz (Fig. 6.16) (48).

a

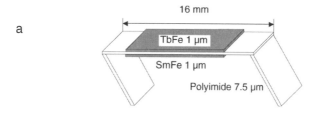

16 mm

TbFe 1 μm

SmFe 1 μm

Polyimide 7.5 μm

b

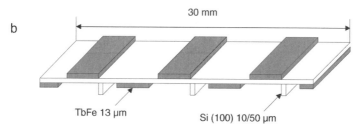

30 mm

TbFe 13 μm

Si (100) 10/50 μm

**FIGURE 6.15.** Schematic view of magnetostrictive linear micromotors: (a) a bimorphous type with a polyimide substrate (45) and (b) a silicon beam with teeth driven by magnetostrictive stripes (42).

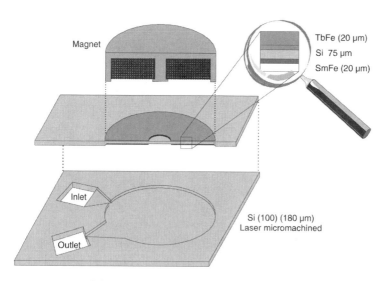

Magnet

TbFe (20 μm)
Si  75 μm
SmFe (20 μm)

Inlet

Outlet

Si (100) (180 μm)
Laser micromachined

**FIGURE 6.16.** Schematic cross section of a magnetostrictive membrane-type micropump using a bimorphous membrane operated with a radial magnetic field.

# REFERENCES

1. W. Bennecke, *Silicon-microactuators: Activation mechanisms and scaling problems, in* "Proc. IEEE," 91CH2817 p. 46, 1991.

2. E. Quandt and H. Holleck, *Materials development for thin film actuators, Microsystems Technol.* **1**, 178–184, (1995).

3. T. Honda, K. I. Arai, and M. Yamaguchi, *Fabrication of actuators using magnetostrictive thin films, in* "Proceedings MEMS 1994," Oiso, Japan, p. 51.

4. G. Flik, M. Schnell, F. Schatz, and M. Hirscher, *Giant Magnetostrictive Thin Film Transducers for Microsystems, in* "Proceedings Actuator 94," Bremen Germany, p. 232.

5. K. I. Arai, W. Sugawara, K. Ishiyama, T. Honda, and M. Yamaguchi, *Fabrication of small flying machines using magnetic thin films, IEEE Trans. Magn.* **31**, 3758, (1995).

6. J. Betz, K. Mackay, J.-C. Peuzin, B. Halstrup, and N. Lhermet, *Torsion based, drift-free magnetostrictive microactuators, in* "Proceedings Actuator '96," Bremen, Germany, p. 283.

7. F. Schatz, M. Hirscher, M. Schnell, G. Flik, and H. Kronmüller, *Magnetic anisotropy and giant magnetostriction of amorphous TbDyFe films, J. Appl. Phys.* **76**, 5380, (1994).

8. E. du Trémolet de Lacheisserie, "*Magnetostriction, Theory and Applications of Magnetoelasticity,*" p.198ff. CRC Press, Boca Raton; 1993.

9. E. Quandt, B. Gerlach, and K. Seemann, *Preparation and applications of magnetostrictive thin films, J. Appl. Phys.* **76**, 7000, (1994).

10. N.H. Duc, K. Mackay, J. Betz, and D. Givord, Giant magnetostriction in amorphous $(Tb_{1-x}Dy_x)Fe_{0.45}Co_{0.55})_y$ films, *J. Appl. Phys.* **79**, 973, (1996).

11. C.-W. Chen, "*Magnetism and Metallurgy of Soft Magnetic Materials,*" Chapter 3. North-Holland, Amsterdam, 1977.

12. T. Honda, Y. Hayashi, K. I. Arai, K. Ishiyama, and M. Yamaguchi, Magnetostriction of sputtered Sm-Fe thin films, *IEEE Trans. Magn.* **29**, 3126 (1993).

13. D. W. Forester, C. Vittoria, J. Schelleng, and P. Lubitz, *Magnetostriction of amorphous $Tb_xFe_{1-x}$ thin films, J. Appl. Phys.* **49**, 1966, (1978).

14. H. Takagi, S. Tsunashima, S. Uchiyama, and T. Fujii, Magnetostriction of amorphous Gd-Fe and Tb-Fe sputtered films, *Jpn. J. Appl. Phys.* **18**, 399, (1979).

15. E. Quandt, Multitarget sputtering of high magnetostrictive Tb-Dy-Fe films, *J. Appl. Phys.* **75**, 5653, (1994).

16. P. J. Grundy, D. G. Lord, and P. I. Williams, Magnetostriction in TbDyFe thin films, *J. Appl. Phys.* **76**, 7003, (1994).

17. J. Huang, C. Prados, J. E. Evetts, and A. Hernando, Giant magnetostriction of amorphous $Tb_xFe_{1-x}$ $(0.10 < x < 0.45)$ thin films and its correlation with perpendicular anisotropy, *Phys. Rev. B* **51**, 297, (1995).

18.  Y. S. Choi, S. R. Lee, S. H. Han, H. J. Kim, and S. H. Lim, The magnetic properties of Tb-Fe-(B) thin films fabricated by rf magnetron sputtering, *J. Alloys. Comp.* **258**, (1997), 155.

19.  Y. Hayashi, T. Honda, K. I. Arai, K. Ishiyama, and M. Yamaguchi, Dependence of magnetostriction of sputtered Tb-Fe films on preparation conditions, *IEEE Trans. Magn.* **29**, 3129, (1993).

20.  E. Quandt, *Giant magnetostrictive thin film materials and applications, J. Alloys Comp.* **258**, 126, (1997).

21.  P. I. Williams and P. J. Grundy, Magnetic and magnetostrictive properties of amorphous rare earth-transition metal alloy films, *J. Phys. D Appl. Phys.* **27**, 897, (1994).

22.  E. Quandt, and K. Seemann, Fabrication and simulation of magnetostrictive thin-film actuators, *Sensors and Actuators,* **A50**, 105, (1995).

23.  E. Quandt, A. Ludwig, J. Betz, K. Mackay, and D. Givord, Giant magnetostrictive spring magnet type multilayers, *J. Appl. Phys.* **81**, 5420, (1997).

24.  E. Quandt, and A. Ludwig, Giant magnetostrictive multilayer thin film transducers, *in* "Proceedings of the MRS Fall meeting 1996", *MRS Symp. Proc.* **459**, 565, (1997).

25.  K. Mackay, personal communication.

26.  K. Mackay, personal communication.

27.  J. P. Hayes, H. V. Snelling, A. G. Jenner, and R. D. Greenough, Thin magnetoelastic films prepared by pulsed TEA $CO_2$ laser ablation deposition, *J. Magn. Magn. Mat.* (1996).

28.  F. Schatz, M. Hirscher, G. Flik, and H. Kronmüller, Magnetic properties of giant magnetostrictive TbDyFe films, *Phys. Stat. Solidi(a)* **137**, 197, (1993).

29.  H. H. Uchida, V. Koeninger, H. Uchida, M. Wada, H. Funakura, Y. Matsumura, T. Kurino, and H. Kaneko, Preparation and characterization of $(Tb,Dy)Fe_2$ giant magnetostrictive thin films for surface acoustic wave devices, *Alloys Compounds* **211/212**, 455, (1994).

30.  H. Uchida, M. Wada, A. Ichikawa, Y. Matsumura, and H. H. Uchida, Effects on the preparation method and condition on the magnetic and giant magnetostrictive properties of the $(Tb,Dy)Fe_2$ thin films, in "Proceedings Actuator 96," Bremen, Germany, p. 275.

31.  R. B. van Dover, E. M. Gyorgy, R. P. Frankenthal, M. Hong, and D. J. Siconolfi, Effect of oxidation on the magnetic properties of unprotected TbFe thin films, *J. Appl. Phys.* **59**, 1291, (1986).

32.  J. Mencik, E. Quandt, and D. Munz, Elastic modulus of TbDyFe films—A comparison of nanoindentation and bending measurements, *Thin Solid Films* **287**, 208, (1996).

33.  Q. Su, J. Morillo, Y. Wen, and M. Wuttig, Young's modulus of amorphous Terfenol-D thin films, *J. Appl. Phys.* **80**, 3604, (1996).

34. E. du Trémolet de Lacheisserie, and J. C. Peuzin, Magnetostriction and internal stresses in thin films: The cantilever method revisited, *J. Magn. Magn. Mat.* **136**, 189, (1994).

35. E. Klokholm and C. V. Jahnes, Comments on "Magnetostriction and Internal Stresses in Thin Films: The Cantilever Method Revisited," *J. Magn. Magn. Mat.* **152**, 226, (1996).

36. E. Klokholm, *IEEE Trans. Magn.* **MAG-12**, 819, (1976).

37. E. Klokholm, Instrument for measuring magnetostriction, *IBM Technical Disclosure Bull.* **19**, 4030, (1977).

38. A. C. Tam and H. Schroeder, Precise measurements of a magnetostriction coefficient of a thin soft-magnetic film deposited on a substrate, *J. Appl. Phys.* **64**, 5422, (1988).

39. V. Koeninger, Y. Matsumura, T. Noguchi, H. H. Uchida, H. Uchida, H. Funakura, H. Kaneko, and T. Kurino, Magnetostriction of (TbDy)Fe2 and TbFe2 thin films, in "Proceedings of the International Symposium on Giant Magnetostrictive Materials and Their Applications," published by AMTDA (Advanced Machining Technology & Development association Tokyo 1992, p. 151.

40. K.I. Arai, M. Yamaguchi, and C.S. Muranaka, Measurement of thin films's magnetostriction with piezoelectric ceramic substrates, *IEEE Trans. Magn.* **25**, 4201, (1989).

41. R. Osiander, S.A. Ecelberger, R.B. Givens, D.K. Wickenden, J.C. Murphy, and T.J. Kistenmacher, A microelectromechanical-based magnetostrictive magnetometer, *Appl. Phys. Lett.* **69**, 2930, (1996).

42. B. Halstrup, J. Betz, K. Mackay, J.-C. Peuzin, and N. Lhermet, Micromachined magnetostrictive actuators, in "Micro Systems Technologies 96," (H. Reichl and A. Heuberger, Eds.), p. 457. VDE-Verlag, Berlin, 1996.

43. T. Honda, K.I. Arai, and M. Yamaguchi, Fabrication of magnetostrictive actuators using rare-earth (Tb,Sm)-Fe thin films, *J. Appl. Phys.* **76**, 6994, (1994).

44. E. Quandt and K. Seemann, Magnetostrictive thin film microflow devices, in "Micro System Technologies 96," (H. Reichl and A. Heuberger, Eds.), p. 451. VDI-Verlag, Berlin, 1996.

45. T. Honda, Y. Hayashi, M. Yamaguchi, and K.I. Arai, Fabrication of thin-film actuators using magnetostriction, *IEEE Trans. Magn. Jpn.* **9**, 27, (1994).

46. E. Quandt and K. Seemann, Magnetostrictive thin film transducers for applications in microsystem technology, in; "Proceedings Actuator 96," Bremen, Germany, p. 279.

47. T. Gerlach and H. Wurmus, Fundamentals of dynamic flux rectification as the basis of valve-less dynamic micropumps, in "Micro Systems Technologies 96," (H. Reichl and A. Heuberger, Eds.), p.445. VDE-Verlag, Berlin, 1996.

48. K. Seemann and E. Quandt, unpublished results.

# APPENDIX A

# Eddy Currents in a Cylindrical Rod with an Axial Magnetizing Field

Göran Engdahl
*Royal Institute of Technology*
*Stockholm, Sweden*

The magnetic field inside a cylindrical rod made of a material with conductivity $\sigma$ magnetized by an externally applied field $\mathbf{H} = (0, 0, H_z)$ $= (0, 0, H_0 e^{jwt})$ along its symmetry axis $z$ (see Fig. 2.19) can be calculated from Maxwell's equations:

$$\nabla \times \mathbf{H} = \mathbf{J} \tag{A.1}$$

$$\nabla \times \mathbf{E} = -\frac{\partial \mathbf{B}}{\partial t} \tag{A.2}$$

$$\mathbf{J} = \sigma \mathbf{E} \tag{A.3}$$

$$\mathbf{B} = \mu_r \mu_0 \mathbf{H} \tag{A.4}$$

where the displacement current has been neglected in Eq. (A.1). The cylindrical symmetry of the problem implies that the magnetic fields do not have any $\phi$ component. Furthermore, the symmetry gives that the $r$ and $z$ components of the magnetic fields also have no $\phi$ dependence. It follows that Eq. (A.1) has only a $\phi$ component, which is

$$\left(\frac{\partial H_z}{\partial r}\right)_\phi = J_\phi \tag{A.5}$$

From Eq. (A.3), $E_\phi = J_\phi / \sigma$, which together with Eq. (A.5) inserted into Eq. (A.2) gives

**345**

$$\left(\frac{1}{r}\frac{\partial}{\partial r}\left(rJ_\phi\right)\right)_z = -\sigma\frac{\partial B_z}{\partial t} \qquad (A.6)$$

By inserting Eq. (A.5) into Eq. (A.6), the following equation for the magnetic fields inside the cylinder is obtained:

$$\left(-\frac{1}{r}\frac{\partial}{\partial r}\left(r\frac{\partial H_z}{\partial r}\right)_\phi\right)_z = -\sigma\frac{\partial B_z}{\partial t} \qquad (A.7)$$

which can also be written as

$$\frac{d^2H}{dr^2} + \frac{1}{r}\frac{dH}{dr} = \sigma\frac{dB}{dt} \qquad (A.8)$$

where $H = H_z$ and $B = B_z$.

For harmonic excitation and $B = \mu_r\mu_0 H$,

$$\frac{d^2H}{dr^2} + \frac{1}{r}\frac{dH}{dr} - j\omega\sigma\mu_r\mu_0 H = 0 \qquad (A.9)$$

Here $\mu_r$ is the relative magnetic permeability of the rod and $\mu_0$ the magnetic permeability of vacuum.

With the substitution

$$u = r\sqrt{-j\omega\sigma\mu_r\mu_0} = r\gamma = r\sqrt{-j}\gamma' \qquad (A.10)$$

one gets

$$\frac{d^2H}{du^2} + \frac{1}{u}\frac{dH}{du} + H = 0 \qquad (A.11)$$

which is recognized as a Bessel equation of order zero.[1] The general solution to Eq. (A.11) is

$$H(u) = AJ_0(u) + BY_0(u) \qquad (A.12)$$

where $J_0$ is the Bessel function of the first kind of order zero and $Y_0$ is the Bessel function of the second kind of order zero. The constants $A$ and $B$ are determined from the boundary conditions. Because the first boundary condition is taken that $H$ must be finite in the center of the rod, $B$ must be zero since $Y_0 \rightarrow \infty$ when $r \rightarrow 0$. In the center of the rod,

$$H(0) = AJ_0(0) = A \qquad (A.13)$$

and

---

[1] M. R. Spiegel, "Mathematical Handbook." McGraw-Hill, New York.

[2] L. Butler, personal communication.

$$H(u) = H(0)J_0(u) \tag{A.14}$$

$J_0$ can be separated into a real and an imaginary component when its argument is complex by means of so-called Kelvin functions.[2] $H(u)$ can then be written as

$$H(u) = H\left(\sqrt{-j}\gamma'r\right) = H(0)J_0\left(\sqrt{-j}\gamma'r\right) = H(0)(\text{ber}(\gamma'r) + j\text{bei}(\gamma'r)) \tag{A.15}$$

The second boundary condition states that the amplitude of the H-field at the surface of the rod in Fig. 2.19 is $H_0$. Therefore,

$$H_0 = H\left(\sqrt{-j}\gamma'r_r\right) = H(0)(\text{ber}(\gamma'r_r) + j\text{bei}(\gamma'r_r)) \tag{A.16}$$

Hence, the field inside the rod is

$$H_{\text{inside}} = H_0 \frac{\text{ber}(\gamma'r) + j\,\text{bei}(\gamma'r)}{\text{ber}(\gamma'r_r) + j\,\text{bei}(\gamma'r_r)} e^{j\omega t} \tag{A.17}$$

when the harmonic time variation is included.

# APPENDIX B

# Impedance of a Straight Wire

Göran Engdahl
*Royal Institute of Technology*
*Stockholm, Sweden*

Assume an axial symmetric straight wire along the z-axis with the geometry according to Fig. B.1 carrying a sinusoidal current with angular frequency $\omega$ along the wire. Therefore $E_z \neq 0$ and $E_r = E_\phi = 0$. The axial and radial components of the magnetic flux density then are zero, i.e., $B_z = B_r = 0$. The rotational symmetry also implies that $E_z$ and $B_\phi$ do not have any $\phi$ dependence.

According to Maxwell's equations,

$$\nabla \times \mathbf{E} = -\frac{\partial \mathbf{B}}{\partial t} \tag{B.1}$$

$$\nabla \times \mathbf{H} = \mathbf{J} \tag{B.2}$$

$$\mathbf{J} = \sigma \mathbf{E} \tag{B.3}$$

$$\mathbf{B} = \mu_r \mu_0 \mathbf{H} \tag{B.4}$$

Taking the curl of the left-hand side of Eq. (B.1) gives

$$\nabla \times \nabla \times \mathbf{E} = -\frac{\partial}{\partial t} \nabla \times \mathbf{B} = -\mu_r \mu_0 \frac{\partial}{\partial t} \nabla \times \mathbf{H} = -j\omega \mu_r \mu_0 \mathbf{J} = -j\omega \mu_r \mu_0 \sigma \mathbf{E} \tag{B.5}$$

$$Eq.(B.4) \qquad Eq.(B.2) \qquad Eq.(B.3)$$

$\nabla \times \mathbf{E}$ can also be expressed as

$$\nabla \times \mathbf{E} = \mathbf{a}_r \overset{=0}{\left(\frac{\partial E_z}{\partial \phi}\right)} \overset{=0}{-\frac{\partial E_\phi}{\partial z}} + \mathbf{a}_\phi \overset{=0}{\left(\frac{\partial E_r}{\partial z}\right)} \overset{\neq 0}{-\frac{\partial E_r}{\partial r}} + \mathbf{a}_z \overset{=0}{\left(\frac{1}{r}\frac{\partial}{\partial r}(rE_\phi)\right)} \overset{=0}{-\frac{1}{r}\frac{\partial E_r}{\partial \phi}}$$

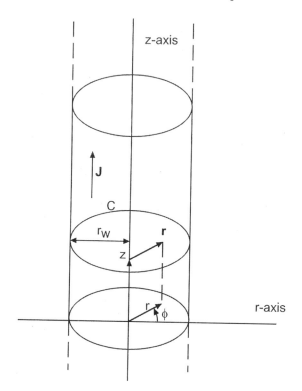

**FIGURE B.1.**   The geometry of a straight wire.

$$\nabla \times \mathbf{E} = -\mathbf{a}_\phi \frac{\partial E_z}{\partial r} \tag{B.6}$$

This then gives

$$\nabla \times \nabla \times \mathbf{E} = \nabla \times \left(0, 0, -\frac{\partial E_z}{\partial r}\right) = -\left(0, 0, \frac{1}{r}\frac{\partial}{\partial r}\left(r\frac{\partial E_z}{\partial r}\right)\right) \tag{B.7}$$

Equation (B.5) and Eq. (B.7) then give

$$-\frac{1}{r}\left(\frac{\partial E_z}{\partial r} + r\frac{\partial^2 E_z}{\partial r^2}\right) = -j\omega\mu_r\mu_0\sigma E_z$$

or

$$\frac{\partial^2 E_z}{\partial r^2} + \frac{1}{r}\frac{\partial E_z}{\partial r} - \gamma^2 E_z = 0 \tag{B.8}$$

where $\gamma = \sqrt{j\gamma'} = \sqrt{j} \cdot \sqrt{\omega \mu_r \mu_0 \sigma}$. The substitution $x = \gamma r$ gives

$$\frac{\partial^2 E_z}{\partial x^2} + \frac{1}{x}\frac{\partial E_z}{\partial r} + E_z = 0$$

which is a modified Bessel equation. Its general solution is

$$E_z = C_1 I_0(\gamma r) + C_2 K_0(\gamma r) \tag{B.9}$$

where $I_0$ and $Y_0$ are modified Bessel functions of zero order and the first and second kind, respectively.

$E_z$ must be finite for $r = 0$, which gives $C_2 = 0$ because

$$\lim_{r \to 0} K_0(r) = \infty$$

Equations (B.3) and (B.9) then give

$$E_z = C_1 I_0(\gamma r) \tag{B.10}$$

$$J_z = \sigma C_1 I_0(\gamma r) \tag{B.11}$$

Equations (B.1) and (B.6) give

$$\frac{\partial B_\phi}{\partial t} = \frac{\partial E_z}{\partial r} \tag{B.12}$$

and

$$j\omega B_\phi = C_1 \frac{d}{dr} I_0(\gamma r) = C_1 \gamma I_1(\gamma r) = C_1 (\sqrt{j\omega\mu\mu\sigma}) I_1(\gamma r) \tag{B.13}$$

where $I_1$ is a modified Bessel function of the first kind and first order.

At the surface of the wire, where $r$ is equal to $r_w$ and the H-field is equal to $H_\phi(r_w)$,

$$H_\phi(r_w) = \frac{B_\phi(r_w)}{\mu_r \mu_0} = C_1 \frac{\sigma}{\gamma} I_1(\gamma r) \tag{B.14}$$

Ampere's law $\oint \mathbf{H} \cdot d\mathbf{l} = I$ integrated around the wire gives

$$\oint_C \mathbf{H} \cdot d\mathbf{l} = 2\pi r_w H_\phi(r_w) = I \tag{B.15}$$

where $I$ is the total current in the wire.

Equations (B.14) and (B.15) give

$$C_1 = \frac{I\gamma}{2\pi r_w \sigma I_1(\gamma r_w)} \tag{B.16}$$

which inserted into Eq. (B.10) gives

$$E_z = \frac{I\gamma I_0(\gamma r)}{2\pi r_w \sigma I_1(\gamma r_w)} \tag{B.17}$$

The voltage $U$ across a length $l_w$ along the wire is

$$U = \int_{z_0}^{z_0+l_w} E_z dl = E_z l_w = \frac{I\gamma l_w I_0(\gamma r)}{2\pi r_w \sigma I_1(\gamma r_w)} \tag{B.18}$$

The impedance $Z_{wire} = U/I$ of the wire of length $l_w$ is

$$Z_{wire} = \frac{U}{I} = R + j\omega L = \frac{l_w \gamma I_0(\gamma r)}{2\pi r_w \sigma I_1(\gamma r)} = \frac{l_w \rho}{\pi r_w^2} \frac{\gamma r_w}{2} \frac{I_0(\gamma r)}{I_1(\gamma r)} \tag{B.19}$$

# APPENDIX C

# Variational Formulation of Electromechanical Coupling

Frank Claeyssen
*Cedrat Recherche*
*Zirst, France*

## THE FINITE ELEMENT ALGORITHM

### Hypotheses

All the basic hypotheses of the retained model are described and discussed. Statement:

1. The phenomena are harmonic.

2. The magnetic phenomena are described in a quasistatic approximation.

3. The sources are directly expressed from a finite number of excitation currents.

4. No magnetic item of infinite permeability, not simply connex, interleaves with a current loop.

5. The material behavior is linear.

6. The system energy is conservative.

Hypothesis 1 is verified for the sonar transducers which are excited by long pulses, including many periods in a sinusoidal steady state. Hypotheses 2 and 3 assume that the role of eddy currents is negligible. The magnetic and piezomagnetic parts of magnetostrictive transducers are designed so that their characteristic frequency $(1, 2)$ is greater than the operating frequency. Under these conditions, the presence of eddy currents introduces a small loss angle in the magnetic field but without

**353**

modifying its module. Hypotheses 2 and 3 will therefore have no significant effects on the desired results.

Hypothesis 4 means that there are no loop-shape infinite permeability circuits interleaving with a current loop. In practice, the magnetic circuits of transducers are designed to focus the magnetic energy into the active material and rare earth–iron materials have particularly low relative permeability levels for ferromagnetics, in the range of 10 times that of vacuum at the most (3). Hence, hypothesis 4 is naturally verified in practice.

Hypothesis 5 is justified for the following reasons:

> Magnetostrictive materials are biased so that they become piezomagnetic and hence linear.

> Mechanical materials are used within their elastic domain; prestress rods in particular are designed with this characteristic taken into account.

> Highly permeable ferromagnetic elements, which are by nature nonlinear with a topology which respects hypothesis 4, contain a very low magnetic energy compared to the energy in the rest of the domain. Consequently, and provided that one ensures that saturation is not reached, the approximation consisting of assigning them with average permeability may give an approximate internal magnetic field but will not affect the external fields. This is especially true as long as the relative permeability is large compared to unity.

It cannot be contested that losses exist in magnetostrictive materials, for example because of eddy currents. However, as long as the quality factors are >5 (as in the case in which the operating frequency is lower than the eddy current characteristic frequency), the structure vibration mode frequencies and the inductive part of the impedance are little affected. The main consequence of hypothesis 6 will be the difficulty of predicting accurately the mode damping and the efficiency, which does not reduce the interest of the model too much.

## Definition of Domains

The elements to be modeled so as to describe the dynamic behavior of a magnetostrictive transducer necessarily fall into at least one of the following three classes:

C1: The element is a magnetic source if its current density $J$ is assumed to not be zero. The set of points such as $J \neq 0$, with a bounded support, defines the domain of the $V_S$ sources, split into a finite number of subdomains such that each subdomain is homeomorphic to a torus.

C2: The element is a magnetic material if its magnetic field $H$ is assumed to not be zero. The set of points such as $H \neq 0$, generally with a nonbounded support, defines the magnetic domain $V_M$.

C3: The element is a mechanical part if its displacement field $u$ is assumed to not be zero. The set of points such as $u \neq 0$, with a bounded support, defines the elastic domain $V_E$.

The source class C1, fully determined because of hypothesis 2, consists of the transducer coils by continuous approximation. It should be noted that if the coils are fitted to a moving part of the transducer, they also fall into category C3 attributed to mechanical parts.

Magnetic material category C2 includes the piezomagnetic and pole items of finite relative permeability greater than unit (the "imperfect couplers") and also the items with a relative permeability equal to unit and vacuum (or the air) surrounding the transducer. However, the pole items of infinite permeability (the "perfect couplers") are excluded from the $V_M$ domain because the magnetic field $H$ and the magnetic energy are zero on account of hypothesis 4 (4). The mechanical element category includes the moving mechanical parts of the transducer as well as its piezomagnetic parts.

These definitions are illustrated by a sample structure (Fig. C.1) consisting of a transducer with two cylindrical magnetostrictive rods in parallel, with each rod surrounded by an axial symmetry-type coil which is not secured to the structure. The current flows through the coils in opposite directions so as to create a magnetic field loop passing through the rods and couplers. The inclusion of a perfect and imperfect coupler was decided on for "pedagogic" purposes. The movement of the rods, induced by the piezomagnetic coupling, drives the couplers plus a head mass and tail mass. The head mass is the radiating face of the transducer under water. Each element of this sample structure was subjected to an a priori analysis according to criteria $J$, $H$ and $u$ (Table C.1) allowing splitting of the element in relation to domains $V_S$, $V_M$, and $V_E$ (Fig. C.2).

Before considering equations, the following definitions and properties are introduced. Domains $V_S$, $V_M$, and $V_E$ are referenced by an origin at zero and a common axis system $x_1$, $x_2$, and $x_3$. They are respectively bounded by surfaces $S_S$, $S_M$, and $S_E$, the normals of which are noted $R$. The

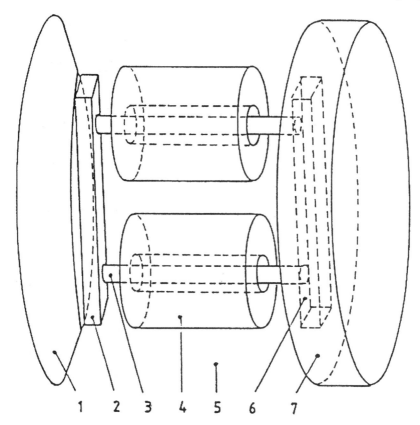

**FIGURE C.1.** Perspective view of the sample structure. 1, Head mass; 2, imperfect coupler; 3, piezomagnetic rod; 4, coil; 5, vacuum; 6, perfect coupler; 7, tail mass.

piezomagnetic materials have a finite relative permeability (often in the order of magnitude of the unit) and are located at the intersection of $V_M$ and $V_E$. The addition of $V_M$ and $V_E$ is denoted $V$ and may be considered as the piezomagnetic domain in the general meaning (by canceling the appropriate terms of the constitutive laws).

## Constructive Laws of a Piezomagnetic Medium

The constitutive laws of a piezomagnetic medium, linking the stresses $T_{ij}$, the strains and the components of induction $B_m$ and that of field $H_n$, are expressed by the following system (5, 6):

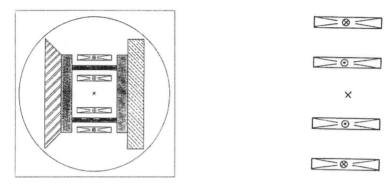

Sample Structure Diagram

Source Domain $V_S$

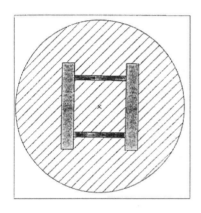

Magnetic Domain $V_M$

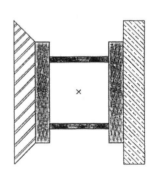

Elastic Domain $V_E$

**FIGURE C.2.** Sample structure diagram and modeling domains $V_S$ , $V_M$, and $V_E$. The white areas are not part of the domain involved.

$$T_{ij} = c_{ijkl}^H \cdot S_{kl} - e_{nij} H_n$$
$$B_m = e_{mkl} \cdot S_{kl} + \mu_{mn}^S H_n$$
(C.1)

where strain $S_{kl}$ is expressed from the components of displacement **u**:

$$S_{kl} = \frac{1}{2} \left( \frac{\partial u_k}{\partial x_l} + \frac{\partial u_l}{\partial x_k} \right)$$
(C.2)

Such laws are generally all that is needed to describe a medium which is

**Table C.1.**   Analysis of the Sample Structure[a]

| Part | Head mass | Imperfect coupler | Piezo magnetic | Coil | Vacuum | Perfect coupler | Tail mass |
|------|-----------|-------------------|----------------|------|--------|-----------------|-----------|
| J    | 0         | 0                 | 0              | *    | 0      | 0               | 0         |
| H    | *         | *                 | *              | *    | *      | 0               | *         |
| u    | *         | *                 | *              | 0    | 0      | *               | *         |

[a] The symbols 0 and * respectively indicate the a priori zero and nonzero values of criteria $J$, $H$, and $u$ for each part.

either purely magnetic or purely elastic, with the transition made by canceling the piezomagnetic tensor $e_{nij}$ and by retaining only the equation involved. In these equations and those that follow the Einstein notation is used for the summations. The letters $i$–$n$ are used to refer to the components 1–3.

## Consideration of Sources

How the sources are taken into account is a factor which is intimately linked to the representative fundamental magnetic variables. It partly determines the formulation of the problem and its numerical resolution. There are currently four methods for processing magnetostatic problems:

1. The vector potential formulation (7)

2. The magnetic field formulation (8)

3. The total scalar potential formulation (9)

4. The reduced scalar potential formulation (10, 11)

The latter method was selected since it has two advantages. First, the state variable, the potential, is a scalar quantity. It reduces the amount of computer memory needed and time requirements in relation to the first two methods. Second, this method is quite simple to implement and does not present algorithmic problems such as those of the third method for selection of the cutting surfaces.

The idea behind scalar potential formulation consists of breaking down magnetic field $H$ into two components:

A rotational component $H^S$ due to source $J$ verifying hypothesis 3 and admitting an analytic solution given by the Biot–Savart law (expression 4):

$$\text{Curl } H^S = J \tag{C.3}$$

$$H^S = \frac{1}{4\pi} \int_{V_s} \frac{J \wedge R}{R^3} \, dv \tag{C.4}$$

An irrotational component $H^M$ due to the magnetization of the material and, consequently, deriving from a scalar potential $\phi$ (Eq. C.5):

$$\begin{aligned} \text{Curl } H^M &= 0 \\ H^M &= -\text{grad } \phi \end{aligned} \tag{C.5}$$

A check can be performed to ensure that the desired field (Eq. C.6) verifies the Ampere theorem (Eq. C.7):

$$H = H^S + H^M \tag{C.6}$$

$$\text{Curl } H = J \tag{C.7}$$

The only constraint imposed by a scalar potential formulation is that a reference potential must be imposed. This can be performed by attributing an arbitrary value to any point of domain $V_M$. In practice, this point is often selected at the external boundaries of $V_M$ and the corresponding potential value is taken as being zero.

This simple and quite general formulation, however, has one disadvantage from the numerical point of view in high-permeability magnetic regions; since the magnetic field strength $H$ is low in this region as a result of the difference between $H^S$ and $H^M$, which nearly balance each other out, its value is not precisely determined by the computation. Although known cures exist for this problem (12, 13) based on the use of the total scalar potential, it was not deemed necessary to apply them in this case. In fact, the solution is precisely searched for especially in magnetostrictive materials having very low permeability levels (12, 13). However, for the magnetic couplers, which are the only high-permeability parts in a magnetostrictive transducer, a qualitative idea of the field is sufficient. A complete reduced scalar potential formulation is therefore quite satisfactory with regard to these objectives.

A final improvement of the formulation remains to be made so that a very specific requirement for the computation of magnetostrictive transducers can be taken into account—that is, the possibility of computing the electric resonance modes (which correspond to the mechanical resonance modes at zero voltage) and to perform harmonic analysis at prescribed voltages. Another similar case is when the magnetostrictive transducer is used as a sensor or a receiver. In these two cases, the values of the currents $I_p$ in the $n_B$ coils are not known a priori, which implies that they have to be explained in expression 4 by using hypothesis 3. Therefore,

$$H^S = \sum_{p-1}^{n_B} {}^\circ H^{S_p} \cdot I_p \qquad (C.8)$$

with

$$ {}^\circ H^{S_p} = \frac{1}{4\pi} \int_{V_s} \frac{J_p^\circ \wedge R}{R^3} \, dv \qquad (C.9)$$

The $J_p^\circ$ is the current density in the coil $p$ corresponding to a coil current equal to unity. Therefore, the fields ${}^\circ H^{S_p}$ can be computed a priori. The total field (Eq. C.8) is then a linear combination of these fields, whose coefficients are the effective currents $I_p$ which can be unknown as explained previously.

The components $H_n$ of the total field $H$ are then written as a function of the fundamental variables $I_p$ and $\phi$ as follows:

$$H_n = H_n^S + H_n^M = {}^\circ H_n^{S_p} \cdot I_p - \frac{\partial \phi}{\partial x_n} \qquad (C.10)$$

Finally, voltage at the terminals of each inductor has to be computed. In the harmonic analysis, it is proportional to the flux $Q_p^V$ seen by the windings:

$$v_p = j\omega Q_p^V \qquad (C.11)$$

This flux $Q_p^V$ is equal to the partial derivative of the magnetic coenergy with respect to the current $I_p$ (14):[16]

$$\delta I_p \cdot Q_p^V = \delta W_M^* \qquad (C.12)$$

## Euler Equations and Boundary Conditions

From an electrical point of view, the solution of Eq. (C.12) which allows the calculation of $Q_p^V$ is the first Euler's equation for the electromechanical problem. However, it is useless in the previous form because Eq. (C.12) is already part of the variational principle having to be derived.

A prescribed current condition, such as the Dirichlet, can only be added:

$$I_p - I'_p = 0 \qquad (C.13)$$

where $I'_p$ designates the prescribed current.

The electromechanical coupling Euler's equations (15) are the Newton's law in $V_E$ and the magnetic flux conservation equation in $V_M$:

$$\frac{\partial T_{ij}}{\partial x_j} + \rho \omega^2 u_i = 0 \qquad (C.14)$$

$$\frac{\partial B_m}{\partial x_m} = 0 \qquad (C.15)$$

Equations (C.14) and (C.15) describe the internal behavior in $V_E$ and $V_M$ and are matched with mechanical boundary conditions on $S_E$ and magnetic boundary conditions on $S_M$. Note that the special interface continuity conditions do not have to be added, provided that all previous relations are admitted in the distribution sense (16).

The mechanical boundary conditions are:

- A Dirichlet-type prescribed displacement field condition on $S_u$, with the border surface defined as follows:

$$u_i - u'_i = 0 \qquad (C.16)$$

where $u'_i$ designates the prescribed displacement. The corresponding force can be computed using the condition

$$F'_i - \int_{S_u} T_{ij} \cdot R_j ds = 0 \qquad (C.17)$$

where $R_j$ are the components of the normal vector on the surface.

- A Neumann-type prescribed normal stress condition on $S_{T_N}$, with the border surface defined as follows:

$$T_{ij} \cdot R_j - T'_{Ni} = 0 \qquad (C.18)$$

where $T'_{Ni}$ designates the prescribed normal stress.
These two conditions are natural and fulfill all possible requirements.

The magnetic boundary conditions are

- A Dirichlet-type prescribed magnetic potential condition on $S_\phi$:

$$\phi - \phi' = 0 \tag{C.19}$$

where $\phi'$ is the prescribed potential which can be used as the reference potential. The corresponding magnetic flux can be computed using the condition

$$Q' - \int_{S_\phi} B_m \cdot R_m ds = 0 \tag{C.20}$$

- A Neumann-type prescribed normal magnetic induction condition on $S_{B_N}$:

$$B_m \cdot R_m - B'_N = 0 \tag{C.21}$$

where $B'_N$ is the induction prescribed normal component.

- A Dirichlet-type prescribed tangential magnetic field condition on $S_{H_T}$

$$H_T - H'_T = 0 \tag{C.22}$$

where $H'_T$ is the magnetic field prescribed tangential component.

In fact, it can be shown that this peak condition at a point M on $S_{H_T}$ is equivalent to a condition on the potential $\phi$ at the same point:

$$\phi = {}^\circ\phi^{S_p} \cdot I_p - \phi^{H'} + \phi^F \tag{C.23}$$

where $\phi^F$ is an unknown floating potential computed at any fixed point A of $S_{H_T}$, and ${}^\circ\phi^{S_p}$ and $\phi^{H'}$ are scalar quantities which can be computed a priori by integration on any path from A to M on $S_{H_T}$ of the tangential components of ${}^\circ H^{S_p}$ and $H'$, respectively.

By assuming that the second term of Eq. (C.23) is a prescribed potential, it becomes clear that the condition (Eq. C.22) can be reduced to a somewhat special Dirichlet condition. A corresponding flux $Q_H$ on $S_{H_T}$ can also be computed using a condition similar to Eq. (C.20). On the surface $S_{H_T}$ therefore, only $\phi^F$ or $Q_H$ is unknown according to the topology of the problem. In the most frequent case, which is that of the surface of the coupler of infinite permeability as incorporated in the sample structure, $S_{H_T}$ is a closed surface for which it can be stated that $Q^H = 0$. In the other situations, $\phi^F$ is either determined or selected as the potential reference. Should several $S_{H_T k}$ surfaces exist, the proposed procedure becomes general by introducing a floating potential $Q_k^F$ and a flux $Q_k^H$ for each $S_{H_T k}$ without forgetting the intervention of the flux conservation relations between the $Q_k^H$ values. Finally, the condition on $S_{H_T}$ is very similar to the Dirichlet conditions. Therefore, only the natural conditions on $S_\phi$ and Neumann conditions on $S_{B_N}$ will be considered so as to simplify the writing of the following equations and to retain the symmetry between elastic and magnetic equations. The condition on the tangential field will only be processed during the final resolution.

Finally, the boundary conditions on the surfaces obtained through reduction of the domain by symmetry are the natural conditions and are part of those presented, and in the "open" magnetic problems the infinite zero-field condition is taken into account by a Dirichlet-type $\phi' = 0$ condition.

## Differential Equations and State Variables

The Euler Eqs. (C.14) and (C.15) are local equations and relate a high number of variables which should be reduced first by replacing the stresses $T_{ij}$ and the induction following the constitutive laws (Eq. C.1) and then by replacing the strains $S_{kl}$ using Eq. (C.2) and the magnetic field $H_n$ using Eq. (C.10):

$$\frac{\partial}{\partial x_i}\left(c_{ijkl}^H \cdot \frac{1}{2}\left(\frac{\partial u_k}{\partial x_l} + \frac{\partial u_l}{\partial x_k}\right) + e_{nij} \cdot \frac{\partial \phi}{\partial x_n} e_{nij} \cdot H_n^{S_p} \cdot I_p\right) + \rho\omega^2 u_i = 0$$

$$\frac{\partial}{\partial x_m}\left(e_{mkl} \cdot \frac{1}{2}\left(\frac{\partial u_k}{\partial x_l} + \frac{\partial u_l}{\partial x_k}\right) + \mu_{mn}^S \cdot \frac{\partial \phi}{\partial x_n} + \mu_{mn}^S \cdot H_n^{S_p} \cdot I_p\right) = 0$$

(C.24)

These differential equations clearly show that the fundamental unknowns or state variables are (i) the components of the mechanical displacement field, $u_k$; (ii) the magnetic potential field $\phi$; and the $n_B$ currents $I_p$.

The Dirichlet conditions Eqs. (C.13), (C.16), and (C.19) have a direct effect on the state variables. The associated Eqs. (C.12), (C.17), and (C.20) enable the derived loads $Q_V$, $F_i'$ and $Q'$ to be expressed as a function of the state variables. The Neumann conditions (Eqs. C.18 and C.21) are also expressed as a function of state variables and prescribed loads $T_N'$ and $B_N'$.

The system (Eq. C.24) constitutes the strong piezomagnetic problem because either its equations are to be processed in the sense of distributions or they implicate the derivability of the displacements and potentials because of the medium discontinuities. This apparent difficulty is overcome by using the Galerkin method (17), which is based on the related weak problem, which no longer requires a differentiability condition for the variables. A variational method having the same advantages and a precise physical interpretation was preferred to this method.

Finally, without performing any approximation, it is assumed that the continuity of the displacements and potentials is sufficient in most of the physical situations to ensure continuity relationships between the different materials (18).

## Functionals Associated with the Variational Principle

Using a "trial-and-error" procedure and taking advantage of studies relevant to magnetostatics (19) and piezoelectricity (20, 21), a quadratic functional $L(u, \phi, I_p)$ is derived with values within IR, its stationary conditions with respect to small variations being equivalent to the previous Euler equations (variational principle):

$$L = L_V + L_{S_E} + L_{S_M} \tag{C.25}$$

$$L_V = \int_V \left( \frac{1}{2} \rho \omega^2 u_i u_i - \frac{1}{2} S_{ij} T_{ij} + \frac{1}{2} H_m B_m \right) dV - I'_p Q_p^V - \int_V \left( I_p - I'_p \right) H_m^{S_p} B_m dV$$

$$L_{S_E} = \int_{S_{T_M}} u_i \cdot T'_{N_i} \cdot ds + \int_{S_u} (u_i - u'_i) \, T_{N_i} ds + u'_i F'_i$$

$$L_{S_M} = \int_{S_{B_N}} \phi \cdot B'_N \cdot ds + \int_{S_\phi} (\phi - \phi') \, B_N ds + \phi' \cdot Q'$$

This expression can be reduced if the space of the allowed functions for $u$, $\phi$, and $I_p$ is restricted to the functions verifying the Dirichlet conditions, Eqs. (C.13), (C.16), and (C.19) and if the associated Eqs. (C.17) and (C.20) are no longer taken into account. Then

$$L' = L'_V + L'_{S_E} + L'_{S_M} \tag{C.26}$$

$$L'_V = \int_V \left( \frac{1}{2} \rho \omega^2 u_i u_i - \frac{1}{2} S_{ij} T_{ij} + \frac{1}{2} H_m B_m \right) dV - I_p Q_p^V$$

$$L'_{S_E} = \int_{S_{T_N}} u_i \cdot T'_{N_i} \cdot ds$$

$$L_{S_M} = \int_{S_{B_N}} \phi \cdot B'_N \cdot ds$$

Equation (C.26) presents advantages in the case of elliptical problems such as noncoupled magnetostatics. Unfortunately, the piezomagnetic problem does not have the ellipticity property and Eq. (C.26) no longer presents a major interest in relation to Eq. (C.25). Moreover, Eqs. (C.25) and (C.26) lead to the same final system to be solved in the case of a finite element approximation, which is the method that is followed. Consequently, the classical discussion between H. F. Tiersten (22) and E. P. Eer Nisse and R. Holland (23) with regard to the equivalent piezoelectric problem will not be undertaken here. Equation (C.26) gives rise to an interpretation using the Hamilton principle, but Eq. (C.25), which contains Eq. (C.26) can be fully proven by differentiation.

## Application of the Variational Principle

The choice of the functional is justified by performing a small variation $\delta$ of the state variables $u$, $\phi$, and I (15). The stationary state of $L$ is then written as

$$\delta L = 0 \tag{C.27}$$

The variational principle is verified if the latter condition implies Euler equations and the problem boundary conditions.

The computation of $\delta L$ and $\delta L'$ can be broken down into several stages, with the first stage being common to both. On account of the symmetries on $[c^H]$ and $[\mu^S]$, the variation of the volume terms is written as

$$\delta \left( \frac{1}{2} S_{ij} T_{ij} - \frac{1}{2} H_m B_m \right) = \delta S_{ij} \cdot T_{ij} - \delta H_m \cdot B_m \tag{C.28}$$

Using Eq. (C.28), the computation of $\delta L'$ is trivial:

$$\delta L' = \delta L'_V + \delta L'_{S_E} + \delta L'_{S_M} = 0 \tag{C.29}$$

$$\delta L'_V = \int_V \left( \rho \omega^2 u_i \, \delta u_i - \delta S_{ij} \, T_{ij} + \delta H_m \cdot B_m \right) dV - \delta I_p Q_p^V$$

$$\delta L'_{S_E} = \int_{S_{T_N}} \delta u_i \cdot T_{N_i} \cdot ds$$

$$\delta L'_{S_M} = \int_{S_{B_N}} \delta\phi \cdot B'_N \cdot ds$$

Equation (C.29) then takes a physical sense:

$$\delta L'_{S_E} = \delta \mathcal{W}_E$$

is the virtual work produced by a small variation of the displacements on the surfaces $S_{T_N}$, and

$$\delta L'_{S_M} = \delta \mathcal{W}_M$$

is the virtual work produced by a small variation of the potentials on the surfaces $S_{B_N}$.

The sum of these works is denoted

$$\delta \mathcal{W} = \delta \mathcal{W}_E + \delta \mathcal{W}_M$$

$\delta L'_V$ comprises a sum of terms. The first one,

$$\int_V \rho\omega^2 \, u_i \, \delta u_i \, dV = \delta E_c$$

is the kinetic energy variation. The second one,

$$\int_V \left( \delta S_{ij} \, T_{ij} - \delta H_m \cdot B_m \right) dv + \delta l_p \, Q_p^V = \delta \mathcal{H}$$

is the variation of the magnetic enthalpy $\mathcal{H}$ which is defined in the following manner in relation to the internal energy $U$ (6, 4):

$$\mathcal{H} = U - \int_V B \cdot H \, dV$$

Then;

$$\delta U = \int_V (T \, \delta S + H \, \delta B) \, dV + Q_p^V \cdot \delta l_p$$

is the sum of the elastic, magnetic, and electric energy variations.

The variational definition of $Q_p^V$ flux appears in Eq. (C.12) for the magnetic enthalpy $\delta \mathcal{H}$ and consequently is completed. The term $\delta L'_V$ is therefore the variation of a Langrangian $L$ equal to the kinetic energy minus the magnetic enthalpy:

$$L = E_c - \mathcal{H}$$

Thus, the variational principle is equivalent to the Hamilton principle (25), also called the virtual work principle:

$$\delta L + \delta \mathcal{W} = 0$$

In this form, the piezomagnetic problem presents a partial analogy with the elastic problem, where the Lagrangian is equal to the kinetic energy minus the internal energy (25). The analogy is total with the piezoelectric problem (26), in which the Lagrangian is equal to the kinetic energy minus the electrical enthalpy (22) by making the electric field and induction $(E, D)$ correspond to the magnetic field and induction $(H, B)$. The application of Neumann's theorem (25, 27) to both problems has shown the uniqueness of the solution. With allowance made for the analogies with these problems, the conclusion of the solution uniqueness can also be drawn from the formulation retained. In addition, the examination of the Lagrangian $L$ shows that the functional is a nondefinite quadratic form. Consequently, the solution for which one is searching is a saddle point of the functional.

Finally, note that the variational principle, presented in its most general thermodynamic form, is applicable to the nonlinear piezomagnetic problem.

The complete proof of the functional $L$ (Eq. C.25) is not be presented because it can easily be accomplished by following the next stages. After differentiating $L$, taking care to disclose the elementary variations of the displacements, potentials, and currents, particularly with the use of Eq. (C.28), it is only necessary to successively apply a partial integration and the Green–Ostrogradski formula to the volume term $\delta L_V$ in the distribution sense and to group the surface terms from this operation together with those of $\delta L_{S_E}$ and $\delta L_{S_M}$. The variation $\delta L$ is then written in the form of a sum of weighted integrals through the variation of one of the state variables. Because all these variations are arbitrary, the variational principle (Eq. C.27) and the fundamental lemma of variations calculus lead to the cancellation of the weighted terms, which restores Euler's Eqs. (C.12)–(C.14) and the problem boundary conditions (Eqs. C.15–C.21).

## RESOLUTION BY FINITE ELEMENT METHOD

Among the many existing numerical methods, the finite element method (28, 29), which is a global method based on space discretization, has the major advantage of being capable of processing heterogeneous geometries of very irregular shapes and being well suited to energy formulations such as those based on a variational principle. It leads to the resolution of a band matrix for which solving algorithms exist providing increasingly better performance. Calculus in the piezoelectric, magnetostatic (7, 19), and structural (29, 30), fields are problems similar to magnetostrictive transducers problems.

For the piezoelectric transducers problem, with which the strongest analogies were recorded, the ATILA code was successfully developed for DCN. Because of all its advantages, the finite element method was applied to the magnetostrictive problem, with the objective of deriving resolution algorithms as close as possible to those used in the ATILA code so as to gain the maximum benefit from the existing structure of this code. Since the principle of the finite element method has been clearly set forth elsewhere (29), only the special features connected with the magnetostrictive problem will be addressed.

## Representation of Solutions

The nodal parameters retained for this application are the three components of displacement $u$ and the magnetic potential $\phi$. At each node $M_i$ of the mesh, these quantities take the values of $U_i$ and $\Phi$, respectively. At any point $M$ of an element $e$, they are expressed with the use of interpolation functions $N$ and $N\phi$ (that can be chosen of different types), respectively, in the following manner:

$$u(M) = \sum_{i \in e} N_i^e(M) \cdot U_i$$

and

$$\phi(M) = \sum_{i \in e} N_{\phi i}^e(M) \cdot \Phi_i$$

Since the continuity of the potentials and displacements is the only property required, polynomials of degree 1 are sufficient for the interpolation functions. Finally, the solution throughout domain V is written by assembling the solutions on each element:

$$u = \sum_i N_i \cdot U_i = [N] \cdot \{U\} \tag{C.30}$$

$$\phi = \sum_i N_{\phi i} \cdot \Phi_i = [N_\phi] \cdot \{\Phi\} \tag{C.31}$$

where

$$N_i = \sum_{e(i)} N_i^e \quad \text{and} \quad N_{\phi i} = \sum_{e(i)} N_{\phi i}^e \tag{C.32}$$

These last summations are performed on all the elements $e(i)$ that contain a node $i$.

## Formulation of Various Tensors

Rather than introducing the approximated expressions of $u$ and $\phi$ directly in the functional, it is advantageous to formulate the magnetic and induction field vector expressions and the strain and stress tensor expressions separately.

According to Eqs.(C.5) and (C.31) the magnetic field $H_M$ is written as

$$H^M = -\operatorname{grad} \phi = -\sum_i B_{\phi i}\Phi_i = -\left[B_\phi\right]\{\Phi\} \qquad \text{(C.33)}$$

where

$$B_{\phi i} = \sum_{e(i)} B_{\phi i}{}^e \quad \text{and} \quad B_{\phi i}{}^e = \operatorname{grad}\left(N^e_{\phi i}\right).$$

The terms $B_{\phi i}$ are expressed from the derivatives of the weighting functions and are determined analytically.

The magnetic field $H$ results from Eq. (C.10):

$$\{H\} = -\left[B_\phi\right]\{\Phi\} + \left[{}^\circ H^S\right]\{I\} \qquad \text{(C.34)}$$

with the writing convention

$$\{H^S\} = {}^\circ H^{S_p} \cdot I_p = \left[{}^\circ H^S\right]\{I\}$$

Likewise, for the strain tensor **S**, the following type of expression is derived with the use of Eq. (C.2):

$$\{S\} = [B] \cdot \{U\} \qquad \text{(C.35)}$$

where

$$B_i = \sum_{e(i)} B_i^e \qquad \text{(C.36)}$$

The terms of $B_i^e$ are also expressed as a function of the derivatives of the weighting functions $N_i^e$ in a slightly more complicated manner however, for $S$ is the vector condensed to six components and resulting from the strain $3 \times 3$ symmetrical tensor *(31)*. The following are stress and induction expressions derived from Eqs. (C.1), (C.30), (C.31), and (C.33)–(C.35):

$$\{T\} = [c^H][B] \cdot \{U\} + [e][B_\phi] \cdot \{\Phi\} - [e][^\circ H^S] \cdot \{I\}$$
$$\{B\} = [e][B] \cdot \{U\} - [\mu^S][B_\phi] \cdot \{\Phi\} + [\mu^S][^\circ H^S] \cdot \{I\}$$

(C.37)

with the essential condensations again for tensors $T$, $c^H$, and $e$.

## Application of the Variational Principle to the Discretized Functional

The discrete problem functional is obtained from the continuous problem functional (Eq. C.25) by replacing the variables by their approximated values:

$$L = -\frac{1}{2}\{U\}^T \cdot ([H_{UU}] - \omega^2[M]) \cdot \{U\} - \frac{1}{2} \cdot \{U\}^T \cdot [K_{U\phi}] \cdot \{\Phi\} - \frac{1}{2} \cdot \{U\}^T \cdot [K_{UI}] \cdot \{I\} + \{U\}^T \cdot \{F\}$$
$$- \frac{1}{2}\{\Phi\}^T \cdot [K_{U\phi}]^T \cdot \{U\} - \frac{1}{2} \cdot \{\Phi\}^T \cdot [K_{\phi\phi}] \cdot \{\Phi\} - \frac{1}{2} \cdot \{\Phi\}^T \cdot [K_{\phi I}] \cdot \{I\} + \{\Phi\}^T \cdot \{Q\}$$
$$- \frac{1}{2}\{I\}^T \cdot [K_{UI}]^T \cdot \{U\} - \frac{1}{2} \cdot \{I\}^T \cdot [K_{\phi I}]^T \cdot \{\Phi\} - \frac{1}{2} \cdot \{I\}^T \cdot [K_{II}] \cdot \{I\} - \{I\}^T \cdot \{Q^V\} \quad \text{(C.38)}$$

In this expression, the integrals disclosing the Dirichlet conditions have disappeared. It is obvious that the application of the condition on the current $I = I'$ cancels the corresponding integral. For the integral on $S_u$, it is necessary to state that the application of the condition $u = u'$ to the nodes on $S_u$ is sufficient because the interpolation functions are such that the values of the displacement field $u$ at any point of $S_u$ only depend on its nodal values. The same applies for the integrals on $S_\phi$.

The matrices introduced are expressed by means of Eqs. (C.32), (C.33), and (C.36) and may give rise to an identification linked with the terms intervening in their computation (Table C.2). The conventional form specific to this type of problem (29), except for the sign (which is of no importance for the form), is recognized in Eq. (C.38):

$$-L = \frac{1}{2}x^T \cdot [A] \cdot \{x\} - \{x\} \cdot \{b\}$$

(C.39)

where $[A]$ is the tangent matrix, $x$ is the vector of unknown quantities, and $b$ is the charge vector:

$$[A] = \begin{bmatrix} [K_{UU}] - \omega^2[M] & [K_{U\phi}] & [K_{UI}] \\ [K_{U\phi}]^T & [K_{\phi\phi}] & [K_{\phi I}] \\ [K_{UI}]^T & [K_\phi]^T & [K_{II}] \end{bmatrix}; \{b\} = \begin{bmatrix} \{F\} \\ \{Q\} \\ \{-Q^V\} \end{bmatrix}; \{x\} = \begin{bmatrix} \{U\} \\ \{\Phi\} \\ \{I\} \end{bmatrix}$$

The application of the variational principle to the functional expressed in Eq. (C.39) is consequently immediate because $A$ is real and symmetrical:

**Table C.2.** Definition of Electromechanical Matrix and Vectors

| | |
|---|---|
| $[M]_{ij} = \int_v N_i^T \cdot \rho \cdot N_j dv$ | Kinematically consistent mass matrix |
| $[K_{UU}]_{ij} = \int_v B_i^T \cdot [c^H] \cdot B_j dv$ | Stiffness matrix |
| $[K_{U\phi}]_{ij} = \int_v B_i^T \cdot [e] \cdot B_{\phi j} dv$ | Piezomagnetic stiffness matrix |
| $[K_{\phi\phi}]_{ij} = -\int_v B_{\phi i}^T \cdot [\mu^s] \cdot B_{\phi j} dv$ | Magnetic stiffness matrix |
| $[K_{UI}]_{ip} = -\int_v B_i^T \cdot [e] \cdot {}^\circ H_p^S dv$ | "Source/structure stiffness" matrix |
| $[K_{\phi I}]_{ip} = \int_v B_{\phi i}^T \cdot [\mu^S] \cdot {}^\circ H_p^s dv$ | "Source/magnetization stiffness" matrix |
| $[K_{II}]_{pq} = -\int_v {}^\circ H_p^s \cdot [\mu^s] \cdot {}^\circ H_q^s dv$ | Inductances in vaccum |
| If $i \in S_{T_N}$, $\{F_i\} = \int_{S_{T_N}} N_i^T \{T_N\} ds$ | Surface $S_{T_N}$ force vector |
| If $i \in S_U$, $\{F_i\} = \{F_i'\}$ | Surface $S_U$ force vector |
| If $i \in V_E$, $\{F_i\} = \{0\}$ | Body force vector |
| If $i \in S_{B_N}$, $\{Q_i\} = \int_{S_{B_N}} N_{\phi i}^T \cdot \{B_N\} ds$ | Surface $S_{B_N}$ flux vector |
| If $i \in S_\phi$, $\{Q_i\} = \{Q_i'\}$ | Surface $S_\phi$ flux vector |
| If $i \in V_M$, $\{Q_i\} = 0$ | Body flux vector |

$$\delta L = 0 = \delta\{x\}^T([A]\{x\} - \{b\})$$

leading to the system resolution:

$$[A]\{x\} = \{b\} \tag{C.40}$$

It is the term $[K_{UU}] - \omega^2[M]$ written as $[KM_{UU}]$ which renders $[A]$ indefinite and therefore the solution is a saddle point of $L$. This conclusion agrees with that derived in chapter 2.

The tangential field condition (Eq. C.33), imposed on the $S_{H_T}$ areas must be introduced in the functional before using the stationarity condition in order to be correctly considered. A few algebrical operations, changing the tangential matrix, make it possible for the equations in relation with the $S_{H_T}$ area consideration to be the same as those of a $S_\phi$ area, where the prescribed potential should be the floating one.

## Reduction of Potentials

The last step before the final resolution consists of eliminating the potential degrees of freedom in the system (Eq. C.40) while noting that they are of two types: the potentials are either known as on the surfaces $S_\phi$ and the associated flux is unknown, or they are unknown as within $V_M$ and the associated flux is known (zero in $V_M$). In this case, the corresponding lines of the system (Eq. C.40) are inverted so as to derive the expression for unknown potentials and to replace it in the computation lines of the forces {F} and the fluxes seen by the windings {$Q^V$}. The final system to be solved is written subsequently to this operation as

$$\begin{bmatrix} [KM_{UU}] & [K_{UI}] \\ [K_{UI}]^T & [K_{II}] \end{bmatrix} \begin{bmatrix} \{U\} \\ \{I\} \end{bmatrix} = \begin{bmatrix} \{F\} \\ \{-Q^V\} \end{bmatrix} \tag{C.41}$$

## Analyses and Resolutions

There are essentially two possible types of analysis:

Modal analysis corresponding to the determination of modes which are specific to the structure (i.e., the forces {F} are zero) under given electrical conditions

Harmonic analysis corresponding to the study of the structural strain versus the frequency inductance

The open-circuit modal analysis, obtained by prescribing zero currents {I} in Eq. (C.41), leads to the computation of the electric antiresonance modes:

$$([K_{UU}] - \omega^2 [M]) \cdot \{U\} = 0 \tag{C.42}$$

The relation

$$\{Q^V\} = -[K_{UI}]^T \cdot \{U\}$$

permits computation of the fluxes seen by the windings.

The short-circuit modal analysis, obtained by prescribing zero fluxes {$Q^V$} in (Eq. C.41), leads to the computation of the electric resonance modes:

$$\left[ [K_{UU}] - [K_{UI}][K_{II}]^{-1}[K_{UI}]^T - \omega^2[M] \right] \cdot \{U\} = 0 \tag{C.43}$$

The relation

$$\{I\} = -[K_{II}]^{-1}[K_{UI}]^T \cdot \{U\}$$

permits the computation of the currents flowing through the windings.

The harmonic analysis, obtained by prescribing the currents and external forces (with circular frequency $\omega$ as a parameter), leads to the computation of displacements:

$$([K_{UU}] - \omega^2[M]) \cdot \{U\} = \{F\} - \left([K_{UI}]^{\mathrm{T}} \cdot \{I\}\right) \tag{C.44}$$

The relation

$$\{-Q^V\} = [K_{UI}]^{\mathrm{T}} \cdot \{U\} + [K_{II}] \cdot \{I\} \tag{C.45}$$

permits computation of the fluxes seen by the windings and thus derivation of the electric impedances.

When, as in practice, the coils are connected in series, the currents (Eq. C.1) are reduced to only one current $I$ and Eq. (C.45) indicates the transducer impedance:

$$Z = -j\omega\left(K_{II} + \{K_{UI}\}^{\mathrm{T}} \cdot \{U\}/I\right) \tag{C.46}$$

where $-\{K_{II}\}$ is a scalar equal to the inductance of the clamped material, and $-\{K_{UI}\}^{\mathrm{T}} \cdot \{U\}/I$ is a scalar equal to the motional inductance.

The analogy with the piezoelectricity in the final system to be solved is found again in the analyses. The resolution algorithms which are open-circuit and short-circuit modal analyses, respectively, for the current problem coincide with those of piezoelectricity for the short-circuit and open-circuit modal analyses (i.e., interchange between current and voltage).

The harmonic analysis, conducted for current in magnetostrictive applications, uses the same algorithms as the harmonic analysis of voltage in piezoelectricity (32).

# REFERENCES

1. R. M. Bozorth, "Ferromagnetism," 2nd ed. Van Nostrand, New York, 1951.

2. J. L. Butler and N. L. Lizza, Eddy current loss factor series for magnetostrictive rods, *J. Acoust. Soc. Am.* **82**, 378, (1987).

3. A. E. Clark, Magnetostrictive rare earth–Fe2 compounds, *in* "Ferromagnetic Materials" (E. P. Wohlfarth, Ed.), North-Holland, Amsterdam, 1980.

4. A. Bossavit, On the condition "h normal to the wall" in magnetic field problems, *Int. J. Num. Meth. Eng.* **24**, 1541–1550, (1987).

5. O. B. Wilson, "An introduction to the Theory and Design of SONAR Transducers," U. S. Government Printing Office, Washington, DC, 1985.

6. D. A. Berlincourt, D. R. Curran and H. Jaffe, Piezoelectric and piezomagnetic materials, *in* "Physical Acoustics, Principles, and Methods," (W.P. Mason, Ed.). Academic Press, New York, 1964.

7.  J. C. Sabonnadiere, *in* "Finite Elements in Electrical and Magnetic Field Problems," (M.V.K. Chari, P. P. Silvester, Eds.): J Wiley, New York. 1980.

8.  A. Bossavit and J. C. Verite, The TRIFOU code: Solving the 3-D eddy-currents problem by using H as state variable, *IEEE Trans. Magn.* **MAG-19**, 2465–2470, (1983).

9.  J. C. Verite, Calculation of multivalued potentials in multiply connected exterior regions, *IEEE Trans. Magn.* **MAG-23**, 1881–1887, (1987).

10. C. Iselin, A scalar integral equation for magnetostatic fields, *in* "Proceedings of the Compumag Conf. Oxford," 15-18, (1976).

11. O. C. Zienkiewicz, J. Lyness, and D. R. Owen, Three-dimensional magnetic field determination using a scalar potential—A finite element solution, *IEEE Trans. Magn.* **MAG-13**(5), 1649–1656, (1977).

12. M. M. Sussman, Remarks on computational magnetostatics, *Int. J. Num. Meth. Eng.* **26**, 987–1000, (1988).

13. J. Simkin and C. W. Trowbridge, On the use of the total scalar potential in the numerical solution of field problems in electromagnetics, *J. Num. Meth. Eng.* **14**, 423–440, (1979).

14. L. Landau and E. Lifchitz, "Electrodynamique des Milieux Continus" MIR, Moscow, 1967.

15. P. M. Morse and H. Feshbach, "Methods of Theoretical Physics." McGraw-Hill, 1953.

16. R. Petit, "L'outil Mathématique," Masson, 1983.

17. B. A. Finlayson, "The Method of Weighted Residuals and Variational Principles," Academic Press, New York, 1972.

18. A. L. Davies and P. Samuels, On the use of the total scalar potential in the numerical solution of field problems in electromagnetics, *Int. J. Num. Meth. Eng.* **26**, 2779–2780, (1988).

19. J. L. Coulomb, Analyse tridimensionnelle des champs électriques et magnétiques par la méthode des éléments finis, doctoral thesis, INPG, Grenoble, France, 1981.

20. D. Boucher, Calcul des modes de vibration de transducteurs piézoélectriques par une méthode Eléments finis—Perturbation, Doctoral thesis, University of Maine, France, 1979.

21. J. N. Decarpigny, Application de la méthode des éléments finis à l'étude de transducteurs piézoélectriques, Doctoral thesis, USTL, Lille, France, 1984.

22. H.F Tiersten, Hamilton's principle for linear piezoelectric media, *Proc. IEEE (Lett.)*, **55**, 1523–1524, (1967).

23. E. P. Eer Nisse R. Holland, On variational techniques for piezoelectric device analysis, *Proc. IEEE (lett.)*, **55**, 1524–1525, (1967).

24. O. B. Wilson, "An Introduction to the Theory and Design of SONAR Transducers," U.S. Government Printing Office, Washington, DC, 1985.

25. A. E. H. Love, "A Treatise on the Mathematical Theory of Elasticity," 4th ed. Dover, New York, 1927.

26. H. Allick and T. J. R. Hughes, Finite element method for piezoelectric vibration, *Int. J. Num. Meth. Eng.* **2**, 31 151–157, (1970).

27. R. D. Mindlin, On the equations of motion of piezoelectric crystals, Problems of continuum mechanics, *Soc. Ind. Appl. Math.*, 282–290, (1961).

28. R. Courant, Variational methods for the solution of problems of equilibrium and vibration, *Bull. Am. Math. Soc.* **49**, 1–23, (1943).

29. O. C. Zienkiewicz, "The Finite Element Method," 3rd ed. MacGraw-Hill, New York, 1979.

30. J. F. Imbert, "Analyse des Structures par Éléments Finis." Cepadues, 1979.

31. J. F. Nye, "Physical Properties of Crystal." Clarendon, Oxford Univ. Press, London, 1957.

32. E. Du Tremolet De Lacheisserie and J. Rouchy, La détermination de la matrice élasto-magnétique et son domaine d'application dans le cas des alliages Terfenol-D, GERDSM Contract Report, France, 1988.

# A Brief Market Inventory

This brief market inventory is supplied to help you get started regarding giant magnetostrictive materials and their applications.

The information is organized under the following headlines:

- Material manufacturers
- Material suppliers
- System suppliers
- Consultants
- Design software

## Material Manufacturers

ETREMA Products, Inc.
2500 North Loop Dr.
Ames, IA 50010-8278
USA
+ 1 515 296-8030 (phone)
+ 1 515 296-7168 (fax)
www.etrema-usa.com

TDK Corporation
Materials Research Centre
Minami-Toba Aza Matsugashita
570-2
Naraita-Shi
Chiba-Prefecture
JAPAN
www.tdk.com

## Material Suppliers

ETREMA Products, Inc.
2500 North Loop Dr.
Ames, IA 50010-8278
USA
+ 1 515 296-8030 (phone)
+ 1 515 296-7168 (fax)
www.etrema-usa.com

ProEngCo AB
 (Representative of ETREMA
 Products)
Solbjersvägen 4
224 68 Lund
SWEDEN
+46 46 189008 (phone)
+46 46 189009 (fax)
www.proengco.se

Marlborough Communications Ltd
(Representative of ETREMA Products)
Dovenby Hall,
Balcombe Road,
Horley, Surrey,
RH6 9UU
ENGLAND
+44-1293-775071 (phone)
+44-1293-820781 (fax)
marlborough.coms@virgin.net

MORITEX Corporation
(Representative of ETREMA Products)
MORITEX Bldg , 3-1-14,
Jingu-Mae,
Shibuya-Ku
Tokyo 150-0001
JAPAN
+81 3 3401 9736 (phone)

Meyer-Industrie-Electronic GmbH
(Representative of ETREMA Products)
Postfach 14 47-48
Gewerbegehiet Lohesch
D-49525 Lengerich
GERMANY
+49 54 81-93 85 (phone)
www.meyle.de

InoTec Solutions Gmbh
(Representative of ETREMA Products)
Gardolostrasse
18-85375 Nuefarhn
GERMANY
+49 8165 708856

Newlands Technology, Ltd
(Representative of ETREMA Products)
Newlands House
Inglemire Lane
Hull, E Yorkshire
HU6 7TQ
ENGLAND
+44 1482 806688 (phone)
+44 1482 806654 (fax)
www.websights.co.uk
/newlands_technology/

**System Suppliers**

ETREMA Products, Inc.
2500 North Loop Dr.
Ames, IA 50010-8278
USA
+ 1 515 296-8030 (phone)
+ 1 515 296-7168 (fax)
www.etrema-usa.com

Bofors SA Marine AB
Box 551
Nettovägen 11
17526 Järfälla
SWEDEN
+46 8 5808 2592 (phone)
+46 8 5803 0226 (fax)

CEDRAT RECHERCHE S. A.
AMA Department
Zirst, 38246 Meylan Cedex
FRANCE
+33 4 76 90 50 45 (phone)
+33 4 76 90 16 09 (fax)
www.cedrat-grenoble.fr

ProEngCo AB
Solbjersvägen 4
224 68 Lund
SWEDEN
+46 46 189008 (phone)
+46 46 189009 (fax)
www.proengco.se

UNACO Systems AB
Björmövägen 20
S-721 31 Västerås
SWEDEN
+46 21 80 55 00 (phone)
+46 21 80 38 45 (fax)
unaco@unaco.se

Newlands Technology, Ltd
Newlands House
Inglemire Lane
Hull, E Yorkshire
HU6 7TQ
ENGLAND
+44 1482 806688 (phone)
+44 1482 806654 (fax)
www.websights.co.uk
/newlands_technology/

Sonic Research Corporation
Texas Office
P.O. Box 224502
Dallas, Texas 75222
USA
+1 214 942-8337 (phone)
+1 214 942-4367 (fax)
www.vline.net/sonic/de-emulsi-
fier.html

AdaptaMat Oy
P.O. Box 547
FIN-02150 Espoo
FINLAND
+358 9 437 5444 (phone)
+358 9 437 5444 (fax)
www.otech.fi/otech/
Adaptamat.html

**Consultants**

ETREMA Products, Inc.
2500 North Loop Dr.
Ames, IA 50010-8278
USA
+ 1 515 296-8030 (phone)
+ 1 515 296-7168 (fax)
www.etrema-usa.com

CEDRAT RECHERCHE S. A.
AMA Department
Zirst, 38246 Meylan Cedex
FRANCE
+33 4 76 90 50 45 (phone)
+33 4 76 90 16 09 (fax)
www.cedrat-grenoble.fr

ProEngCo AB
Solbjervägen 4
224 68 Lund
SWEDEN
+46 46 189008 (phone)
+46 46 189009 (fax)
www.proengco.se

Newlands Technology, Ltd
Newlands House
Inglemire Lane
Hull, E Yorkshire
HU6 7TQ
ENGLAND
+44 (0)1482 806688 (phone)
+44 (0)1482 806654 (fax)
www.websights.co.uk
/newlands_technology/

Image Acoustics Inc.
97 Elm St.
Cohasset, MA 02025-1805
USA
+1 781 383-2002 (phone)

HYMAG
Rospiggsvägen 4
S-183 63 Täby
SWEDEN
+46 8 7327057 (phone)
+46 8 7327057 (fax)
goran.engdahl@taby.mail.
telia.com

Clark Associates
10421 Floral Drive
Hyattsville, MD 20783
USA
+1 301 434 6325 (phone)

ZIP-Aktorik
Altenkesseler Str. 17/D2
D-66115 Saarbrücken
GERMANY
+49 681 302 6039 (phone)
+49 681 302 6031 (fax)
www.zip.uni-sb.de

**Design Software**

CEDRAT RECHERCHE S. A.
AMA Department
Zirst, 38246 Meylan Cedex
FRANCE
+33 4 76 90 50 45 (phone)
+33 4 76 90 16 09 (fax)
www.cedrat-grenoble.fr

Magnetic Materials Modelling Laboratory
Kungl. Tekniska Högskolan
Teknikringen 33
S-100 44 Stockholm
SWEDEN
+46 8 790 77 60 (phone)
+46 8 20 52 68 (phone)
www.ekc.kth.se

The cited contact information is not claimed to be complete. In fact, it is merely a first entrance into the technology of giant magnetostrictive materials

# INDEX